SCHAUM'S OUTLINE OF

Modern Introductory DIFFERENTIAL EQUATIONS

with

Laplace Transforms • Numerical Methods
Matrix Methods • Eigenvalue Problems

by

RICHARD BRONSON, Ph.D.

Chairman, Department of Mathematics and Computer Science
Fairleigh Dickinson University

SCHAUM'S OUTLINE SERIES

McGRAW-HILL BOOK COMPANY

New York, St. Louis, San Francisco, Düsseldorf, Johannesburg, Kuala Lumpur, London, Mexico, Montreal, New Delhi, Panama, São Paulo, Singapore, Sydney, and Toronto

To Ignace and Gwendolyn Bronson
Samuel and Rose Feldschuh

07-008009-7

12 13 14 15 SH SH 8 7 6 5 4 3

Library of Congress Cataloging in Publication Data

Bronson, Richard.
Schaum's outline of modern introductory differential equations.

(Schaum's outline series)
1. Differential equations. I. Title.
II. Title: Modern introductory differential equations.
[QA372.B855] 515'.35 73-19601

ISBN 0-07-008009-7

Preface

Over the past twenty years there have been significant developments in the field of differential equations. The advent of high-speed computers has made solutions by numerical techniques feasible and has resulted in a host of new methods. The systems approach favored in many present-day engineering problems lends itself to both matrix methods and Laplace transforms.

This book outlines, with many solved problems, both the classical theory of differential equations and the more modern techniques currently available. The only prerequisite for any of the topics treated is calculus. As a supplement to standard textbooks, or as a textbook in its own right, it should prove useful for undergraduate courses and for independent study.

Chapters 1 through 21 and Chapters 37 through 39 cover the classical material, including separable and exact equations, solutions of linear equations with constant coefficients by the characteristic equation method, variation of parameters and the method of undetermined coefficients, infinite series solutions, and boundary-value and Sturm-Liouville problems. In contrast, Chapters 22 through 36 deal with the newer techniques currently in vogue, in particular, Laplace transforms, matrix methods, and numerical techniques. This last subject, because of its great practical importance, has been developed more completely than is usual at this level.

Each chapter of the book is divided into three parts. The first outlines the salient points, drawing attention to potential difficulties and pointing out subtleties that could be easily overlooked. The second part consists of completely worked-out problems which clarify the material presented in the first part and which, on occasion, also expand on that development. Finally, there is a section of problems with answers through which the student can test his understanding of the material.

I should like to thank the many individuals who helped make this book a reality. The valuable suggestions by Joseph Klein and Jack Mieses for Chapters 22 through 27, and those of Mabel Dukeshire, are all warmly acknowledged. Particular thanks are due Raymond Raggi who programmed most of the numerical methods and David Beckwith of the Schaum's staff for his splendid editing. Finally, my greatest debt is to my wife Evelyn who besides doing most of the typing contributed substantially to the editing and proofreading phases of this project.

RICHARD BRONSON

Fairleigh Dickinson University
October 1973

CONTENTS

CONTENTS

CONTENTS

CONTENTS

Chapter 1

Basic Concepts

1.1 ORDINARY DIFFERENTIAL EQUATIONS

A *differential equation* is an equation involving an unknown function and its derivatives.

Example 1.1. The following are differential equations involving the unknown function y.

$$\frac{dy}{dx} = 5x + 3 \tag{1.1}$$

$$e^y \frac{d^2y}{dx^2} + 2\left(\frac{dy}{dx}\right)^2 = 1 \tag{1.2}$$

$$4\frac{d^3y}{dx^3} + (\sin x)\frac{d^2y}{dx^2} + 5xy = 0 \tag{1.3}$$

$$\left(\frac{d^2y}{dx^2}\right)^3 + 3y\left(\frac{dy}{dx}\right)^7 + y^3\left(\frac{dy}{dx}\right)^2 = 5x \tag{1.4}$$

$$\frac{\partial^2 y}{\partial t^2} - 4\frac{\partial^2 y}{\partial x^2} = 0 \tag{1.5}$$

A differential equation is an *ordinary differential equation* if the unknown function depends on only one independent variable. If the unknown function depends on two or more independent variables, the differential equation is a *partial differential equation.*

Example 1.2. Equations (*1.1*) through (*1.4*) are examples of ordinary differential equations, since the unknown function y depends solely on the variable x. Equation (*1.5*) is a partial differential equation, since y depends on both the independent variables t and x.

In this book we will be concerned solely with ordinary differential equations.

1.2 ORDER AND DEGREE

The *order* of a differential equation is the order of the highest derivative appearing in the equation.

Example 1.3. Equation (*1.1*) is a first-order differential equation; (*1.2*), (*1.4*), and (*1.5*) are second-order differential equations. (Note in (*1.4*) that the order of the highest *derivative* appearing in the equation is two.) Equation (*1.3*) is a third-order differential equation.

The *degree* of a differential equation that can be written as a polynomial in the unknown function and its derivatives is the power to which the highest-order derivative is raised.

Example 1.4. Equation (*1.4*) is a differential equation of degree three, since the highest-order derivative, in this case the second, is raised to the third power. Equations (*1.1*) and (*1.3*) are examples of first-degree differential equations.

Not every equation can be classified by degree. For instance, *(1.2)* has no degree, as it cannot be written as a *polynomial* in the unknown function and its derivatives (because of the term e^y).

1.3 LINEAR DIFFERENTIAL EQUATIONS

An nth-order ordinary differential equation in the unknown function y and the independent variable x is *linear* if it has the form

$$b_n(x)\frac{d^ny}{dx^n} + b_{n-1}(x)\frac{d^{n-1}y}{dx^{n-1}} + \cdots + b_1(x)\frac{dy}{dx} + b_0(x)y = g(x) \qquad (1.6)$$

The functions $b_j(x)$ $(j = 0, 1, 2, \ldots, n)$ and $g(x)$ are presumed known and depend only on the variable x. Differential equations that cannot be put into the form *(1.6)* are *nonlinear.*

Example 1.5. Equation *(1.1)* is a first-order linear equation, with $b_1(x) = 1$, $b_0(x) = 0$, and $g(x) = 5x + 3$. Equation *(1.3)* is a third-order linear equation, with $b_3(x) = 4$, $b_2(x) = \sin x$, $b_1(x) = 0$, $b_0(x) = 5x$, and $g(x) = 0$. Equations *(1.2)* and *(1.4)* are nonlinear.

1.4 NOTATION

The expressions y', y'', y''', $y^{(4)}$, $\ldots$, $y^{(n)}$ are often used to represent, respectively, the first, second, third, fourth, $\ldots$, nth derivatives of y with respect to the independent variable under consideration. Thus, y'' represents d^2y/dx^2 if the independent variable is x, but represents d^2y/dp^2 if the independent variable is p. If the independent variable is time, usually denoted by t, primes are often replaced by dots. Thus, $\dot{y}$, $\ddot{y}$, and $\dddot{y}$ represent dy/dt, d^2y/dt^2, and d^3y/dt^3, respectively.

Observe that parentheses are used in $y^{(n)}$ to distinguish it from the nth power, y^n.

Solved Problems

In the following problems, classify each differential equation as to order, degree (if appropriate), and linearity. Determine the unknown function and the independent variable.

1.1. $y''' - 5xy' = e^x + 1.$

Third-order: the highest-order derivative is the third. *First-degree*: the equation has the form required in Section 1.2, and the third derivative is raised to the first power. *Linear*: $b_3(x) = 1$, $b_2(x) = 0$, $b_1(x) = -5x$, $b_0(x) = 0$, $g(x) = e^x + 1$. The unknown function is y; the independent variable is x.

1.2. $t\ddot{y} + t^2\dot{y} - (\sin t)\sqrt{y} = t^2 - t + 1.$

Second-order: the highest-order derivative is the second. *No degree*: because of the term $\sqrt{y}$, the equation cannot be written as a polynomial in y and its derivatives. *Nonlinear*: the equation cannot be put into the form *(1.6)*. The unknown function is y; the independent variable is t.

1.3. $s^2\dfrac{d^2t}{ds^2} + st\dfrac{dt}{ds} = s.$

Second-order. First-degree: the equation is a polynomial in the unknown function t and its derivatives (with coefficients in s), and the second derivative is raised to the first power. *Nonlinear*: $b_1 = st$, which depends on both s and t. The unknown function is t; the independent variable is s.

1.4. $5\left(\frac{d^4b}{dp^4}\right)^5 + 7\left(\frac{db}{dp}\right)^{10} + b^7 - b^5 = p.$

Fourth-order. Fifth-degree: the equation has the form required in Section 1.2, and the fourth derivative is raised to the fifth power. *Nonlinear.* The unknown function is b; the independent variable is p.

1.5. $y\frac{d^2x}{dy^2} = y^2 + 1.$

Second-order. First-degree. Linear: $b_2(y) = y$, $b_1(y) = 0$, $b_0(y) = 0$, and $g(y) = y^2 + 1$. The unknown function is x; the independent variable is y.

Supplementary Problems

For the following differential equations, determine (*a*) order, (*b*) degree (if appropriate), (*c*) linearity, (*d*) unknown function, and (*e*) independent variable.

1.6. $(y'')^2 - 3yy' + xy = 0.$

1.7. $x^4y^{(4)} + xy''' = e^x.$

1.8. $t^2\ddot{s} - t\dot{s} = 1 - \sin t.$

1.9. $y^{(4)} + xy''' + x^2y'' - xy' + \sin y = 0.$

1.10. $\frac{d^nx}{dy^n} = y^2 + 1.$

1.11. $\left(\frac{d^2r}{dy^2}\right)^2 + \frac{d^2r}{dy^2} + y\frac{dr}{dy} = 0.$

1.12. $\left(\frac{d^2y}{dx^2}\right)^{3/2} + y = x.$

1.13. $\frac{d^7b}{dp^7} = 3p.$

1.14. $\left(\frac{db}{dp}\right)^7 = 3p.$

Answers to Supplementary Problems

1.6.	(*a*) 2	(*b*) 2	(*c*) nonlinear	(*d*) y	(*e*) x
1.7.	(*a*) 4	(*b*) 1	(*c*) linear	(*d*) y	(*e*) x
1.8.	(*a*) 2	(*b*) 1	(*c*) linear	(*d*) s	(*e*) t
1.9.	(*a*) 4	(*b*) none	(*c*) nonlinear	(*d*) y	(*e*) x
1.10.	(*a*) n	(*b*) 1	(*c*) linear	(*d*) x	(*e*) y
1.11.	(*a*) 2	(*b*) 2	(*c*) nonlinear	(*d*) r	(*e*) y
1.12.	(*a*) 2	(*b*) none	(*c*) nonlinear	(*d*) y	(*e*) x
1.13.	(*a*) 7	(*b*) 1	(*c*) linear	(*d*) b	(*e*) p
1.14	(*a*) 1	(*b*) 7	(*c*) nonlinear	(*d*) b	(*e*) p

Chapter 2

Solutions

2.1 DEFINITION OF SOLUTION

A *solution* of a differential equation in the unknown function y and the independent variable x on the interval $\mathcal{J}$ is a function $y(x)$ that satisfies the differential equation identically for all x in $\mathcal{J}$.

Example 2.1. Is $y(x) = c_1 \sin 2x + c_2 \cos 2x$, where c_1 and c_2 are arbitrary constants, a solution of $y'' + 4y = 0$?

Differentiating y, we find:

$$y' = 2c_1 \cos 2x - 2c_2 \sin 2x \qquad y'' = -4c_1 \sin 2x - 4c_2 \cos 2x$$

Hence,

$$\begin{aligned} y'' + 4y &= (-4c_1 \sin 2x - 4c_2 \cos 2x) + 4(c_1 \sin 2x + c_2 \cos 2x) \\ &= (-4c_1 + 4c_1) \sin 2x + (-4c_2 + 4c_2) \cos 2x \\ &= 0 \end{aligned}$$

Thus, $y = c_1 \sin 2x + c_2 \cos 2x$ satisfies the differential equation for all values of x and is a solution on the interval $(-\infty, \infty)$.

Example 2.2. Determine whether $y = x^2 - 1$ is a solution of $(y')^4 + y^2 = -1$.

Note that the left side of the differential equation must be nonnegative for every real function $y(x)$ and any x, since it is the sum of terms raised to the second and fourth powers, while the right side of the equation is negative. Since no function $y(x)$ will satisfy this equation, the given differential equation has no solution.

We see that some differential equations have infinitely many solutions (Example 2.1), whereas other differential equations have no solutions (Example 2.2). It is also possible that a differential equation has exactly one solution. Consider $(y')^4 + y^2 = 0$, which for reasons identical to those given in Example 2.2 has only the solution $y \equiv 0$.

2.2 PARTICULAR AND GENERAL SOLUTIONS

A *particular solution* of a differential equation is any one solution. The *general solution* of a differential equation is the set of all solutions.

Example 2.3. The general solution to the differential equation in Example 2.1 can be shown to be (see Chapters 11 and 12) $y = c_1 \sin 2x + c_2 \cos 2x$. That is, every particular solution of the differential equation has this general form. A few particular solutions are: (*a*) $y = 5 \sin 2x - 3 \cos 2x$ (choose $c_1 = 5$ and $c_2 = -3$), (*b*) $y = \sin 2x$ (choose $c_1 = 1$ and $c_2 = 0$), and (*c*) $y \equiv 0$ (choose $c_1 = c_2 = 0$).

The general solution of a differential equation cannot always be expressed by a single formula. As an example consider the differential equation $y' + y^2 = 0$, which has two particular solutions $y = \dfrac{1}{x}$ and $y \equiv 0$. Linear differential equations are special in this regard and their general solutions are discussed in Chapter 11.

2.3 INITIAL-VALUE PROBLEMS. BOUNDARY-VALUE PROBLEMS

A differential equation along with subsidiary conditions on the unknown function and its derivatives, all given at the same value of the independent variable, constitutes an *initial-value problem.* The subsidiary conditions are *initial conditions.* If the subsidiary conditions are given at more than one value of the independent variable, the problem is a *boundary-value problem* and the conditions are *boundary conditions.*

Example 2.4. The problem $y'' + 2y' = e^x$; $y(\pi) = 1$, $y'(\pi) = 2$ is an initial-value problem, since the two subsidiary conditions are both given at $x = \pi$. The problem $y'' + 2y' = e^x$; $y(0) = 1$, $y(1) = 1$ is a boundary-value problem, since the two subsidiary conditions are given at the different values $x = 0$ and $x = 1$.

A *solution* to an initial-value or boundary-value problem is a function $y(x)$ that both solves the differential equation (in the sense of Section 2.1) and satisfies all given subsidiary conditions.

Example 2.5. Determine whether any of the functions (*a*) $y_1 = \sin 2x$, (*b*) $y_2(x) = x$, or (*c*) $y_3(x) = \frac{1}{2}\sin 2x$ is a solution to the initial-value problem $y'' + 4y = 0$; $y(0) = 0$, $y'(0) = 1$. (*a*) $y_1(x)$ is a solution to the differential equation and satisfies the first initial condition $y(0) = 0$. However, $y_1(x)$ does not satisfy the second initial condition ($y_1'(x) = 2\cos 2x$; $y_1'(0) = 2\cos 0 = 2 \neq 1$); hence it is not a solution to the initial-value problem. (*b*) $y_2(x)$ satisfies both initial conditions but does not satisfy the differential equation; hence $y_2(x)$ is not a solution. (*c*) $y_3(x)$ satisfies the differential equation and both initial conditions; therefore, it is a solution to the initial-value problem.

Solved Problems

2.1. Determine whether $y(x) = 2e^{-x} + xe^{-x}$ is a solution of $y'' + 2y' + y = 0$.

Differentiating $y(x)$, it follows that

$$y'(x) = -2e^{-x} + e^{-x} - xe^{-x} = -e^{-x} - xe^{-x}$$

$$y''(x) = e^{-x} - e^{-x} + xe^{-x} = xe^{-x}$$

Substituting these values into the differential equation, we obtain

$$y'' + 2y' + y = xe^{-x} + 2(-e^{-x} - xe^{-x}) + (2e^{-x} + xe^{-x}) = 0$$

Thus, $y(x)$ is a solution.

2.2. Is $y(x) \equiv 1$ a solution of $y'' + 2y' + y = x$?

From $y(x) \equiv 1$ it follows that $y'(x) \equiv 0$ and $y''(x) \equiv 0$. Substituting these values into the differential equation, we obtain

$$y'' + 2y' + y = 0 + 2(0) + 1 = 1 \neq x$$

Thus, $y(x) \equiv 1$ is not a solution.

2.3. Show that $y = \ln x$ is a solution of $xy'' + y' = 0$ on $\mathcal{I} = (0, \infty)$ but is not a solution on $\mathcal{I} = (-\infty, \infty)$.

On $(0, \infty)$ we have $y' = 1/x$ and $y'' = -1/x^2$. Substituting these values into the differential equation, we obtain

$$xy'' + y' = x\left(-\frac{1}{x^2}\right) + \frac{1}{x} = 0$$

Thus, $y = \ln x$ is a solution on $(0, \infty)$.

Note that $y = \ln x$ could not be a solution on $(-\infty, \infty)$, since the logarithm is undefined for negative numbers and zero.

2.4. Show that $y = 1/(x^2 - 1)$ is a solution of $y' + 2xy^2 = 0$ on $\mathcal{J} = (-1, 1)$ but not on any larger interval containing $\mathcal{J}$.

On $(-1, 1)$, $y = 1/(x^2-1)$ and its derivative $y' = -2x/(x^2-1)^2$ are well-defined functions. Substituting these values into the differential equation, we have

$$y' + 2xy^2 = -\frac{2x}{(x^2-1)^2} + 2x\left[\frac{1}{x^2-1}\right]^2 = 0$$

Thus, $y = 1/(x^2-1)$ is a solution on $\mathcal{J} = (-1, 1)$.

Note, however, that $1/(x^2-1)$ is not defined at $x = \pm 1$ and therefore could not be a solution on any interval containing either of these two points.

2.5. Find the solution to the initial-value problem $y' + y = 0$; $y(3) = 2$, if the general solution to the differential equation is known to be (see Chapter 8) $y(x) = c_1 e^{-x}$, where c_1 is an arbitrary constant.

Since $y(x)$ is a solution of the differential equation for every value of c_1, we seek that value of c_1 which will also satisfy the initial condition. Note that $y(3) = c_1 e^{-3}$. To satisfy the initial condition $y(3) = 2$, it is sufficient to choose c_1 so that $c_1 e^{-3} = 2$, that is, to choose $c_1 = 2e^3$. Substituting this value for c_1 into $y(x)$, we obtain $y(x) = 2e^3 e^{-x} = 2e^{3-x}$ as the solution of the initial-value problem.

2.6. Find a solution to the initial-value problem $y'' + 4y = 0$; $y(0) = 0$, $y'(0) = 1$, if the general solution to the differential equation is known to be (see Chapter 12) $y(x) = c_1 \sin 2x + c_2 \cos 2x$.

Since $y(x)$ is a solution of the differential equation for all values of c_1 and c_2 (see Example 2.1), we seek those values of c_1 and c_2 that will also satisfy the initial conditions. Note that $y(0) = c_1 \sin 0 + c_2 \cos 0 = c_2$. To satisfy the first initial condition, $y(0) = 0$, we choose $c_2 = 0$. Furthermore, $y'(x) = 2c_1 \cos 2x - 2c_2 \sin 2x$; thus, $y'(0) = 2c_1 \cos 0 - 2c_2 \sin 0 = 2c_1$. To satisfy the second initial condition, $y'(0) = 1$, we choose $2c_1 = 1$, or $c_1 = \frac{1}{2}$. Substituting these values of c_1 and c_2 into $y(x)$, we obtain $y(x) = \frac{1}{2}\sin 2x$ as the solution of the initial-value problem. (See Example 2.5.)

2.7. Find a solution to the boundary-value problem $y'' + 4y = 0$; $y\left(\frac{\pi}{8}\right) = 0$, $y\left(\frac{\pi}{6}\right) = 1$, if the general solution to the differential equation is $y(x) = c_1 \sin 2x + c_2 \cos 2x$.

Note that

$$y\left(\frac{\pi}{8}\right) = c_1 \sin\left(\frac{\pi}{4}\right) + c_2 \cos\left(\frac{\pi}{4}\right) = c_1(\tfrac{1}{2}\sqrt{2}) + c_2(\tfrac{1}{2}\sqrt{2})$$

To satisfy the condition $y\left(\frac{\pi}{8}\right) = 0$, we require

$$c_1(\tfrac{1}{2}\sqrt{2}) + c_2(\tfrac{1}{2}\sqrt{2}) = 0$$

Furthermore,

$$y\left(\frac{\pi}{6}\right) = c_1 \sin\left(\frac{\pi}{3}\right) + c_2 \cos\left(\frac{\pi}{3}\right) = c_1(\tfrac{1}{2}\sqrt{3}) + c_2(\tfrac{1}{2})$$

To satisfy the second condition, $y\left(\frac{\pi}{6}\right) = 1$, we require

$$\tfrac{1}{2}\sqrt{3}\,c_1 + \tfrac{1}{2}c_2 = 1 \tag{2}$$

Solving (1) and (2) simultaneously, we find

$$c_1 = -c_2 = \frac{2}{\sqrt{3}-1}$$

Substituting these values into $y(x)$, we obtain

$$y(x) = \frac{2}{\sqrt{3}-1}(\sin 2x - \cos 2x)$$

as the solution of the boundary-value problem.

2.8. Find a solution to the boundary-value problem $y''+4y=0$; $y(0)=1$, $y(\pi/2)=2$, if the general solution to the differential equation is known to be $y(x)=c_1\sin 2x + c_2\cos 2x$.

Since $y(0) = c_1\sin 0 + c_2\cos 0 = c_2$, we must choose $c_2 = 1$ to satisfy the condition $y(0) = 1$. Since $y\left(\frac{\pi}{2}\right) = c_1\sin\pi + c_2\cos\pi = -c_2$, we must choose $c_2 = -2$ to satisfy the second condition, $y\left(\frac{\pi}{2}\right) = 2$. Thus, to satisfy both boundary conditions simultaneously, we must require c_2 to equal both one and minus two, which is impossible. Therefore, there does not exist a solution to this problem.

2.9. Determine c_1 and c_2 so that $y(x) = c_1\sin 2x + c_2\cos 2x + 1$ will satisfy the conditions $y\left(\frac{\pi}{8}\right) = 0$ and $y'\left(\frac{\pi}{8}\right) = \sqrt{2}$.

Note that

$$y\left(\frac{\pi}{8}\right) = c_1\sin\left(\frac{\pi}{4}\right) + c_2\cos\left(\frac{\pi}{4}\right) + 1 = c_1(\tfrac{1}{2}\sqrt{2}) + c_2(\tfrac{1}{2}\sqrt{2}) + 1$$

To satisfy the condition $y\left(\frac{\pi}{8}\right) = 0$, we require $c_1(\tfrac{1}{2}\sqrt{2}) + c_2(\tfrac{1}{2}\sqrt{2}) + 1 = 0$, or equivalently,

$$c_1 + c_2 = -\sqrt{2} \tag{1}$$

Since $y'(x) = 2c_1\cos 2x - 2c_2\sin 2x$,

$$\begin{aligned} y'\left(\frac{\pi}{8}\right) &= 2c_1\cos\left(\frac{\pi}{4}\right) - 2c_2\sin\left(\frac{\pi}{4}\right) \\ &= 2c_1(\tfrac{1}{2}\sqrt{2}) - 2c_2(\tfrac{1}{2}\sqrt{2}) = \sqrt{2}\,c_1 - \sqrt{2}\,c_2 \end{aligned}$$

To satisfy the condition $y'\left(\frac{\pi}{8}\right) = \sqrt{2}$, we require $\sqrt{2}\,c_1 - \sqrt{2}\,c_2 = \sqrt{2}$, or equivalently,

$$c_1 - c_2 = 1 \tag{2}$$

Solving (1) and (2) simultaneously, we obtain $c_1 = -\tfrac{1}{2}(\sqrt{2}-1)$ and $c_2 = -\tfrac{1}{2}(\sqrt{2}+1)$.

2.10. Determine c_1 and c_2 so that $y(x) = c_1e^{2x} + c_2e^{x} + 2\sin x$ will satisfy the conditions $y(0) = 0$ and $y'(0) = 1$.

Because $\sin 0 = 0$, $y(0) = c_1 + c_2$. To satisfy the condition $y(0) = 0$, we require

$$c_1 + c_2 = 0 \tag{1}$$

From

$$y'(x) = 2c_1e^{2x} + c_2e^x + 2\cos x$$

we have $y'(0) = 2c_1 + c_2 + 2$. To satisfy the condition $y'(0) = 1$, we require $2c_1 + c_2 + 2 = 1$, or

$$2c_1 + c_2 = -1 \tag{2}$$

Solving (1) and (2) simultaneously, we obtain $c_1 = -1$ and $c_2 = 1$.

Supplementary Problems

2.11. Which of the following functions are solutions of the differential equation $y'' - y = 0$? (a) e^x, (b) $\sin x$, (c) $4e^{-x}$, (d) 0, (e) $\frac{1}{2}x^2 + 1$.

2.12. Which of the following functions are solutions of the differential equation $y'' - 4y' + 4y = e^x$? (a) e^x, (b) e^{2x}, (c) $e^{2x} + e^x$, (d) $xe^{2x} + e^x$, (e) $e^{2x} + xe^x$.

In Problems 2.13–2.22, find c_1 and c_2 so that $y(x) = c_1 \sin x + c_2 \cos x$ will satisfy the given conditions. Determine whether the given conditions are initial conditions or boundary conditions.

2.13. $y(0) = 1,\ y'(0) = 2.$

2.14. $y(0) = 2,\ y'(0) = 1.$

2.15. $y\left(\frac{\pi}{2}\right) = 1,\ y'\left(\frac{\pi}{2}\right) = 2.$

2.16. $y(0) = 1,\ y\left(\frac{\pi}{2}\right) = 1.$

2.17. $y'(0) = 1,\ y'\left(\frac{\pi}{2}\right) = 1.$

2.18. $y(0) = 1,\ y'(\pi) = 1.$

2.19. $y(0) = 1,\ y(\pi) = 2.$

2.20. $y(0) = 0,\ y'(0) = 0.$

2.21. $y\left(\frac{\pi}{4}\right) = 0,\ y\left(\frac{\pi}{6}\right) = 1.$

2.22. $y(0) = 0,\ y'\left(\frac{\pi}{2}\right) = 1.$

In Problems 2.23–2.27, find values of c_1 and c_2 so that the given functions will satisfy the prescribed initial conditions.

2.23. $y(x) = c_1e^x + c_2e^{-x} + 4\sin x;\ y(0) = 1,\ y'(0) = -1.$

2.24. $y(x) = c_1x + c_2 + x^2 - 1;\ y(1) = 1,\ y'(1) = 2.$

2.25. $y(x) = c_1e^x + c_2e^{2x} + 3e^{3x};\ y(0) = 0,\ y'(0) = 0.$

2.26. $y(x) = c_1 \sin x + c_2 \cos x + 1;\ y(\pi) = 0,\ y'(\pi) = 0.$

2.27. $y(x) = c_1e^x + c_2xe^x + x^2e^x;\ y(1) = 1,\ y'(1) = -1.$

Answers to Supplementary Problems

2.11. (*a*), (*c*), (*d*)

2.12. (*a*), (*c*), (*d*)

2.13. $c_1 = 2,\ c_2 = 1$; initial conditions

2.14. $c_1 = 1,\ c_2 = 2$; initial conditions

2.15. $c_1 = 1,\ c_2 = -2$; initial conditions

2.16. $c_1 = c_2 = 1$; boundary conditions

2.17. $c_1 = 1,\ c_2 = -1$; boundary conditions

2.18. $c_1 = -1,\ c_2 = 1$; boundary conditions

2.19. no values; boundary conditions

2.20. $c_1 = c_2 = 0$; initial conditions

2.21. $c_1 = \dfrac{-2}{\sqrt{3}-1},\ c_2 = \dfrac{2}{\sqrt{3}-1}$; boundary conditions

2.22. no values; boundary conditions

2.23. $c_1 = -2,\ c_2 = 3$

2.24. $c_1 = 0,\ c_2 = 1$

2.25. $c_1 = 3,\ c_2 = -6$

2.26. $c_1 = 0,\ c_2 = 1$

2.27. $c_1 = 1 + \dfrac{3}{e},\ c_2 = -2 - \dfrac{2}{e}$

Chapter 3

Classification of First-Order Differential Equations

3.1 STANDARD FORM AND DIFFERENTIAL FORM

The *standard form* of a first-order differential equation is

$$y' = f(x, y) \tag{3.1}$$

Example 3.1. For the differential equation $y' = -y + \sin x$, we have $f(x, y) = -y + \sin x$. For $y' = 3yx^2/(x^3 + y^4)$, $f(x, y) = 3yx^2/(x^3 + y^4)$. The differential equation $e^x y' + e^{2x} y = \sin x$ is not in standard form; it can be put into such form, however, by algebraically solving for y'. Thus, $e^x y' = -e^{2x} y + \sin x$, or $y' = -e^x y + e^{-x} \sin x$, and $f(x, y) = -e^x y + e^{-x} \sin x$.

Theorem 3.1. If $f(x, y)$ and $\dfrac{\partial f(x, y)}{\partial y}$ are continuous in a rectangle $\mathcal{R}$: $|x - x_0| \leq a$, $|y - y_0| \leq b$, then there exists an interval about x_0 in which the initial-value problem $y' = f(x, y)$; $y(x_0) = y_0$ has a unique solution.

The function $f(x, y)$ given in (*3.1*) can always be written as the quotient of two other functions $M(x, y)$ and $-N(x, y)$. (The minus sign is only for convenience.) Thus, recalling that $y' = dy/dx$, we can rewrite (*3.1*) as $dy/dx = M(x, y)/-N(x, y)$, which is equivalent to the *differential form*

$$M(x, y)\,dx + N(x, y)\,dy = 0 \tag{3.2}$$

Example 3.2. Given any function $f(x, y)$, there are infinitely many ways to decompose it into a quotient of two other functions. For example, if $f(x, y) = \dfrac{x + y}{y^2}$, then four decompositions are:

(*a*) $M(x, y) = x + y$, $N(x, y) = -y^2$, and

$$\frac{M(x, y)}{-N(x, y)} = \frac{x + y}{-(-y^2)} = \frac{x + y}{y^2}$$

(*b*) $M(x, y) = -1$, $N(x, y) = \dfrac{y^2}{x + y}$, and

$$\frac{M(x, y)}{-N(x, y)} = \frac{-1}{-y^2/(x + y)} = \frac{x + y}{y^2}$$

(*c*) $M(x, y) = \dfrac{x + y}{2}$, $N(x, y) = \dfrac{-y^2}{2}$, and

$$\frac{M(x, y)}{-N(x, y)} = \frac{(x + y)/2}{-(-y^2/2)} = \frac{x + y}{y^2}$$

(*d*) $M(x, y) = \dfrac{-x - y}{x^2}$, $N(x, y) = \dfrac{y^2}{x^2}$, and

$$\frac{M(x, y)}{-N(x, y)} = \frac{(-x - y)/x^2}{-y^2/x^2} = \frac{x + y}{y^2}$$

In the remainder of this chapter we employ the standard or differential form to define four categories of differential equations. It is important to emphasize that most first-order differential equations do *not* fall into, and can *not* be transformed into, any of these categories. For such differential equations, there are generally no analytic techniques available that will yield the solution. The best one can do is to use numerical techniques (see Chapters 32–35) to obtain approximate solutions.

3.2 LINEAR EQUATIONS

Consider a differential equation in standard form (*3.1*). If $f(x,y)$ can be written as $f(x,y) = -p(x)y + q(x)$ (that is, as a function of x times y, plus another function of x), the differential equation is *linear*. First-order linear differential equations can always be expressed as

$$y' + p(x)y = q(x) \tag{3.3}$$

Linear equations are solved in Chapter 8. Equation (*3.3*) is a special case of (*1.6*), page 2, with $n = 1$, $p(x) = b_0(x)/b_1(x)$, and $q(x) = g(x)/b_1(x)$.

3.3 HOMOGENEOUS EQUATIONS

A differential equation in standard form (*3.1*) is *homogeneous* if

$$f(tx, ty) = f(x,y) \tag{3.4}$$

for every real number t. Homogeneous equations are solved in Chapter 5.

Note: In the general framework of differential equations, the word "homogeneous" has an entirely different meaning (see Section 10.1). Only in the context of first-order differential equations does "homogeneous" have the meaning defined above.

3.4 SEPARABLE EQUATIONS

Consider a differential equation in differential form (*3.2*). If $M(x,y) = A(x)$ (a function only of x) and $N(x,y) = B(y)$ (a function only of y), the differential equation is *separable*, or has its *variables separated*. Separable equations are solved in Chapter 4.

3.5 EXACT EQUATIONS

A differential equation in differential form (*3.2*) is *exact* if

$$\frac{\partial M(x,y)}{\partial y} = \frac{\partial N(x,y)}{\partial x} \tag{3.5}$$

Exact equations are solved in Chapter 6 (where a more precise definition of exactness is given).

Solved Problems

3.1. Determine if the following differential equations are linear:

(*a*) $y' = (\sin x)y + e^x$ (*b*) $y' = x \sin y + e^x$ (*c*) $y' = 5$ (*d*) $y' = y^2 + x$

(*a*) The equation is linear; here $p(x) = -\sin x$ and $q(x) = e^x$.

(*b*) The equation is not linear, because of the term $\sin y$.

(*c*) The equation is linear; here $p(x) = 0$ and $q(x) = 5$.

(*d*) The equation is not linear, because of the term y^2.

3.2. Determine if the following differential equations are homogeneous:

(a) $y' = \dfrac{y+x}{x}$ (b) $y' = \dfrac{y^2}{x}$ (c) $y' = \dfrac{2xye^{x/y}}{x^2 + y^2 \sin\dfrac{x}{y}}$ (d) $y' = \dfrac{x^2+y}{x^3}$

(a) The equation is homogeneous, since

$$f(tx, ty) = \frac{ty+tx}{tx} = \frac{t(y+x)}{tx} = \frac{y+x}{x} = f(x,y)$$

(b) The equation is not homogeneous, since

$$f(tx, ty) = \frac{(ty)^2}{tx} = \frac{t^2y^2}{tx} = t\frac{y^2}{x} \neq f(x,y)$$

(c) The equation is homogeneous, since

$$f(tx, ty) = \frac{2(tx)(ty)e^{tx/ty}}{(tx)^2 + (ty)^2 \sin\dfrac{tx}{ty}} = \frac{t^2 2xy\, e^{x/y}}{t^2x^2 + t^2y^2 \sin\dfrac{x}{y}}$$

$$= \frac{2xy\, e^{x/y}}{x^2 + y^2 \sin\dfrac{x}{y}} = f(x,y)$$

(d) The equation is not homogeneous, since

$$f(tx, ty) = \frac{(tx)^2 + ty}{(tx)^3} = \frac{t^2x^2 + ty}{t^3x^3} = \frac{tx^2 + y}{t^2x^3} \neq f(x,y)$$

3.3. Determine if the following differential equations are separable:

(a) $\sin x\, dx + y^2\, dy = 0$ (c) $(1+xy)\, dx + y\, dy = 0$

(b) $xy^2\, dx - x^2y^2\, dy = 0$

(a) The differential equation is separable; here $M(x,y) = A(x) = \sin x$ and $N(x,y) = B(y) = y^2$.

(b) The equation is not separable in its present form, since $M(x,y) = xy^2$ is not a function of x alone. But if we divide both sides of the equation by x^2y^2, we obtain the equation $(1/x)\, dx + (-1)\, dy = 0$, which is separable. Here, $A(x) = 1/x$ and $B(y) = -1$.

(c) The equation is not separable, since $M(x,y) = 1 + xy$, which is not a function of x alone.

3.4. Determine whether the following differential equations are exact:

(a) $3x^2y\, dx + (y + x^3)\, dy = 0$ (b) $xy\, dx + y^2\, dy = 0$

(a) The equation is exact; here $M(x,y) = 3x^2y$, $N(x,y) = y + x^3$, and $\dfrac{\partial M}{\partial y} = \dfrac{\partial N}{\partial x} = 3x^2$.

(b) The equation is not exact. Here $M(x,y) = xy$ and $N(x,y) = y^2$; hence $\dfrac{\partial M}{\partial y} = x$, $\dfrac{\partial N}{\partial x} = 0$, and $\dfrac{\partial M}{\partial y} \neq \dfrac{\partial N}{\partial x}$.

3.5. Prove that a separable equation is always exact.

For a separable differential equation, $M(x,y) = A(x)$ and $N(x,y) = B(y)$. Thus,

$$\frac{\partial M(x,y)}{\partial y} = \frac{\partial A(x)}{\partial y} = 0 \quad \text{and} \quad \frac{\partial N(x,y)}{\partial x} = \frac{\partial B(y)}{\partial x} = 0$$

Since $\dfrac{\partial M}{\partial y} = \dfrac{\partial N}{\partial x}$, the differential equation is exact.

3.6. The initial-value problem $y' = 2\sqrt{|y|};\ y(0) = 0$ has the two solutions $y = x|x|$ and $y \equiv 0$. Does this result violate Theorem 3.1?

No. Here, $f(x, y) = 2\sqrt{|y|}$ and, therefore, $\dfrac{\partial f}{\partial y}$ does not exist at the origin.

Supplementary Problems

The following first-order differential equations are given in both standard and differential forms. Determine whether the equations in standard form are linear and/or homogeneous and whether the equations in differential form, as given, are separable and/or exact.

3.7. $y' = xy;\quad xy\,dx - dy = 0.$

3.8. $y' = xy;\quad x\,dx - \dfrac{1}{y}dy = 0.$

3.9. $y' = xy + 1;\quad (xy + 1)\,dx - dy = 0.$

3.10. $y' = \dfrac{x^2}{y^2};\quad \dfrac{x^2}{y^2}dx - dy = 0.$

3.11. $y' = \dfrac{x^2}{y^2};\quad -x^2\,dx + y^2\,dy = 0.$

3.12. $y' = -\dfrac{2y}{x};\quad 2xy\,dx + x^2\,dy = 0.$

3.13. $y' = \dfrac{xy^2}{x^2y + y^3};\quad xy^2\,dx - (x^2y + y^3)\,dy = 0.$

3.14. $y' = \dfrac{-xy^2}{x^2y + y^2};\quad xy^2\,dx + (x^2y + y^2)\,dy = 0.$

Answers to Supplementary Problems

3.7. linear

3.8. linear, separable, and exact

3.9. linear

3.10. homogeneous

3.11. homogeneous, separable, and exact

3.12. linear, homogeneous, and exact

3.13. homogeneous

3.14. exact

Chapter 4

Separable First-Order Differential Equations

4.1 GENERAL SOLUTION

Consider the first-order separable differential equation (see Section 3.4):

$$A(x)\,dx + B(y)\,dy = 0 \tag{4.1}$$

As shown in Problem 4.8, the solution to (4.1) is

$$\int A(x)\,dx + \int B(y)\,dy = c \tag{4.2}$$

where c represents an arbitrary constant.

The integrals obtained in (4.2) may be, for all practical purposes, impossible to evaluate. In such case, numerical techniques (see Chapters 32–35) may have to be used to obtain an approximate solution. Even if the indicated integrations in (4.2) can be performed, it may not be algebraically possible to solve for y explicitly in terms of x. In that case, the solution is left in implicit form. (See Problem 4.4.)

4.2 INITIAL-VALUE PROBLEM

The solution to the initial-value problem

$$A(x)\,dx + B(y)\,dy = 0; \qquad y(x_0) = y_0 \tag{4.3}$$

can be obtained, as usual, by first using (4.2) to solve the differential equation and then applying the initial condition directly to evaluate c. (See Problem 4.5.)

Alternatively, the solution to (4.3) can be obtained from

$$\int_{x_0}^{x} A(x)\,dx + \int_{y_0}^{y} B(y)\,dy = 0 \tag{4.4}$$

Equation (4.4), however, may not determine the solution of (4.3) *uniquely*; that is, (4.4) may have many solutions, of which only one will satisfy the initial-value problem. (See Problem 4.6.)

Solved Problems

4.1. Solve $x\,dx - y^2\,dy = 0$.

For this differential equation, $A(x) = x$ and $B(y) = -y^2$. Substituting these values into (4.2), we have $\int x\,dx + \int (-y^2)\,dy = c$, which, after the indicated integrations are performed, becomes $x^2/2 - y^3/3 = c$. Solving for y explicitly, we obtain the solution as

$$y = (\tfrac{3}{2}x^2 + k)^{1/3}, \qquad k = -3c$$

4.2. Solve $y' = y^2x^3$.

We first rewrite this equation in the differential form (see Section 3.1) $x^3\,dx - (1/y^2)\,dy = 0$. Then $A(x) = x^3$ and $B(y) = -1/y^2$. Substituting these values into (*4.2*), we have

$$\int x^3\,dx + \int (-1/y^2)\,dy = c$$

or, by performing the indicated integrations, $x^4/4 + 1/y = c$. Solving explicitly for y, we obtain the solution as

$$y = \frac{-4}{x^4 + k}, \qquad k = -4c$$

4.3. Solve $y' = 5y$.

First rewrite this equation in the differential form $5\,dx - (1/y)\,dy = 0$. Then $A(x) = 5$ and $B(y) = -1/y$. Substituting these results into (*4.2*), we obtain the solution implicitly as

$$\int 5\,dx + \int (-1/y)\,dy = c$$

or, by evaluating, as $5x - \ln|y| = c$.

To solve for y explicitly, we first rewrite the solution as $\ln|y| = 5x - c$ and then take the exponential of both sides. Thus, $e^{\ln|y|} = e^{5x-c}$. Noting that $e^{\ln|y|} = |y|$, we obtain $|y| = e^{5x}e^{-c}$, or $y = \pm e^{-c}e^{5x}$. The solution is given explicitly by $y = ke^{5x}$, $k = \pm e^{-c}$.

Note that the presence of the term $(-1/y)$ in the differential form of the differential equation requires the restriction $y \neq 0$ in our derivation of the solution. This restriction is equivalent to the restriction $k \neq 0$, since $y = ke^{5x}$. However, by inspection, $y \equiv 0$ is a solution of the differential equation as originally given. Thus, $y = ke^{5x}$ is the solution for all k.

The differential equation as originally given is also linear. See Problem 8.5 for an alternate method of solution.

4.4. Solve $y' = \dfrac{x+1}{y^4+1}$.

This equation, in differential form, is $(x+1)\,dx + (-y^4-1)\,dy = 0$. Thus, $A(x) = x+1$ and $B(y) = -y^4 - 1$. The solution is given by (*4.2*) implicitly as $\int (x+1)\,dx + \int (-y^4-1)\,dy = c$, or, by evaluating, as

$$\frac{x^2}{2} + x - \frac{y^5}{5} - y = c$$

Since it is impossible algebraically to solve this equation explicitly for y, the solution must be left in its present implicit form.

4.5. Solve $e^x\,dx - y\,dy = 0$; $y(0) = 1$.

The solution to the differential equation is given by (*4.2*) as $\int e^x\,dx + \int (-y)\,dy = c$, or, by evaluating, as $y^2 = 2e^x + k$, $k = -2c$. Applying the initial condition, we obtain $(1)^2 = 2e^0 + k$, $1 = 2 + k$, or $k = -1$. Thus, the solution to the initial-value problem is

$$y^2 = 2e^x - 1 \qquad \text{or} \qquad y = \sqrt{2e^x - 1}$$

(Note that we cannot choose the negative square root, since then $y(0) = -1$, which violates the initial condition.)

To insure that y remains real, we must restrict x so that $2e^x - 1 \geqq 0$. To guarantee that y' exists (note that $y'(x) = dy/dx = e^x/y$), we must restrict x so that $2e^x - 1 \neq 0$. Together these conditions imply that $2e^x - 1 > 0$, or $x > \ln\frac{1}{2}$.

4.6. Use (*4.4*) to solve Problem 4.5.

For this problem, $x_0 = 0$, $y_0 = 1$, $A(x) = e^x$, and $B(y) = -y$. Substituting these values into (*4.4*), we obtain

$$\int_0^x e^x\,dx + \int_1^y (-y)\,dy = 0$$

Evaluating these integrals, we have

$$e^x\Big|_0^x + \left(\frac{-y^2}{2}\right)\Big|_1^y = 0 \quad \text{or} \quad e^x - e^0 + \left(\frac{-y^2}{2}\right) - (-\tfrac{1}{2}) = 0$$

Thus, $y^2 = 2e^x - 1$, and, as in Problem 4.5, $y = \sqrt{2e^x - 1}$, $x > \ln \frac{1}{2}$.

4.7. Solve $x \cos x \, dx + (1 - 6y^5)\, dy = 0$; $y(\pi) = 0$.

Here, $x_0 = \pi$, $y_0 = 0$, $A(x) = x \cos x$, and $B(y) = 1 - 6y^5$. Substituting these values into (4.4), we obtain

$$\int_\pi^x x \cos x \, dx + \int_0^y (1 - 6y^5)\, dy = 0$$

Evaluating these integrals (the first one by integration by parts), we find

$$x \sin x\Big|_\pi^x + \cos x\Big|_\pi^x + (y - y^6)\Big|_0^y = 0$$

or

$$x \sin x + \cos x + 1 = y^6 - y$$

Since we cannot solve this last equation for y explicitly, we must be content with the solution in its present implicit form.

4.8. Prove that every solution of (4.1) satisfies (4.2).

Rewrite (4.1) as $A(x) + B(y)y' = 0$. If $y(x)$ is a solution, it must satisfy this equation identically in x; hence,

$$A(x) + B[y(x)]\, y'(x) = 0$$

Integrating both sides of this last equation with respect to x, we obtain

$$\int A(x)\, dx + \int B[y(x)]\, y'(x)\, dx = c$$

In the second integral, make the change of variables $y = y(x)$, hence $dy = y'(x)\, dx$. The result of this substitution is (4.2).

4.9. Prove that every solution of system (4.3) is a solution of (4.4).

Following the same reasoning as in Problem 4.8, except now integrating from $x = x_0$ to $x = x$, we obtain

$$\int_{x_0}^x A(x)\, dx + \int_{x_0}^x B[y(x)]\, y'(x)\, dx = 0$$

The substitution $y = y(x)$ again gives the desired result. Note that as x varies from x_0 to x, y will vary from $y(x_0) = y_0$ to $y(x) = y$.

Supplementary Problems

4.10. Solve $x\, dx + y\, dy = 0$.

4.11. Solve $\frac{1}{x}\, dx - \frac{1}{y}\, dy = 0$.

4.12. Solve $\frac{1}{x} dx + dy = 0$.

4.13. Solve $x\,dx + \frac{1}{y}dy = 0$.

4.14. Solve $(x^2+1)\,dx + (y^2+y)\,dy = 0$.

4.15. Solve $\sin x\,dx + y\,dy = 0;\ y(0) = -2$.

4.16. Solve $(x^2+1)\,dx + \frac{1}{y}dy = 0;\ y(-1) = 1$.

4.17. Solve $xe^{x^2}\,dx + (y^5-1)\,dy = 0;\ y(0) = 0$.

4.18. Rewrite the following differential equations in differential forms which are separable and then solve:

(a) $y' = \frac{y}{x^2}$ (b) $y' = \frac{xe^x}{2y}$ (c) $y' = \frac{x^2y - y}{y+1};\ y(3) = -1$

Answers to Supplementary Problems

4.10. $y = \pm\sqrt{k - x^2},\ k = 2c$

4.11. $y = kx,\ k = \pm e^{-c}$

4.12. $y = \ln\left|\frac{k}{x}\right|,\ c = \ln|k|$

4.13. $y = ke^{-x^2/2},\ k = \pm e^c$

4.14. $2x^3 + 6x + 2y^3 + 3y^2 = k,\ k = 6c$

4.15. $y = -\sqrt{2 + 2\cos x}$

4.16. $y = e^{-\frac{1}{3}(x^3+3x+4)}$

4.17. $\frac{1}{2}e^{x^2} + \frac{1}{6}y^6 - y = \frac{1}{2}$

4.18. (a) $\frac{1}{x^2}dx - \frac{1}{y}dy = 0;\ y = ke^{-1/x},\ k = \pm e^{-c}$

(b) $xe^x\,dx - 2y\,dy = 0;\ y = \pm\sqrt{xe^x - e^x - c}$

(c) $(x^2-1)\,dx - \left(1 + \frac{1}{y}\right)dy = 0;\ \frac{x^3}{3} - x - y - \ln|y| = 7$

Chapter 5

Homogeneous First-Order Differential Equations

5.1 FIRST METHOD OF SOLUTION

In the homogeneous differential equation

$$\frac{dy}{dx} = f(x, y) \tag{5.1}$$

where from Section 3.3 $f(tx, ty) = f(x, y)$, substitute

$$y = xv \tag{5.2}$$

and the corresponding derivative

$$\frac{dy}{dx} = v + x\frac{dv}{dx} \tag{5.3}$$

After simplifying, the resulting differential equation will be one with variables (v and x) separable, which can be solved by the methods given in Chapter 4. The required solution to (*5.1*) then is obtained by back-substitution (see the Solved Problems).

5.2 ALTERNATE METHOD OF SOLUTION

Rewrite the differential equation as

$$\frac{dx}{dy} = \frac{1}{f(x, y)} \tag{5.4}$$

and then substitute

$$x = yu \tag{5.5}$$

and the corresponding derivative

$$\frac{dx}{dy} = u + y\frac{du}{dy} \tag{5.6}$$

into (*5.4*). After simplifying, the resulting differential equation will be one with variables (this time, u and y) separable, which can be solved by the methods given in Chapter 4. The required solution to (*5.4*) then is obtained by back-substitution.

Since either method of solution involves solving an associated separable differential equation, the discussion in Section 4.1 remains relevant here. Ordinarily, it is immaterial which method of solution is used (see Problems 5.2 and 5.3). Occasionally, however, one of the substitutions (*5.2*) or (*5.5*) is definitely superior to the other one. In such cases, the better substitution is usually apparent from the form of the differential equation itself. (See Problem 5.7.) Problem 5.8 provides insight into both methods. For an alternate definition of homogeneous equations see Problems 5.9 and 5.10.

Solved Problems

5.1. Solve $y' = \dfrac{y+x}{x}$.

Substituting (*5.2*) and (*5.3*) into the differential equation, we obtain

$$v + x\frac{dv}{dx} = \frac{xv + x}{x}$$

which can be algebraically simplified to

$$x\frac{dv}{dx} = 1 \quad \text{or} \quad \frac{1}{x}dx - dv = 0$$

This last equation is separable; its solution is (Section 4.1) $v = \ln|x| - c$, or

$$v = \ln|kx| \tag{1}$$

where we have set $c = -\ln|k|$ and have noted that $\ln|x| + \ln|k| = \ln|kx|$. Finally, substituting $v = y/x$, see (*5.2*), into (*1*), we obtain the solution to the given differential equation as $y = x\ln|kx|$.

5.2. Solve $y' = \dfrac{2y^4 + x^4}{xy^3}$.

Substituting (*5.2*) and (*5.3*) into the differential equation, we obtain

$$v + x\frac{dv}{dx} = \frac{2(xv)^4 + x^4}{x(xv)^3}$$

which can be algebraically simplified to

$$x\frac{dv}{dx} = \frac{v^4+1}{v^3} \quad \text{or} \quad \frac{1}{x}dx - \frac{v^3}{v^4+1}dv = 0$$

This last equation is separable; its solution is (Section 4.1) $\ln|x| - \frac{1}{4}\ln(v^4+1) = c$, or

$$v^4 + 1 = (kx)^4 \tag{1}$$

where we have set $c = -\ln|k|$ and then used the identities

$$\ln|x| + \ln|k| = \ln|kx| \quad \text{and} \quad 4\ln|kx| = \ln(kx)^4$$

Finally, substituting $v = y/x$, see (*5.2*), into (*1*), we obtain

$$y^4 = c_1x^8 - x^4 \qquad (c_1 = k^4) \tag{2}$$

5.3. Solve the differential equation of Problem 5.2 by using (*5.4*)–(*5.6*).

We first rewrite the differential equation as

$$\frac{dx}{dy} = \frac{xy^3}{2y^4 + x^4}$$

Then substituting (*5.5*) and (*5.6*) into this new differential equation, we obtain

$$u + y\frac{du}{dy} = \frac{(yu)y^3}{2y^4 + (yu)^4}$$

which can be algebraically simplified to

$$y\frac{du}{dy} = -\frac{u+u^5}{2+u^4} \quad \text{or} \quad \frac{1}{y}dy + \frac{2+u^4}{u+u^5}du = 0$$

This last equation is separable; using partial fractions

$$\frac{2+u^4}{u+u^5} = \frac{2+u^4}{u(1+u^4)} = \frac{2}{u} - \frac{u^3}{1+u^4}$$

we obtain $\ln|y| + 2\ln|u| - \frac{1}{4}\ln(1+u^4) = c$, which can be rewritten as

$$ky^4u^8 = 1 + u^4 \tag{1}$$

where $c = -\frac{1}{4}\ln|k|$. Substituting $u = x/y$, see (5.6), into (1), we once again have (2) of Problem 5.2.

5.4. Solve $y' = \dfrac{2xy}{x^2 - y^2}$.

Substituting (5.2) and (5.3) into the differential equation, we get

$$v + x\frac{dv}{dx} = \frac{2x(xv)}{x^2-(xv)^2}$$

which can be algebraically simplified to

$$x\frac{dv}{dx} = -\frac{v(v^2+1)}{v^2-1}$$

or

$$\frac{1}{x}dx + \frac{v^2-1}{v(v^2+1)}dv = 0 \tag{1}$$

Using partial fractions, we can expand (1) to

$$\frac{1}{x}dx + \left(-\frac{1}{v} + \frac{2v}{v^2+1}\right)dv = 0$$

The solution to this separable equation is $\ln|x| - \ln|v| + \ln(v^2+1) = c$, which can be simplified to

$$x(v^2+1) = kv \qquad (c = \ln|k|) \tag{2}$$

Substituting $v = y/x$ into (2), we find the solution of the given differential equation is $x^2 + y^2 = ky$.

5.5. Solve $y' = \dfrac{x^2+y^2}{xy}$.

Substituting (5.2) and (5.3) into the differential equation, we obtain

$$v + x\frac{dv}{dx} = \frac{x^2+(xv)^2}{x(xv)}$$

which can be algebraically simplified to

$$x\frac{dv}{dx} = \frac{1}{v} \quad \text{or} \quad \frac{1}{x}dx - v\,dv = 0$$

The solution to this separable equation is $\ln|x| - v^2/2 = c$, or equivalently

$$v^2 = \ln x^2 + k \qquad (k = -2c) \tag{1}$$

Substituting $v = y/x$ into (1), we find that the solution to the given differential equation is

$$y^2 = x^2\ln x^2 + kx^2$$

5.6. Solve $y' = \dfrac{x^2+y^2}{xy}$; $y(1) = -2$.

The solution to the differential equation is given in Problem 5.5 as $y^2 = x^2\ln x^2 + kx^2$. Applying the initial condition, we obtain $(-2)^2 = (1)^2\ln(1)^2 + k(1)^2$, or $k = 4$. (Recall that $\ln 1 = 0$.) Thus, the solution to the initial-value problem is

$$y^2 = x^2 \ln x^2 + 4x^2 \qquad \text{or} \qquad y = -\sqrt{x^2 \ln x^2 + 4x^2}$$

The negative square root is taken, consistent with the initial condition.

5.7. Solve $y' = \dfrac{2xy\, e^{(x/y)^2}}{y^2 + y^2 e^{(x/y)^2} + 2x^2 e^{(x/y)^2}}$.

Noting the (x/y)-term in the exponential, we try the substitution $u = x/y$, which is an equivalent form of (5.5). Rewriting the differential equation as

$$\frac{dx}{dy} = \frac{y^2 + y^2 e^{(x/y)^2} + 2x^2 e^{(x/y)^2}}{2xy\, e^{(x/y)^2}}$$

we have upon using substitutions (5.5) and (5.6) and simplifying,

$$y\frac{du}{dy} = \frac{1+e^{u^2}}{2ue^{u^2}} \qquad \text{or} \qquad \frac{1}{y}\,dy - \frac{2ue^{u^2}}{1+e^{u^2}}\,du = 0$$

This equation is separable; its solution is

$$\ln|y| - \ln(1+e^{u^2}) = c$$

which can be rewritten as

$$y = k(1+e^{u^2}) \qquad (c = \ln|k|) \tag{1}$$

Substituting $u = x/y$ into (1), we obtain the solution of the given differential equation as

$$y = k[1 + e^{(x/y)^2}]$$

5.8. Prove that if $y' = f(x,y)$ is homogeneous, then the differential equation can be rewritten as $y' = g(y/x)$, where $g(y/x)$ depends only on the quotient y/x.

By (3.4), page 11, we have that $f(x,y) = f(tx,ty)$. Since this equation is valid for all t, it must be true, in particular, for $t = 1/x$. Thus, $f(x,y) = f(1, y/x)$. If we now define $g(y/x) = f(1,y/x)$, we then have $y' = f(x,y) = f(1,y/x) = g(y/x)$ as required.

Note that this form suggests the substitution $v = y/x$ which is equivalent to (5.2). If, in the above, we had set $t = 1/y$, then $f(x,y) = f(x/y, 1) = h(x/y)$, which suggests the alternate substitution (5.5).

5.9. A function $g(x,y)$ is *homogeneous of degree n* if $g(tx,ty) = t^n g(x,y)$ for all t. Determine whether the following functions are homogeneous, and, if so, find their degree: (a) $xy+y^2$, (b) $x + y\sin(y/x)^2$, (c) $x^3 + xy^2e^{x/y}$, and (d) $x + xy$.

(a) $(tx)(ty) + (ty)^2 = t^2(xy+y^2)$; homogeneous of degree two.

(b) $tx + ty\sin\left(\dfrac{ty}{tx}\right)^2 = t\left[x + y\sin\left(\dfrac{y}{x}\right)^2\right]$; homogeneous of degree one.

(c) $(tx)^3 + (tx)(ty)^2 e^{tx/ty} = t^3(x^3 + xy^2 e^{x/y})$; homogeneous of degree three.

(d) $tx + (tx)(ty) = tx + t^2xy$; not homogeneous.

5.10. An alternate definition of a homogeneous differential equation is as follows: A differential equation $M(x,y)\,dx + N(x,y)\,dy = 0$ is *homogeneous* if both $M(x,y)$ and $N(x,y)$ are homogeneous of the same degree (see Problem 5.9). Show that this definition implies the definition given in Section 3.3.

If $M(x,y)$ and $N(x,y)$ are homogeneous of degree n, then

$$f(tx,ty) = \frac{M(tx,ty)}{-N(tx,ty)} = \frac{t^nM(x,y)}{-t^nN(x,y)} = \frac{M(x,y)}{-N(x,y)} = f(x,y)$$

Supplementary Problems

In the following problems, determine whether the given differential equations are homogeneous and, if so, solve them.

5.11. $y' = \dfrac{y - x}{x}$.

5.12. $y' = \dfrac{2y + x}{x}$.

5.13. $y' = \dfrac{x^2 + 2y^2}{xy}$.

5.14. $y' = \dfrac{2x + y^2}{xy}$.

5.15. $y' = \dfrac{x^2 + y^2}{2xy}$.

5.16. $y' = \dfrac{2xy}{y^2 - x^2}$.

5.17. $y' = \dfrac{y}{x + \sqrt{xy}}$.

5.18. $y' = \dfrac{y^2}{xy + (xy^2)^{1/3}}$.

5.19. $y' = \dfrac{x^4 + 3x^2y^2 + y^4}{x^3y}$.

Answers to Supplementary Problems

5.11. $y = x \ln |k/x|$

5.12. $y = kx^2 - x$

5.13. $y^2 = kx^4 - x^2$

5.14. not homogeneous

5.15. $y^2 = x^2 - kx$

5.16. $3yx^2 - y^3 = k$

5.17. $-2\sqrt{x/y} + \ln |y| = c$

5.18. not homogeneous

5.19. $y^2 = -x^2\left(1 + \dfrac{1}{\ln |kx^2|}\right)$

Chapter 6

Exact First-Order Differential Equations

6.1 DEFINITION

A differential equation

$$M(x,y)\,dx + N(x,y)\,dy = 0 \tag{6.1}$$

is *exact* if there exists a function $g(x,y)$ such that

$$dg(x,y) = M(x,y)\,dx + N(x,y)\,dy \tag{6.2}$$

Test for exactness: If $M(x,y)$ and $N(x,y)$ are continuous functions and have continuous first partial derivatives on some rectangle of the xy-plane, then (*6.1*) is exact if and only if

$$\frac{\partial M(x,y)}{\partial y} = \frac{\partial N(x,y)}{\partial x} \tag{6.3}$$

Example 6.1. For the differential equation $2xy\,dx + (1+x^2)\,dy = 0$, we have $M(x,y) = 2xy$ and $N(x,y) = 1+x^2$. Since $\dfrac{\partial M}{\partial y} = \dfrac{\partial N}{\partial x} = 2x$, the differential equation is exact.

6.2 METHOD OF SOLUTION

To solve (*6.1*), assuming that it is exact, first solve the equations

$$\frac{\partial g(x,y)}{\partial x} = M(x,y) \tag{6.4}$$

$$\frac{\partial g(x,y)}{\partial y} = N(x,y) \tag{6.5}$$

for $g(x,y)$ (see the Solved Problems). The solution to (*6.1*) is then given implicitly by

$$g(x,y) = c \tag{6.6}$$

where c represents an arbitrary constant.

Equation (*6.6*) is immediate from (*6.1*) and (*6.2*). In fact, if (*6.2*) is substituted into (*6.1*), we obtain $dg(x,y(x)) = 0$. Integrating this equation (note that we can write 0 as $0\,dx$), we have $\int dg(x,y(x)) = \int 0\,dx$, which, in turn, implies (*6.6*).

Solved Problems

6.1. Solve $2xy\,dx + (1+x^2)\,dy = 0$.

This equation is exact (see Example 6.1). We now determine a function $g(x,y)$ that satisfies *(6.4)* and *(6.5)*. Substituting $M(x,y) = 2xy$ into *(6.4)*, we obtain $\dfrac{\partial g}{\partial x} = 2xy$. Integrating both sides of this equation with respect to x, we find

$$\int \frac{\partial g}{\partial x}\,dx = \int 2xy\,dx$$

or

$$g(x,y) = x^2y + h(y) \tag{1}$$

Note that when integrating with respect to x, the constant (*with respect to* x) of integration can depend on y.

We now determine $h(y)$. Differentiating *(1)* with respect to y, we obtain $\dfrac{\partial g}{\partial y} = x^2 + h'(y)$. Substituting this equation along with $N(x,y) = 1 + x^2$ into *(6.5)*, we have

$$x^2 + h'(y) = 1 + x^2 \qquad \text{or} \qquad h'(y) = 1$$

Integrating this last equation with respect to y, we obtain $h(y) = y + c_1$ (c_1 = constant). Substituting this expression into *(1)* yields

$$g(x,y) = x^2y + y + c_1$$

The solution to the differential equation, which is given implicitly by *(6.6)* as $g(x,y) = c$, is

$$x^2y + y = c_2 \qquad (c_2 = c - c_1)$$

Solving for y explicitly, we obtain the solution as $y = \dfrac{c_2}{x^2+1}$.

6.2. Solve $(x + \sin y)\,dx + (x\cos y - 2y)\,dy = 0$.

Here $M(x,y) = x + \sin y$ and $N(x,y) = x\cos y - 2y$. Thus, $\dfrac{\partial M}{\partial y} = \dfrac{\partial N}{\partial x} = \cos y$, and the differential equation is exact. We now seek a function $g(x,y)$ that satisfies *(6.4)* and *(6.5)*. Substituting $M(x,y)$ into *(6.4)*, we obtain $\dfrac{\partial g}{\partial x} = x + \sin y$. Integrating both sides of this equation with respect to x, we find

$$\int \frac{\partial g}{\partial x}\,dx = \int (x + \sin y)\,dx$$

or

$$g(x,y) = \tfrac{1}{2}x^2 + x\sin y + h(y) \tag{1}$$

To find $h(y)$, we differentiate *(1)* with respect to y, yielding $\dfrac{\partial g}{\partial y} = x\cos y + h'(y)$, and then substitute this result along with $N(x,y) = x\cos y - 2y$ into *(6.5)*. Thus we find

$$x\cos y + h'(y) = x\cos y - 2y \qquad \text{or} \qquad h'(y) = -2y$$

from which it follows that $h(y) = -y^2 + c_1$. Substituting this $h(y)$ into *(1)*, we obtain

$$g(x,y) = \tfrac{1}{2}x^2 + x\sin y - y^2 + c_1$$

The solution of the differential equation is given implicitly by *(6.6)* as

$$\tfrac{1}{2}x^2 + x\sin y - y^2 = c_2 \qquad (c_2 = c - c_1)$$

6.3. Solve $(xy + x^2)\,dx + (-1)\,dy = 0$.

Here, $M(x,y) = xy + x^2$ and $N(x,y) = -1$; hence $\dfrac{\partial M}{\partial y} = x$ and $\dfrac{\partial N}{\partial x} = 0$. Since $\dfrac{\partial M}{\partial y} \neq \dfrac{\partial N}{\partial x}$, the equation is *not* exact and the method given in this chapter is not applicable. (However, the equation can be rewritten as a linear equation, to which the method of Chapter 8 applies.)

6.4. Solve $y' = \dfrac{2 + ye^{xy}}{2y - xe^{xy}}$.

Rewriting this equation in differential form, we obtain

$$(2 + ye^{xy})\,dx + (xe^{xy} - 2y)\,dy = 0$$

Here, $M(x,y) = 2 + ye^{xy}$ and $N(x,y) = xe^{xy} - 2y$ and, since $\dfrac{\partial M}{\partial y} = \dfrac{\partial N}{\partial x} = e^{xy} + xye^{xy}$, the differential equation is exact. Substituting $M(x,y)$ into *(6.4)*, we find $\dfrac{\partial g}{\partial x} = 2 + ye^{xy}$; then integrating with respect to x, we obtain

$$\int \frac{\partial g}{\partial x}\,dx = \int [2 + ye^{xy}]\,dx$$

or

$$g(x,y) = 2x + e^{xy} + h(y) \tag{1}$$

To find $h(y)$, first differentiate *(1)* with respect to y, obtaining $\dfrac{\partial g}{\partial y} = xe^{xy} + h'(y)$; then substitute this result along with $N(x,y)$ into *(6.5)* to obtain

$$xe^{xy} + h'(y) = xe^{xy} - 2y \qquad \text{or} \qquad h'(y) = -2y$$

It follows that $h(y) = -y^2 + c_1$. Substituting this $h(y)$ into *(1)*, we obtain

$$g(x,y) = 2x + e^{xy} - y^2 + c_1$$

The solution to the differential equation is given implicitly by *(6.6)* as

$$2x + e^{xy} - y^2 = c_2 \qquad (c_2 = c - c_1)$$

6.5. Solve $y' = \dfrac{-2xy}{1 + x^2}$; $y(2) = -5$.

The solution to the differential equation (written in differential form) is given in Problem 6.1 as $x^2y + y = c_2$. Using the initial condition, we obtain $(2)^2(-5) + (-5) = c_2$, or $c_2 = -25$. The solution to the initial-value problem is therefore $x^2y + y = -25$ or $y = -25/(x^2 + 1)$.

Supplementary Problems

Test the following differential equations for exactness and solve all exact equations.

6.6. $(2xy + x)\,dx + (x^2 + y)\,dy = 0$.

6.7. $(y + 2xy^3)\,dx + (1 + 3x^2y^2 + x)\,dy = 0$.

6.8. $ye^{xy}\,dx + xe^{xy}\,dy = 0$.

6.9. $xe^{xy}\,dx + ye^{xy}\,dy = 0$.

6.10. $3x^2y^2\,dx + (2x^3y + 4y^3)\,dy = 0$.

6.11. $y\,dx + x\,dy = 0$.

6.12. $(x - y)\,dx + (x + y)\,dy = 0$.

6.13. $(y \sin x + xy \cos x)\,dx + (x \sin x + 1)\,dy = 0$.

Answers to Supplementary Problems

6.6. $yx^2 + \frac{1}{2}x^2 + \frac{1}{2}y^2 = c_2$

6.7. $xy + x^2y^3 + y = c_2$

6.8. $e^{xy} = c_2$

6.9. not exact

6.10. $x^3y^2 + y^4 = c_2$

6.11. $xy = c_2$

6.12. not exact

6.13. $xy \sin x + y = c_2$

Chapter 7

Integrating Factors

7.1 WHAT IS AN INTEGRATING FACTOR?

In general, the differential equation

$$M(x,y)\,dx + N(x,y)\,dy = 0 \tag{7.1}$$

is not exact. Occasionally, however, it is possible to transform (*7.1*) into an exact differential equation by a judicious multiplication.

Definition: A function $I(x,y)$ is an *integrating factor* for (*7.1*) if the equation

$$I(x,y)[M(x,y)\,dx + N(x,y)\,dy] = 0 \tag{7.2}$$

is exact.

Example 7.1. Determine whether $-1/x^2$ is an integrating factor for $y\,dx - x\,dy = 0$.

Multiplying the given differential equation by $-1/x^2$, we obtain

$$\frac{-1}{x^2}(y\,dx - x\,dy) = 0 \qquad \text{or} \qquad \frac{-y}{x^2}\,dx + \frac{1}{x}\,dy = 0$$

This last equation is exact; hence $-1/x^2$ is an integrating factor.

Example 7.2. Determine whether $-1/xy$ is an integrating factor for $y\,dx - x\,dy = 0$.

Multiplying the given differential equation by $-1/xy$, we obtain

$$\frac{-1}{xy}(y\,dx - x\,dy) = 0 \qquad \text{or} \qquad -\frac{1}{x}\,dx + \frac{1}{y}\,dy = 0$$

This last equation is exact; hence $-1/xy$ is an integrating factor.

Comparing with Example 7.1, we see that a differential equation can have more than one integrating factor.

7.2 SOLUTION BY USE OF AN INTEGRATING FACTOR

If $I(x,y)$ is an integrating factor for (*7.1*), then (*7.2*) is exact and can be solved either by the method of Section 6.2 or, often, by direct integration. The solution of (*7.2*) is also the solution of (*7.1*).

7.3 FINDING AN INTEGRATING FACTOR

It follows from the test for exactness given in Section 6.1 that an integrating factor is a solution to a certain partial differential equation. That equation, however, is usually more difficult to solve than the original differential equation under consideration. Consequently, integrating factors are generally obtained by inspection. The whole success of the method thus depends on the user's ability to recognize or to guess that a particular group of terms composes an exact differential $dh(x, y)$. In this connection, Table 7-1 may prove helpful.

Table 7-1

Group of Terms	Integrating Factor $I(x, y)$	Exact Differential $dh(x, y)$
$y\,dx - x\,dy$	$-\dfrac{1}{x^2}$	$\dfrac{x\,dy - y\,dx}{x^2} = d\left(\dfrac{y}{x}\right)$
$y\,dx - x\,dy$	$\dfrac{1}{y^2}$	$\dfrac{y\,dx - x\,dy}{y^2} = d\left(\dfrac{x}{y}\right)$
$y\,dx - x\,dy$	$-\dfrac{1}{xy}$	$\dfrac{x\,dy - y\,dx}{xy} = d\left(\ln \dfrac{y}{x}\right)$
$y\,dx - x\,dy$	$-\dfrac{1}{x^2 + y^2}$	$\dfrac{x\,dy - y\,dx}{x^2 + y^2} = d\left(\arctan \dfrac{y}{x}\right)$
$y\,dx + x\,dy$	$\dfrac{1}{xy}$	$\dfrac{y\,dx + x\,dy}{xy} = d(\ln xy)$
$y\,dx + x\,dy$	$\dfrac{1}{(xy)^n}, \quad n > 1$	$\dfrac{y\,dx + x\,dy}{(xy)^n} = d\left[\dfrac{-1}{(n-1)(xy)^{n-1}}\right]$
$y\,dy + x\,dx$	$\dfrac{1}{x^2 + y^2}$	$\dfrac{y\,dy + x\,dx}{x^2 + y^2} = d[\frac{1}{2} \ln (x^2 + y^2)]$
$y\,dy + x\,dx$	$\dfrac{1}{(x^2 + y^2)^n}, \quad n > 1$	$\dfrac{y\,dy + x\,dx}{(x^2 + y^2)^n} = d\left[\dfrac{-1}{2(n-1)(x^2 + y^2)^{n-1}}\right]$
$ay\,dx + bx\,dy$ (a, b constants)	$x^{a-1}y^{b-1}$	$x^{a-1}y^{b-1}(ay\,dx + bx\,dy) = d(x^a y^b)$

Sometimes, an integrating factor becomes apparent if the terms of the differential equation are strategically regrouped. (See Problems 7.3–7.5.)

Integrating factors are known if $M(x, y)$ and $N(x, y)$ in (*7.1*) obey certain conditions:

(a) If $\dfrac{1}{N}\left(\dfrac{\partial M}{\partial y} - \dfrac{\partial N}{\partial x}\right) \equiv g(x)$, a function of x alone, then

$$I(x, y) = e^{\int g(x)\,dx} \tag{7.3}$$

(See Problem 7.6.)

(b) If $\dfrac{1}{M}\left(\dfrac{\partial M}{\partial y} - \dfrac{\partial N}{\partial x}\right) \equiv h(y)$, a function of y alone, then

$$I(x, y) = e^{-\int h(y)\,dy} \tag{7.4}$$

(See Problem 7.8.)

(c) If $M = yf(xy)$ and $N = xg(xy)$, then

$$I(x, y) = \frac{1}{xM - yN} \tag{7.5}$$

(See Problem 7.7.)

Solved Problems

7.1. Solve $y\,dx - x\,dy = 0$.

Using the integrating factor $I(x,y) = -1/x^2$ (see Example 7.1 or Table 7-1), we can rewrite the differential equation as

$$\frac{x\,dy - y\,dx}{x^2} = 0 \tag{1}$$

Since (1) is exact, it can be solved by the method of Section 6.2. Alternatively, we note from Table 7-1 that (1) can be rewritten as $d(y/x) = 0$. Hence, by direct integration, we have $y/x = c$, or $y = cx$, as the solution.

7.2. Solve the differential equation of Problem 7.1 by using a different integrating factor.

Using the integrating factor $I(x,y) = -1/xy$ (see Example 7.2 or Table 7-1), we can rewrite the differential equation as

$$\frac{x\,dy - y\,dx}{xy} = 0 \tag{1}$$

Since (1) is exact, it can be solved by the method of Section 6.2. Alternatively, we note from Table 7-1 that (1) can be rewritten as $d[\ln(y/x)] = 0$. Then, by direct integration, $\ln(y/x) = c_1$. Taking the exponential of both sides, we find $\frac{y}{x} = e^{c_1}$, or finally,

$$y = cx \qquad (c = e^{c_1})$$

7.3. Solve $(y^2 - y)\,dx + x\,dy = 0$.

No integrating factor is immediately apparent. Note, however, that if terms are strategically regrouped, the differential equation can be rewritten as

$$-(y\,dx - x\,dy) + y^2\,dx = 0 \tag{1}$$

The group of terms in parentheses has many integrating factors (see Table 7-1). Trying each integrating factor separately, we find that the only one that makes the entire equation exact is $I(x,y) = 1/y^2$. Using this integrating factor, we can rewrite (1) as

$$-\frac{y\,dx - x\,dy}{y^2} + 1\,dx = 0 \tag{2}$$

Since (2) is exact, it can be solved by the method of Section 6.2. Alternatively, we note from Table 7-1 that (2) can be rewritten as $-d(x/y) + 1\,dx = 0$, or as $d(x/y) = 1\,dx$. Integrating, we obtain the solution

$$\frac{x}{y} = x + c \qquad \text{or} \qquad y = \frac{x}{x+c}$$

7.4. Solve $(y - xy^2)\,dx + (x + x^2y^2)\,dy = 0$.

No integrating factor is immediately apparent. Note, however, that the differential equation can be rewritten as

$$(y\,dx + x\,dy) + (-xy^2\,dx + x^2y^2\,dy) = 0 \tag{1}$$

The first group of terms has many integrating factors (see Table 7-1). One of these factors, namely $I(x,y) = 1/(xy)^2$, is an integrating factor for the entire equation. Multiplying (1) by $1/(xy)^2$, we find

$$\frac{y\,dx + x\,dy}{(xy)^2} + \frac{-xy^2\,dx + x^2y^2\,dy}{(xy)^2} = 0$$

or equivalently,

$$\frac{y\,dx + x\,dy}{(xy)^2} = \frac{1}{x}\,dx - 1\,dy \tag{2}$$

From Table 7-1,

$$\frac{y\,dx + x\,dy}{(xy)^2} = d\left(\frac{-1}{xy}\right)$$

so that (*2*) can be rewritten as

$$d\left(\frac{-1}{xy}\right) = \frac{1}{x}\,dx - 1\,dy$$

Integrating both sides of this last equation, we find

$$\frac{-1}{xy} = \ln|x| - y + c$$

which is the solution in implicit form.

7.5. Solve $y' = \dfrac{3yx^2}{x^3 + 2y^4}$.

Rewriting the equation in differential form, we have

$$(3yx^2)\,dx + (-x^3 - 2y^4)\,dy = 0$$

No integrating factor is immediately apparent. We can, however, rearrange this equation as

$$x^2(3y\,dx - x\,dy) - 2y^4\,dy = 0 \tag{1}$$

The group in parentheses is of the form $ay\,dx + bx\,dy$, where $a = 3$ and $b = -1$, which has an integrating factor x^2y^{-2}. Since the expression in parentheses is already multiplied by x^2, we try an integrating factor of the form $I(x,y) = y^{-2}$. Multiplying (*1*) by y^{-2}, we have

$$x^2y^{-2}(3y\,dx - x\,dy) - 2y^2\,dy = 0$$

which can be simplified (see Table 7-1) to

$$d(x^3y^{-1}) = 2y^2\,dy \tag{2}$$

Integrating both sides of (*2*), we obtain

$$x^3y^{-1} = \frac{2}{3}y^3 + c$$

as the solution in implicit form.

7.6. Solve $y' = 2xy - x$.

Rewriting this equation in differential form, we have

$$(-2xy + x)\,dx + dy = 0 \tag{1}$$

No integrating factor is immediately apparent. Note, however, that for this equation $M(x,y) = -2xy + x$ and $N(x,y) = 1$; so that

$$\frac{1}{N}\left(\frac{\partial M}{\partial y} - \frac{\partial N}{\partial x}\right) = \frac{(-2x) - (0)}{1} = -2x$$

is a function of x alone. By (*7.3*), therefore, we have $I(x,y) = e^{\int -2x\,dx} = e^{-x^2}$ as an integrating factor. Multiplying (*1*) by e^{-x^2}, we obtain

$$(-2xye^{-x^2} + xe^{-x^2})\,dx + e^{-x^2}\,dy = 0 \tag{2}$$

which is exact. Solving (*2*) by the method of Section 6.2, we obtain the solution as

$$y = ce^{x^2} + \frac{1}{2}$$

Note that the given differential equation is also linear. (In particular, see Problem 8.2.)

7.7. Solve $y' = \dfrac{xy^2 - y}{x}$.

Rewriting this equation in differential form, we have

$$y(1-xy)\,dx + x(1)\,dy = 0 \tag{1}$$

From (7.5), we choose

$$I(x,y) = \frac{1}{x[y(1-xy)] - yx} = \frac{-1}{(xy)^2}$$

Multiplying (1) by $I(x,y)$, we obtain

$$\frac{xy-1}{x^2y}\,dx - \frac{1}{xy^2}\,dy = 0$$

which is exact. Using the method of Section 6.2, we find that the solution is $y = -1/(x \ln|cx|)$.

Notice that we could have rewritten (1) as

$$(y\,dx + x\,dy) - xy^2\,dx = 0$$

which also suggests (see Table 7-1) the integrating factor $I(x,y) = 1/(xy)^2$.

7.8. Solve $y^2\,dx + xy\,dy = 0$.

Here $M(x,y) = y^2$ and $N(x,y) = xy$; hence,

$$\frac{1}{M}\left(\frac{\partial M}{\partial y} - \frac{\partial N}{\partial x}\right) = \frac{2y-y}{y^2} = \frac{1}{y}$$

a function of y alone. From (7.4), $I(x,y) = e^{-\int (1/y)\,dy} = e^{-\ln y} = 1/y$. Multiplying the given differential equation by $I(x,y) = 1/y$, we obtain the exact equation $y\,dx + x\,dy = 0$, which has the solution $y = c/x$.

An alternate method would be first to divide the given differential equation by xy^2 and then to note that the resulting equation is separable.

Supplementary Problems

In Problems 7.9–7.23, find an appropriate integrating factor for each differential equation and solve.

7.9. $(y+1)\,dx - x\,dy = 0$.

7.10. $y\,dx + (1-x)\,dy = 0$.

7.11. $(x^2 + y + y^2)\,dx - x\,dy = 0$.

7.12. $(y + x^3y^3)\,dx + x\,dy = 0$.

7.13. $(y + x^4y^2)\,dx + x\,dy = 0$.

7.14. $(3x^2y - x^2)\,dx + dy = 0$.

7.15. $dx - 2xy\,dy = 0$.

7.16. $2xy\,dx + y^2\,dy = 0$.

7.17. $y\,dx + 3x\,dy = 0$.

7.18. $\left(2xy^2 + \dfrac{x}{y^2}\right)dx + 4x^2y\,dy = 0$.

7.19. $xy^2\,dx + (x^2y^2 + x^2y)\,dy = 0$.

7.20. $xy^2\,dx + x^2y\,dy = 0$.

7.21. $(y + x^3 + xy^2)\,dx - x\,dy = 0$.

7.22. $(x^3y^2 - y)\,dx + (x^2y^4 - x)\,dy = 0$.

7.23. $3x^2y^2\,dx + (2x^3y + x^3y^4)\,dy = 0$.

Answers to Supplementary Problems

7.9. $I(x,y) = \dfrac{-1}{x^2};\quad y = cx - 1$

7.10. $I(x,y) = \dfrac{1}{y^2};\quad cy = x - 1$

7.11. $I(x,y) = -\dfrac{1}{x^2+y^2};\quad y = x\tan(x+c)$

7.12. $I(x,y) = \dfrac{1}{(xy)^3};\quad \dfrac{1}{y^2} = 2x^2(x-c)$

7.13. $I(x,y) = \dfrac{1}{(xy)^2};\quad \dfrac{1}{y} = \frac{1}{3}x^4 - cx$

7.14. $I(x,y) = e^{x^3};\quad y = ce^{-x^3} + \dfrac{1}{3}$

7.15. $I(x,y) = e^{-y^2};\quad y^2 = \ln|kx|$

7.16. $I(x,y) = \dfrac{1}{y};\quad y^2 = 2(c - x^2)$

7.17. $I(x,y) = y^2;\quad y^3 = \dfrac{c}{x}$

7.18. $I(x,y) = y^2;\quad x^2y^4 + \dfrac{x^2}{2} = c$

7.19. $I(x,y) = \dfrac{1}{(xy)^2};\quad \ln|xy| = c - y$

7.20. $I(x,y) = 1$ (the equation is exact); $\frac{1}{2}x^2y^2 = c$

7.21. $I(x,y) = -\dfrac{1}{x^2+y^2};\quad y = x\tan(\frac{1}{2}x^2 + c)$

7.22. $I(x,y) = \dfrac{1}{(xy)^2};\quad 3x^3y + 2xy^4 + kxy = -6$

7.23. $I(x,y) = e^{y^3/3};\quad x^3y^2e^{y^3/3} = c$

Chapter 8

Linear First-Order Differential Equations

8.1 AN INTEGRATING FACTOR

The first-order linear differential equation was defined by (*3.3*) as

$$y' + p(x)y = q(x) \tag{8.1}$$

An integrating factor for (*8.1*) is given by

$$I(x,y) = e^{\int p(x)\,dx} \tag{8.2}$$

Note that for a linear equation the integrating factor does not depend on y. We recall that integrating factors were defined originally for equations in differential form (see Section 7.1). However, it is more convenient to work with linear differential equations in the form (*8.1*).

Example 8.1. Find an integrating factor for $y' - 2xy = x$.

Here, $p(x) = -2x$. Thus

$$\int p(x)\,dx = \int (-2x)\,dx = -x^2$$

and

$$I(x,y) = e^{\int p(x)\,dx} = e^{-x^2}$$

If we rewrite the given equation in differential form and then multiply by $I(x,y)$, we obtain

$$(-2xye^{-x^2} - xe^{-x^2})\,dx + e^{-x^2}\,dy = 0$$

which is exact. Hence, $I(x,y) = e^{-x^2}$ *is* an integrating factor as defined in Section 7.1.

8.2 METHOD OF SOLUTION

Multiply (*8.1*) by $I(x,y)$ as defined by (*8.2*). The left side of the resulting equation will equal $\dfrac{d[yI(x,y)]}{dx}$. Direct integration of this resulting equation then gives the solution of (*8.1*) (see Problems 8.1–8.5).

Solved Problems

8.1. Solve $y' - 3y = 6$.

Here $p(x) = -3$. Solving for the integrating factor, we have

$$\int p(x)\,dx = \int -3\,dx = -3x \qquad \text{from which} \qquad I(x,y) = e^{-3x}$$

Multiplying the differential equation by $I(x, y)$, we obtain

$$e^{-3x}y' - 3e^{-3x}y = 6e^{-3x} \qquad \text{or} \qquad \frac{d}{dx}(ye^{-3x}) = 6e^{-3x}$$

Integrating both sides of this last equation with respect to x, we have

$$\int \frac{d}{dx}(ye^{-3x})\,dx = \int 6e^{-3x}\,dx$$

$$ye^{-3x} = -2e^{-3x} + c$$

$$y = ce^{3x} - 2$$

8.2. Solve $y' - 2xy = x$.

From Example 8.1, $I(x, y) = e^{-x^2}$. Multiplying the differential equation by $I(x, y)$, we obtain

$$e^{-x^2}y' - 2xe^{-x^2}y = xe^{-x^2} \qquad \text{or} \qquad \frac{d}{dx}[ye^{-x^2}] = xe^{-x^2}$$

Integrating both sides of this last equation with respect to x, we find that

$$\int \frac{d}{dx}(ye^{-x^2})\,dx = \int xe^{-x^2}\,dx$$

$$ye^{-x^2} = -\frac{1}{2}e^{-x^2} + c$$

$$y = ce^{x^2} - \frac{1}{2}$$

8.3. Solve $y' + (4/x)y = x^4$.

Here $p(x) = 4/x$; hence

$$\int p(x)\,dx = \int \frac{4}{x}\,dx = 4\ln|x| = \ln x^4$$

whence

$$I(x, y) = e^{\int p(x)\,dx} = e^{\ln x^4} = x^4$$

Multiplying the differential equation by $I(x, y)$, we find

$$x^4y' + 4x^3y = x^8 \qquad \text{or} \qquad \frac{d}{dx}(yx^4) = x^8$$

Integrating both sides of this last equation with respect to x, we obtain

$$yx^4 = \tfrac{1}{9}x^9 + c \qquad \text{or} \qquad y = \frac{c}{x^4} + \tfrac{1}{9}x^5$$

8.4. Solve $y' + y = \sin x$.

Here $p(x) = 1$; hence $I(x, y) = e^{\int 1\,dx} = e^x$. Multiplying the differential equation by $I(x, y)$, we obtain

$$e^xy' + e^xy = e^x \sin x \qquad \text{or} \qquad \frac{d}{dx}(ye^x) = e^x \sin x$$

Integrating both sides of the last equation with respect to x (to integrate the right side, we use integration by parts twice), we find

$$ye^x = \tfrac{1}{2}e^x(\sin x - \cos x) + c \qquad \text{or} \qquad y = ce^{-x} + \tfrac{1}{2}\sin x - \tfrac{1}{2}\cos x$$

8.5. Solve $y' - 5y = 0$.

Here $p(x) = -5$ and $I(x, y) = e^{\int(-5)\,dx} = e^{-5x}$. Multiplying the differential equation by $I(x, y)$, we obtain

$$e^{-5x}y' - 5e^{-5x}y = 0 \quad \text{or} \quad \frac{d}{dx}(ye^{-5x}) = 0$$

Integrating, we obtain $ye^{-5x} = c$, or $y = ce^{5x}$.

Note that the differential equation is also separable, and can be solved by the method of Section 4.1. (See Problem 4.3.)

8.6. Solve the initial-value problem $y' + y = \sin x;\ y(\pi) = 1$.

From Problem 8.4 the solution to the differential equation is

$$y = ce^{-x} + \frac{1}{2}\sin x - \frac{1}{2}\cos x$$

Applying the initial condition directly, we obtain

$$1 = ce^{-\pi} + \frac{1}{2} \quad \text{or} \quad c = \frac{1}{2}e^{\pi}$$

Thus $$y = \frac{1}{2}e^{\pi}e^{-x} + \frac{1}{2}\sin x - \frac{1}{2}\cos x = \frac{1}{2}(e^{\pi-x} + \sin x - \cos x)$$

8.7. Solve $y' + xy = xy^2$.

This equation is not linear. It is, however, a special case of the *Bernoulli differential equation*, $y' + p(x)y = q(x)y^n$, where n is any real number. In our equation, $p(x) = x$, $q(x) = x$, and $n = 2$. To solve the Bernoulli equation, make the substitution $z = y^{1-n}$. The resulting differential equation will be linear and hence solvable by the method of this chapter.

Since $n = 2$ in the given equation, we make the substitution $z = y^{1-2} = y^{-1}$, from which follow

$$y = \frac{1}{z} \quad \text{and} \quad y' = -\frac{z'}{z^2}$$

Substituting these equations into the differential equation, we obtain

$$-\frac{z'}{z^2} + \frac{x}{z} = \frac{x}{z^2} \quad \text{or} \quad z' - xz = -x$$

This last equation is linear. Solving it by the method of this chapter, we find $z = ce^{x^2/2} + 1$. The solution of the original differential equation is then

$$y = \frac{1}{z} = \frac{1}{ce^{x^2/2} + 1}$$

8.8. Solve $y' - \frac{3}{x}y = x^4y^{1/3}$.

This is a Bernoulli differential equation with $p(x) = -\frac{3}{x}$, $q(x) = x^4$, and $n = \frac{1}{3}$. We make the substitution $z = y^{1-(1/3)} = y^{2/3}$. Thus, $y = z^{3/2}$ and $y' = \frac{3}{2}z^{1/2}z'$. Substituting these values into the differential equation, we obtain

$$\frac{3}{2}z^{1/2}z' - \frac{3}{x}z^{3/2} = x^4z^{1/2} \quad \text{or} \quad z' - \frac{2}{x}z = \frac{2}{3}x^4$$

The solution to this last equation, which is linear, is $z = cx^2 + \frac{2}{9}x^5$. Since $z = y^{2/3}$, the solution of the original problem is given implicitly by $y^{2/3} = cx^2 + \frac{2}{9}x^5$, or explicitly by $y = \pm(cx^2 + \frac{2}{9}x^5)^{3/2}$.

8.9. Find the general form of the solution of (*8.1*).

Multiplying (*8.1*) by (*8.2*), we have

$$e^{\int p(x)\,dx}y' + e^{\int p(x)\,dx}p(x)y = e^{\int p(x)\,dx}q(x) \tag{1}$$

Since

$$\frac{d}{dx}[e^{\int p(x)\,dx}] = e^{\int p(x)\,dx}p(x)$$

it follows from the product rule of differentiation that the left side of (*1*) equals $\frac{d}{dx}[e^{\int p(x)\,dx}y]$. Thus, (*1*) can be rewritten as

$$\frac{d}{dx}[e^{\int p(x)\,dx}y] = e^{\int p(x)\,dx}q(x) \tag{2}$$

Integrating both sides of (*2*) with respect to x, we have

$$\int \frac{d}{dx}[e^{\int p(x)\,dx}y]\,dx = \int e^{\int p(x)\,dx}q(x)\,dx$$

or,

$$e^{\int p(x)\,dx}y + c_1 = \int e^{\int p(x)\,dx}q(x)\,dx \tag{3}$$

Finally, setting $c_1 = -c$ and solving (*3*) for y, we obtain

$$y = ce^{-\int p(x)\,dx} + e^{-\int p(x)\,dx}\int e^{\int p(x)\,dx}q(x)\,dx \tag{4}$$

Supplementary Problems

Solve the following differential equations and initial-value problems.

8.10. $y' - 7y = e^x$.

8.11. $y' - 7y = 14x$.

8.12. $y' - 7y = \sin 2x$.

8.13. $y' + x^2y = x^2$.

8.14. $y' + \frac{2}{x}y = x;\ \ y(1) = 0$.

8.15. $y' + 6xy = 0;\ \ y(\pi) = 5$.

8.16. $y' - \frac{3}{x^2}y = \frac{1}{x^2}$.

8.17. $y' = \cos x$.

8.18. $y' + 2xy = 2x^3;\ \ y(0) = 1$.

8.19. $y' + y = y^2$.

8.20. $y' + xy = 6x\sqrt{y}$.

8.21. $y' + \frac{2}{x}y = -x^9y^5;\ \ y(-1) = 2$.

Answers to Supplementary Problems

8.10. $y = ce^{7x} - \frac{1}{6}e^x$

8.11. $y = ce^{7x} - 2x - \frac{2}{7}$

8.12. $y = ce^{7x} - \frac{2}{53}\cos 2x - \frac{7}{53}\sin 2x$

8.13. $y = ce^{-x^3/3} + 1$

8.14. $y = \frac{1}{4}(-x^{-2} + x^2)$

8.15. $y = 5e^{-3(x^2-\pi^2)}$

8.16. $y = ce^{-3/x} - \frac{1}{3}$

8.17. $y = c + \sin x$

8.18. $y = 2e^{-x^2} + x^2 - 1$

8.19. $\frac{1}{y} = ce^x + 1$

8.20. $y = (ce^{-x^2/4} + 6)^2$

8.21. $\frac{1}{y^4} = -\frac{31}{16}x^8 + 2x^{10}$

Chapter 9

Applications of First-Order Differential Equations

9.1 PROBLEMS IN COOLING

Newton's law of cooling states that *the time rate of change of the temperature of a body is proportional to the temperature difference between the body and its surrounding medium.* Let T denote the temperature of the body and let T_m denote the temperature of the surrounding medium. Then the time rate of change of the temperature of the body is dT/dt, and Newton's law of cooling can be formulated as $dT/dt = -k(T - T_m)$, or as

$$\frac{dT}{dt} + kT = kT_m \tag{9.1}$$

where k is a positive constant of proportionality. (See Problems 9.1, 9.2, and 9.3.) Once k is chosen positive, the minus sign is required in Newton's law to make dT/dt negative in a cooling process. Note that in such a process, T is greater than T_m; thus $T - T_m$ is positive.

9.2 GROWTH AND DECAY PROBLEMS

Let $N(t)$ denote the amount of substance (or population) that is either growing or decaying. If we assume that dN/dt, the time rate of change of this amount of substance, is proportional to the amount of substance present, then $dN/dt = kN$, or

$$\frac{dN}{dt} - kN = 0 \tag{9.2}$$

where k is the constant of proportionality. (See Problems 9.4, 9.5, and 9.6.)

We are assuming that $N(t)$ is a differentiable, hence continuous, function of time. For population problems, where $N(t)$ is actually discrete and integer-valued, this assumption is incorrect. Nonetheless, (*9.2*) still provides a good approximation to the physical laws governing such a system. (See Problem 9.6.)

9.3 FALLING BODIES WITH AIR RESISTANCE

Consider a vertically falling body of mass m that is being influenced only by gravity g and an air resistance that is proportional to the velocity of the body. Assume that both gravity and mass remain constant and, for convenience, choose the downward direction as the positive direction.

Newton's second law of motion: The net force acting on a body is equal to the time rate of change of the momentum of the body; or, for constant mass,

$$F = m\frac{dv}{dt} \tag{9.3}$$

where F is the net force on the body and v is the velocity of the body, both at time t.

For the problem at hand, there are two forces acting on the body: (1) the force due to gravity given by the weight w of the body, which equals mg and (2) the force due to air resistance given by $-kv$, where $k \geq 0$ is a constant of proportionality. The minus sign is required because this force opposes the velocity; that is, it acts in the upward, or negative, direction (see Fig. 9-1). The net force F on the body is, therefore, $F = mg - kv$. Substituting this result into (*9.3*), we obtain $mg - kv = m\frac{dv}{dt}$, or

$$\frac{dv}{dt} + \frac{k}{m}v = g \tag{9.4}$$

as the equation of motion for the body.

If air resistance is negligible or nonexistent, then $k = 0$ and (*9.4*) simplifies to

$$\frac{dv}{dt} = g \tag{9.5}$$

(See Problem 9.7.) When $k > 0$, the limiting velocity v_l is defined by

$$v_l = \frac{mg}{k} \tag{9.6}$$

Caution: Equations (*9.4*), (*9.5*), and (*9.6*) are valid only if the given conditions are satisfied. These equations are not valid if, for example, air resistance is not proportional to velocity but to the velocity squared, or if the upward direction is taken to be the positive direction. (See Problems 9.9 and 9.10.)

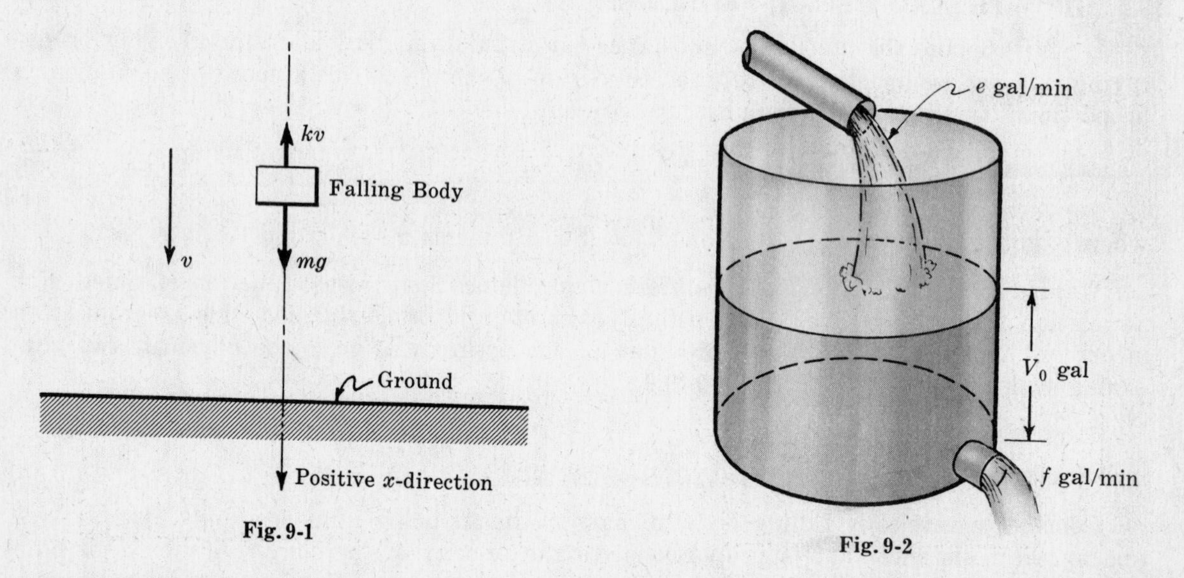

Fig. 9-1

Fig. 9-2

9.4 DILUTION PROBLEMS

Consider a tank which initially holds V_0 gallons of brine that contains a lb of salt. Another brine solution, containing b lb of salt per gallon, is poured into the tank at the rate of e gal/min while, simultaneously, the well-stirred solution leaves the tank at the rate of f gal/min (Fig. 9-2). The problem is to find the amount of salt in the tank at any time t.

Let Q denote the amount (in pounds) of salt in the tank at any time. The time rate of change of Q, dQ/dt, equals the rate at which salt enters the tank minus the rate at which salt leaves the tank. Salt enters the tank at the rate of be lb/min. To determine the rate at which salt leaves the tank, we first calculate the volume of brine in the tank at any time t, which is the initial volume V_0 plus the volume of brine added et minus the volume of brine removed ft. Thus, the volume of brine at any time is

$$V_0 + et - ft \tag{9.7}$$

The concentration of salt in the tank at any time is $Q/(V_0 + et - ft)$, from which it follows that salt leaves the tank at the rate of $f[Q/(V_0 + et - ft)]$ lb/min. Thus, $dQ/dt = be - f[Q/(V_0 + et - ft)]$ or

$$\frac{dQ}{dt} + \frac{f}{V_0 + (e - f)t}Q = be \tag{9.8}$$

(See Problems 9.11–9.13.)

9.5 ELECTRICAL CIRCUITS

The basic equation governing the amount of current I (in amperes) in a simple RL circuit (Fig. 9-3) consisting of a resistance R (in ohms), an inductor L (in henries), and an electromotive force (abbreviated emf) E (in volts) is

$$\frac{dI}{dt} + \frac{R}{L}I = \frac{E}{L} \tag{9.9}$$

For an RC circuit consisting of a resistance, a capacitance C (in farads), an emf, and no inductance (Fig. 9-4), the equation governing the amount of electrical charge q (in coulombs) on the capacitor is

$$\frac{dq}{dt} + \frac{1}{RC}q = \frac{E}{R} \tag{9.10}$$

The relationship between q and I is

$$I = \frac{dq}{dt} \tag{9.11}$$

(See Problems 9.14–9.17.) For more complex circuits see Chapter 17.

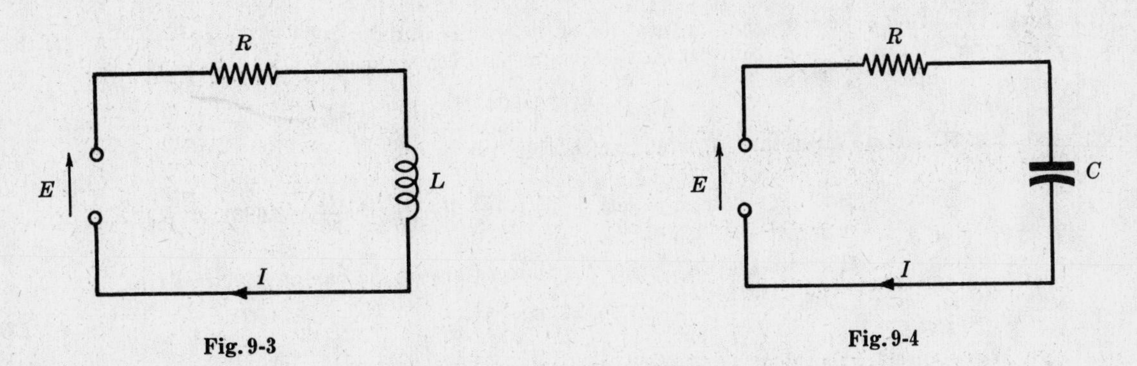

Fig. 9-3

Fig. 9-4

9.6 ORTHOGONAL TRAJECTORIES

Consider a one-parameter family of curves in the xy-plane defined by

$$F(x, y, c) = 0 \tag{9.12}$$

where c denotes the parameter. The problem is to find another one-parameter family of curves, called the *orthogonal trajectories* of the family (9.12) and given analytically by

$$G(x, y, k) = 0 \tag{9.13}$$

such that every curve in this new family (*9.13*) intersects at right angles every curve in the original family (*9.12*).

We first implicitly differentiate (*9.12*) with respect to x, then eliminate c between this derived equation and (*9.12*). This gives an equation connecting x, y, and y', which we solve for y' to obtain a differential equation of the form

$$\frac{dy}{dx} = f(x,y) \tag{9.14}$$

The orthogonal trajectories of (*9.12*) are the solutions of

$$\frac{dy}{dx} = -\frac{1}{f(x,y)} \tag{9.15}$$

(See Problems 9.18–9.20.)

For many families of curves, one cannot explicitly solve for dy/dx and obtain a differential equation of the form (*9.14*). We do not consider such curves in this book.

Solved Problems

9.1. A metal bar at a tempertaure of 100° F is placed in a room at a constant temperature of 0° F. If after 20 minutes the temperature of the bar is 50° F, find (*a*) the time it will take the bar to reach a temperature of 25° F and (*b*) the temperature of the bar after 10 minutes.

Use (*9.1*) with $T_m = 0$: the medium here is the room which is being held at a constant temperature of 0° F. Thus we have $\frac{dT}{dt} + kT = 0$, a linear equation whose solution is (see Chapter 8)

$$T = ce^{-kt} \tag{1}$$

Since $T = 100$ at $t = 0$ (the temperature of the bar is initially 100° F), it follows from (*1*) that $100 = ce^{-k(0)}$ or $100 = c$. Substituting this value into (*1*), we obtain

$$T = 100e^{-kt} \tag{2}$$

At $t = 20$, we are given that $T = 50$; hence, from (*2*),

$$50 = 100e^{-20k} \qquad \text{from which} \qquad k = \frac{-1}{20}\ln\frac{50}{100} = \frac{-1}{20}(-0.693) = 0.035$$

Substituting this value into (*2*), we obtain the temperature of the bar at any time t as

$$T = 100e^{-0.035t} \tag{3}$$

(*a*) We require t when $T = 25$. Substituting $T = 25$ into (*3*), we have

$$25 = 100e^{-0.035t} \qquad \text{or} \qquad -0.035t = \ln\frac{1}{4}$$

Solving, we find that $t = 39.6$ min.

(*b*) We require T when $t = 10$. Substituting $t = 10$ into (*3*) and then solving for T, we find that

$$T = 100e^{(-0.035)(10)} = 100(0.705) = 70.5^\circ \text{ F}$$

It should be noted that since Newton's law is valid only for small temperature differences, the above calculations represent only a first approximation to the physical situation.

9.2. **A body at a temperature of 50° F is placed outdoors where the temperature is 100° F. If after 5 minutes the temperature of the body is 60° F, find (*a*) how long it will take the body to reach a temperature of 75° F and (*b*) the temperature of the body after 20 minutes.**

Using (*9.1*) with $T_m = 100$ (the surrounding medium is the outside air), we have $\frac{dT}{dt} + kT = 100k$. This differential equation is linear, and its solution is (see Chapter 8)

$$T = ce^{-kt} + 100 \qquad (1)$$

Since $T = 50$ when $t = 0$, it follows from (*1*) that $50 = ce^{-k(0)} + 100$, or $c = -50$. Substituting this value into (*1*), we obtain

$$T = -50e^{-kt} + 100 \qquad (2)$$

At $t = 5$, we are given that $T = 60$; hence, from (*2*), $60 = -50e^{-5k} + 100$. Solving for k, we obtain

$$-40 = -50e^{-5k} \quad \text{or} \quad k = \frac{-1}{5}\ln\frac{40}{50} = \frac{-1}{5}(-0.223) = 0.045$$

Substituting this value into (*2*), we obtain the temperature of the body at any time t as

$$T = -50e^{-0.045t} + 100 \qquad (3)$$

(*a*) We require t when $T = 75$. Substituting $T = 75$ into (*3*), we have

$$75 = -50e^{-0.045t} + 100 \quad \text{or} \quad e^{-0.045t} = \frac{1}{2}$$

Solving for t, we find

$$-0.045t = \ln\frac{1}{2}, \quad \text{or} \quad t = 15.4 \text{ min}$$

(*b*) We require T when $t = 20$. Substituting $t = 20$ into (*3*) and then solving for T, we find

$$T = -50e^{(-0.045)(20)} + 100 = -50(0.41) + 100 = 79.5° \text{ F}$$

9.3. **A body at an unknown temperature is placed in a room which is held at a constant temperature of 30° F. If after 10 minutes the temperature of the body is 0° F and after 20 minutes the temperature of the body is 15° F, find the unknown initial temperature.**

From (*9.1*), $\frac{dT}{dt} + kT = 30k$. Solving, we obtain

$$T = ce^{-kt} + 30 \qquad (1)$$

At $t = 10$, we are given that $T = 0$. Hence, from (*1*),

$$0 = ce^{-10k} + 30 \quad \text{or} \quad ce^{-10k} = -30 \qquad (2)$$

At $t = 20$, we are given that $T = 15$. Hence, from (*1*) again,

$$15 = ce^{-20k} + 30 \quad \text{or} \quad ce^{-20k} = -15 \qquad (3)$$

Solving (*2*) and (*3*) for k and c, we find

$$k = \frac{1}{10}\ln 2 = 0.069 \quad \text{and} \quad c = -30e^{10k} = -30(2) = -60$$

Substituting these values into (*1*), we have for the temperature of the body at any time t

$$T = -60e^{-0.069t} + 30 \qquad (4)$$

Since we require T at the initial time $t = 0$, it follows from (*4*) that

$$T = -60e^{(-0.069)(0)} + 30 = -60 + 30 = -30° \text{ F}$$

9.4. A certain radioactive material is known to decay at a rate proportional to the amount present. If initially there is 50 milligrams of the material present and after two hours it is observed that the material has lost 10% of its original mass, find (*a*) an expression for the mass of the material remaining at any time t, (*b*) the mass of the material after four hours, and (*c*) the time at which the material has decayed to one half of its initial mass.

(*a*) Let N denote the amount of material present at time t. Then from (*9.2*), $\frac{dN}{dt} - kN = 0$. This differential equation is linear; its solution is

$$N = ce^{kt} \tag{1}$$

At $t = 0$, we are given that $N = 50$. Therefore, from (*1*), $50 = ce^{k(0)}$, or $c = 50$. Thus,

$$N = 50e^{kt} \tag{2}$$

At $t = 2$, 10% of the original mass of 50 mg, or 5 mg, has decayed. Hence, at $t = 2$, $N = 50 - 5 = 45$. Substituting these values into (*2*) and solving for k, we have

$$45 = 50e^{2k} \qquad \text{or} \qquad k = \frac{1}{2}\ln\frac{45}{50} = -0.053$$

Substituting this value into (*2*), we obtain the amount of mass present at any time t as

$$N = 50e^{-0.053t} \tag{3}$$

where t is measured in hours.

(*b*) We require N at $t = 4$. Substituting $t = 4$ into (*3*) and then solving for N, we find that

$$N = 50e^{(-0.053)(4)} = 50(0.809) = 40.5 \text{ mg}$$

(*c*) We require t when $N = 50/2 = 25$. Substituting $N = 25$ into (*3*) and solving for t, we find

$$25 = 50e^{-0.053t} \qquad \text{or} \qquad -0.053\,t = \ln\frac{1}{2} \qquad \text{or} \qquad t = 13 \text{ hours}$$

The time required to reduce a decaying material to one half its original mass is called the *half-life* of the material. For this problem, the half-life is 13 hours.

9.5. A bacteria culture is known to grow at a rate proportional to the amount present. After one hour, 1000 strands of the bacteria are observed in the culture; and after four hours, 3000 strands. Find (*a*) an expression for the number of strands of the bacteria present in the culture at any time t and (*b*) the number of strands of the bacteria originally in the culture.

(*a*) From (*9.2*), $\frac{dN}{dt} - kN = 0$. The solution to this differential equation is

$$N = ce^{kt} \tag{1}$$

At $t = 1$, $N = 1000$; hence,

$$1000 = ce^{k} \tag{2}$$

At $t = 4$, $N = 3000$; hence,

$$3000 = ce^{4k} \tag{3}$$

Solving (*2*) and (*3*) for k and c, we find

$$k = \frac{1}{3}\ln 3 = 0.366 \qquad \text{and} \qquad c = 1000e^{-0.366} = 694$$

Substituting these values of k and c into (*1*), we obtain

$$N = 694e^{0.366t} \tag{4}$$

as an expression for the amount of the bacteria present at any time t.

(*b*) We require N at $t = 0$. Substituting $t = 0$ into (*4*), we obtain $N = 694e^{(0.366)(0)} = 694$.

9.6. The population of a certain country is known to increase at a rate proportional to the number of people presently living in the country. If after two years the population has doubled, and after three years the population is 20,000, find the number of people initially living in the country.

Let N denote the number of people living in the country at any time t, and let N_0 denote the number of people initially living in the country. Then, from (9.2), $\frac{dN}{dt} - kN = 0$, which has the solution

$$N = ce^{kt} \tag{1}$$

At $t = 0$, $N = N_0$; hence, it follows from (1) that $N_0 = ce^{k(0)}$, or that $c = N_0$. Thus,

$$N = N_0 e^{kt} \tag{2}$$

At $t = 2$, $N = 2N_0$. Substituting these values into (2), we have

$$2N_0 = N_0 e^{2k} \qquad \text{from which} \qquad k = \frac{1}{2}\ln 2 = 0.347$$

Substituting this value into (2) gives

$$N = N_0 e^{0.347t} \tag{3}$$

At $t = 3$, $N = 20{,}000$. Substituting these values into (3), we obtain

$$20{,}000 = N_0 e^{(0.347)(3)} = N_0(2.832) \qquad \text{or} \qquad N_0 = 7062$$

9.7. A body of mass 5 slugs is dropped from a height of 100 feet with zero velocity. Assuming no air resistance, find (*a*) an expression for the velocity of the body at any time t, (*b*) an expression for the position of the body at any time t, and (*c*) the time required to reach the ground.

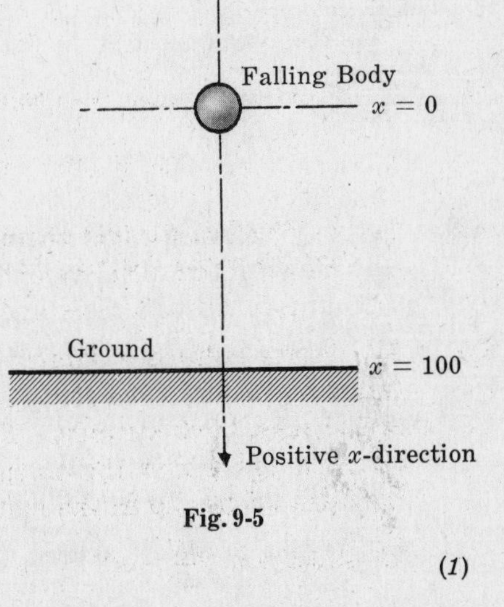

Fig. 9-5

(*a*) Choose the coordinate system as in Fig. 9-5. Then, since there is no air resistance, (9.5) applies: $\frac{dv}{dt} = g$. This differential equation is linear or, in differential form, separable; its solution is $v = gt + c$. When $t = 0$, $v = 0$ (initially the body has zero velocity); hence $0 = g(0) + c$, or $c = 0$. Thus, $v = gt$ or, assuming $g = 32\ \text{ft/sec}^2$,

$$v = 32t \tag{1}$$

(*b*) Recall that velocity is the time rate of change of displacement, designated here by x. Hence, $v = dx/dt$, and (1) becomes $\frac{dx}{dt} = 32t$. This differential equation is also both linear and separable; its solution is

$$x = 16t^2 + c_1 \tag{2}$$

But at $t = 0$, $x = 0$ (see Fig. 9-5). Thus, $0 = (16)(0)^2 + c_1$, or $c_1 = 0$. Substituting this value into (2), we have

$$x = 16t^2 \tag{3}$$

(*c*) We require t when $x = 100$. From (3), $t = \sqrt{(100)/(16)} = 2.5$ sec.

9.8. A body weighing 64 lb is dropped from a height of 100 ft with an initial velocity of 10 ft/sec. Assume that the air resistance is proportional to the velocity of the body. If the limiting velocity is known to be 128 ft/sec, find (*a*) an expression for the velocity of the body at any time t and (b) an expression for the position of the body at any time t.

(a) Locate the coordinate system as in Fig. 9-5. Here $w = 64$ lb. Since $w = mg$, it follows that $mg = 64$, or $m = 2$ slugs. Given that $v_l = 128$ ft/sec, it follows from (9.6) that $128 = 64/k$, or $k = \frac{1}{2}$. Substituting these values into (9.4), we obtain the linear differential equation

$$\frac{dv}{dt} + \frac{1}{4}v = 32$$

which has the solution

$$v = ce^{-t/4} + 128 \tag{1}$$

At $t = 0$, we are given that $v = 10$. Substituting these values into (1), we have $10 = ce^0 + 128$, or $c = -118$. The velocity at any time t is given by

$$v = -118e^{-t/4} + 128 \tag{2}$$

(b) Since $v = dx/dt$, where x is displacement, (2) can be rewritten as

$$\frac{dx}{dt} = -118e^{-t/4} + 128$$

This last equation, in differential form, is separable; its solution is

$$x = 472e^{-t/4} + 128t + c_1 \tag{3}$$

At $t = 0$, we have $x = 0$ (see Fig. 9-5). Thus, (3) gives

$$0 = 472e^0 + (128)(0) + c_1 \quad \text{or} \quad c_1 = -472$$

The displacement at any time t is then given by

$$x = 472e^{-t/4} + 128t - 472$$

9.9. A body of mass m is thrown vertically into the air with an initial velocity v_0. If the body encounters an air resistance proportional to its velocity, find (a) the equation of motion in the coordinate system of Fig. 9-6, (b) an expression for the velocity of the body at any time t, and (c) the time at which the body reaches its maximum height.

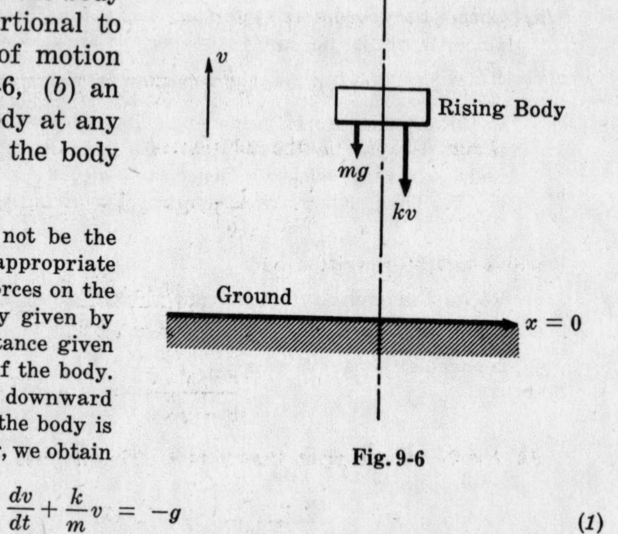

Fig. 9-6

(a) In this coordinate system, (9.4) may not be the equation of motion. To derive the appropriate equation, we note that there are two forces on the body: (1) the force due to the gravity given by mg and (2) the force due to air resistance given by kv, which will impede the velocity of the body. Since both of these forces act in the downward or negative direction, the net force on the body is $-mg - kv$. Using (9.3) and rearranging, we obtain

$$\frac{dv}{dt} + \frac{k}{m}v = -g \tag{1}$$

as the equation of motion.

(b) Equation (1) is a linear differential equation, and its solution is $v = ce^{-(k/m)t} - mg/k$. At $t = 0$, $v = v_0$; hence $v_0 = ce^{-(k/m)0} - (mg/k)$, or $c = v_0 + (mg/k)$. The velocity of the body at any time t is

$$v = \left(v_0 + \frac{mg}{k}\right)e^{-(k/m)t} - \frac{mg}{k} \tag{2}$$

(c) The body reaches its maximum height when $v = 0$. Thus, we require t when $v = 0$. Substituting $v = 0$ into (2) and solving for t, we find

$$0 = \left(v_0 + \frac{mg}{k}\right)e^{-(k/m)t} - \frac{mg}{k}$$

$$e^{-(k/m)t} = \frac{1}{1 + \dfrac{v_0 k}{mg}}$$

$$-(k/m)t = \ln\left(\frac{1}{1 + \dfrac{v_0 k}{mg}}\right)$$

$$t = \frac{m}{k}\ln\left(1 + \frac{v_0 k}{mg}\right)$$

9.10. A body of mass 2 slugs is dropped with no initial velocity and encounters an air resistance that is proportional to the square of its velocity. Find an expression for the velocity of the body at any time t.

The force due to air resistance is $-kv^2$, so that Newton's second law of motion becomes

$$m\frac{dv}{dt} = mg - kv^2 \quad \text{or} \quad 2\frac{dv}{dt} = 64 - kv^2$$

Rewriting this equation in differential form, we have

$$\frac{2}{64 - kv^2}\,dv - dt = 0 \tag{1}$$

which is separable. By partial fractions,

$$\frac{2}{64 - kv^2} = \frac{2}{(8 - \sqrt{k}\,v)(8 + \sqrt{k}\,v)} = \frac{\frac{1}{8}}{8 - \sqrt{k}\,v} + \frac{\frac{1}{8}}{8 + \sqrt{k}\,v}$$

hence (1) can be rewritten as

$$\frac{1}{8}\left(\frac{1}{8 - \sqrt{k}\,v} + \frac{1}{8 + \sqrt{k}\,v}\right)dv - dt = 0$$

From Chapter 4, the solution is

$$\frac{1}{8}\left[-\frac{1}{\sqrt{k}}\ln|8 - \sqrt{k}\,v| + \frac{1}{\sqrt{k}}\ln|8 + \sqrt{k}\,v|\right] - t = c$$

which can be rewritten as

$$\ln\left|\frac{8 + \sqrt{k}\,v}{8 - \sqrt{k}\,v}\right| = 8\sqrt{k}\,t + 8\sqrt{k}\,c$$

or

$$\frac{8 + \sqrt{k}\,v}{8 - \sqrt{k}\,v} = c_1 e^{8\sqrt{k}\,t} \qquad (c_1 = \pm e^{8\sqrt{k}\,c})$$

At $t = 0$, we are given that $v = 0$. This implies $c_1 = 1$, and the velocity is given by

$$\frac{8 + \sqrt{k}\,v}{8 - \sqrt{k}\,v} = e^{8\sqrt{k}\,t} \quad \text{or} \quad v = \frac{8}{\sqrt{k}}\tanh 4\sqrt{k}\,t$$

Note that without additional information, we cannot obtain a numerical value for the constant k.

9.11. A tank initially holds 100 gallons of a brine solution containing 20 lb of salt. At $t = 0$, fresh water is poured into the tank at the rate of 5 gal/min, while the well-stirred mixture leaves the tank at the same rate. Find the amount of salt in the tank at any time t.

Here, $V_0 = 100$, $a = 20$, $b = 0$, and $e = f = 5$. Equation (9.8) becomes $\frac{dQ}{dt} + \frac{1}{20}Q = 0$. The solution of this linear equation is

$$Q = ce^{-t/20} \tag{1}$$

At $t = 0$, we are given that $Q = a = 20$. Substituting these values into (1), we find that $c = 20$, so that (1) can be rewritten as $Q = 20e^{-t/20}$.

Note that as $t \to \infty$, $Q \to 0$ as it should, since only fresh water is being added.

9.12. A tank initially holds 100 gallons of a brine solution containing 1 lb of salt. At $t = 0$ another brine solution containing 1 lb of salt per gallon is poured into the tank at the rate of 3 gal/min, while the well-stirred mixture leaves the tank at the same rate. Find (a) the amount of salt in the tank at any time t and (b) the time at which the mixture in the tank contains 2 lb of salt.

(a) Here $V_0 = 100$, $a = 1$, $b = 1$, and $e = f = 3$; hence, (9.8) becomes $\frac{dQ}{dt} + 0.03\,Q = 3$. The solution to this linear differential equation is

$$Q = ce^{-0.03t} + 100 \tag{1}$$

At $t = 0$, $Q = a = 1$. Substituting these values into (1), we find $1 = ce^0 + 100$, or $c = -99$. Then (1) can be rewritten as

$$Q = -99e^{-0.03t} + 100 \tag{2}$$

(b) We require t when $Q = 2$. Substituting $Q = 2$ into (2), we obtain

$$2 = -99e^{-0.03t} + 100 \qquad \text{or} \qquad e^{-0.03t} = \frac{98}{99}$$

from which

$$t = -\frac{1}{0.03}\ln\frac{98}{99} = 0.338 \text{ min}$$

9.13. A 50-gallon tank initially contains 10 gallons of fresh water. At $t = 0$, a brine solution containing 1 lb of salt per gallon is poured into the tank at the rate of 4 gal/min, while the well-stirred mixture leaves the tank at the rate of 2 gal/min. Find (a) the amount of time required for overflow to occur and (b) the amount of salt in the tank at the moment of overflow.

(a) Here $a = 0$, $b = 1$, $e = 4$, $f = 2$, and $V_0 = 10$. The volume of brine in the tank at any time t is given by (9.7) as $V_0 + et - ft = 10 + 2t$. We require t when $10 + 2t = 50$; hence, $t = 20$ min.

(b) For this problem, (9.8) becomes

$$\frac{dQ}{dt} + \frac{2}{10 + 2t}Q = 4$$

This is a linear equation; its solution [see Chapter 8, with $p(t) = 2/(10 + 2t)$ and $q(t) = 4$] can be written as

$$Q = \frac{40t + 4t^2 + c}{10 + 2t} \tag{1}$$

At $t = 0$, $Q = a = 0$. Substituting these values into (1), we find that $c = 0$. We require Q at the moment of overflow, which from part (a) is $t = 20$. Thus,

$$Q = \frac{40(20) + 4(20)^2}{10 + 2(20)} = 48 \text{ lb}$$

9.14. An RL circuit has an emf of 5 volts, a resistance of 50 ohms, an inductance of 1 henry, and no initial current. Find the current in the circuit at any time t.

Here $E = 5$, $R = 50$, and $L = 1$; hence (*9.9*) becomes $\frac{dI}{dt} + 50I = 5$. This equation is linear; its solution is

$$I = ce^{-50t} + \frac{1}{10}$$

At $t = 0$, $I = 0$; thus, $0 = ce^{-50(0)} + \frac{1}{10}$, or $c = -\frac{1}{10}$. The current at any time t is then

$$I = -\frac{1}{10}e^{-50t} + \frac{1}{10} \tag{1}$$

The quantity $-\frac{1}{10}e^{-50t}$ in (*1*) is called the *transient current,* since this quantity goes to zero ("dies out") as $t \to \infty$. The quantity $\frac{1}{10}$ in (*1*) is called the *steady-state current.* As $t \to \infty$, the current I approaches the value of the steady-state current.

9.15. An RL circuit has an emf given (in volts) by $3 \sin 2t$, a resistance of 10 ohms, an inductance of 0.5 henry, and an initial current of 6 amperes. Find the current in the circuit at any time t.

Here, $E = 3 \sin 2t$, $R = 10$, and $L = 0.5$; hence (*9.9*) becomes $\frac{dI}{dt} + 20I = 6 \sin 2t$. This equation is linear, with solution (see Chapter 8) $\int d(Ie^{20t}) = \int 6e^{20t} \sin 2t\, dt$. Carrying out the integrations (the second integral requires two integrations by parts), we obtain

$$I = ce^{-20t} + \frac{30}{101}\sin 2t - \frac{3}{101}\cos 2t$$

At $t = 0$, $I = 6$; hence,

$$6 = ce^{-20(0)} + \frac{30}{101}\sin 2(0) - \frac{3}{101}\cos 2(0) \qquad \text{or} \qquad 6 = c - \frac{3}{101}$$

whence $c = \frac{609}{101}$. The current at any time t is

$$I = \frac{609}{101}e^{-20t} + \frac{30}{101}\sin 2t - \frac{3}{101}\cos 2t$$

As in Problem 9.14, the current is the sum of a transient current, here $\frac{609}{101}e^{-20t}$, and a steady-state current, $\frac{30}{101}\sin 2t - \frac{3}{101}\cos 2t$.

9.16. Rewrite the steady-state current of Problem 9.15 in the form $A \sin(2t - \phi)$. The angle ϕ is called the *phase angle.*

Since $A \sin(2t - \phi) = A(\sin 2t \cos\phi - \cos 2t \sin\phi)$, we require

$$I_s = \frac{30}{101}\sin 2t - \frac{3}{101}\cos 2t = A\cos\phi \sin 2t - A\sin\phi\cos 2t$$

Thus, $A\cos\phi = \frac{30}{101}$ and $A\sin\phi = \frac{3}{101}$. It now follows that

$$\left(\frac{30}{101}\right)^2 + \left(\frac{3}{101}\right)^2 = A^2\cos^2\phi + A^2\sin^2\phi = A^2(\cos^2\phi + \sin^2\phi) = A^2$$

and

$$\tan\phi = \frac{A\sin\phi}{A\cos\phi} = \left(\frac{3}{101}\right)\Big/\left(\frac{30}{101}\right) = \frac{1}{10}$$

Consequently, I_s has the required form if

$$A = \sqrt{\frac{909}{(101)^2}} = \frac{3}{\sqrt{101}} \qquad \text{and} \qquad \phi = \arctan\frac{1}{10}$$

9.17. An RC circuit has an emf given (in volts) by $400 \cos 2t$, a resistance of 100 ohms, and a capacitance of 10^{-2} farad. Initially there is no charge on the capacitor. Find the current in the circuit at any time t.

We first find the charge q and then use (*9.11*) to obtain the current. Here, $E = 400 \cos 2t$, $R = 100$, and $C = 10^{-2}$; hence (*9.10*) becomes $\dfrac{dq}{dt} + q = 4 \cos 2t$. This equation is linear, and its solution is (two integrations by parts are required)

$$q = ce^{-t} + \frac{8}{5}\sin 2t + \frac{4}{5}\cos 2t$$

At $t = 0$, $q = 0$; hence,

$$0 = ce^{-(0)} + \frac{8}{5}\sin 2(0) + \frac{4}{5}\cos 2(0) \quad \text{or} \quad c = -\frac{4}{5}$$

Thus
$$q = -\frac{4}{5}e^{-t} + \frac{8}{5}\sin 2t + \frac{4}{5}\cos 2t$$

and using (*9.11*), we obtain

$$I = \frac{dq}{dt} = \frac{4}{5}e^{-t} + \frac{16}{5}\cos 2t - \frac{8}{5}\sin 2t$$

9.18. Find the orthogonal trajectories of the family of curves $x^2 + y^2 = c^2$.

The family, which is given by (*9.12*) with $F(x, y, c) = x^2 + y^2 - c^2$, consists of circles with centers at the origin and radii c. Implicitly differentiating the given equation with respect to x, we obtain

$$2x + 2yy' = 0 \quad \text{or} \quad \frac{dy}{dx} = -\frac{x}{y}$$

Here $f(x, y) = -x/y$, so that (*9.15*) becomes

$$\frac{dy}{dx} = \frac{y}{x}$$

This equation is linear (and, in differential form, separable); its solution is

$$y = kx \tag{1}$$

which represents the orthogonal trajectories.

In Fig. 9-7 some members of the family of circles are shown in solid lines and some members of the family (*1*), which are straight lines through the origin, are shown in dashed lines. Observe that each straight line intersects each circle at right angles.

9.19. Find the orthogonal trajectories of the family of curves $y = cx^2$.

The family, which is given by (*9.12*) with $F(x, y, c) = y - cx^2$, consists of parabolas symmetric about the y-axis with vertices at the origin. Implicitly differentiating the given equation with respect to x, we obtain $\dfrac{dy}{dx} = 2cx$. To eliminate c, we observe, from the given equation, that $c = y/x^2$; hence, $\dfrac{dy}{dx} = \dfrac{2y}{x}$. Here $f(x, y) = 2y/x$, so (*9.15*) becomes

$$\frac{dy}{dx} = \frac{-x}{2y} \quad \text{or} \quad x\,dx + 2y\,dy = 0$$

The solution of this separable equation is $\frac{1}{2}x^2 + y^2 = k$.

These orthogonal trajectories are ellipses. Some members of this family, along with some members of the original family of parabolas, are shown in Fig. 9-8. Note that each ellipse intersects each parabola at right angles.

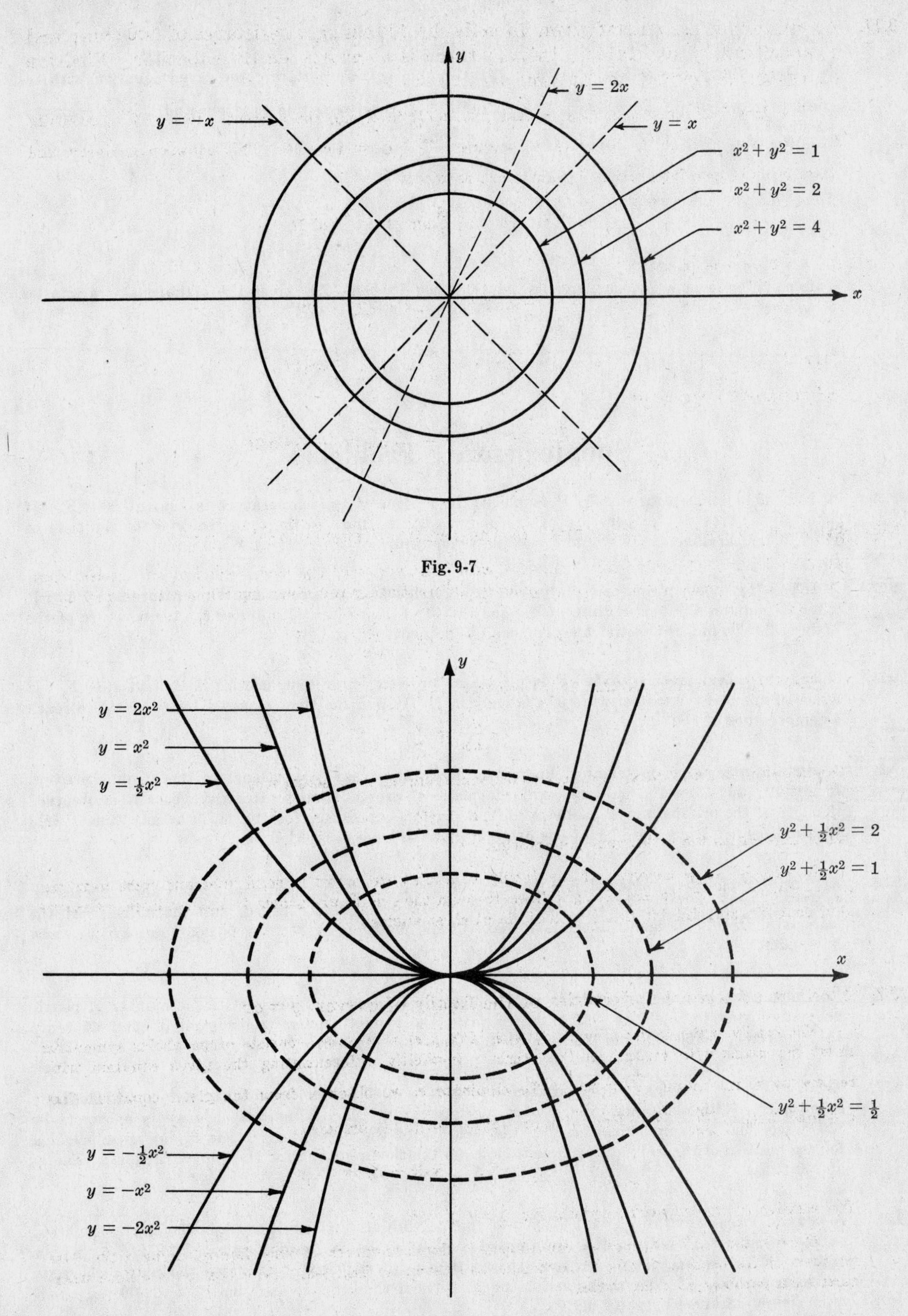

y
x
$y = 2x$
$y = -x$
$y = x$
$x^2 + y^2 = 1$
$x^2 + y^2 = 2$
$x^2 + y^2 = 4$

Fig. 9-7

y
x
$y = 2x^2$
$y = x^2$
$y = \frac{1}{2}x^2$
$y^2 + \frac{1}{2}x^2 = 2$
$y^2 + \frac{1}{2}x^2 = 1$
$y^2 + \frac{1}{2}x^2 = \frac{1}{2}$
$y = -\frac{1}{2}x^2$
$y = -x^2$
$y = -2x^2$

Fig. 9-8

9.20. Find the orthogonal trajectories of the family of curves $x^2 + y^2 = cx$.

Here, $F(x, y, c) = x^2 + y^2 - cx$ in (*9.12*). Implicitly differentiating the given equation with respect to x, we obtain $2x + 2y\frac{dy}{dx} = c$. Eliminating c between this equation and $x^2 + y^2 - cx = 0$, we find

$$\frac{dy}{dx} = \frac{y^2 - x^2}{2xy}$$

Here $f(x, y) = (y^2 - x^2)/2xy$, so (*9.15*) becomes

$$\frac{dy}{dx} = \frac{2xy}{x^2 - y^2}$$

This equation is homogeneous, and its solution (see Problem 5.4) gives the orthogonal trajectories as $x^2 + y^2 = ky$.

Supplementary Problems

9.21. A body at a temperature of 0° F is placed in a room whose temperature is kept at 100° F. If after 10 minutes the temperature of the body is 25° F, find (*a*) the time required for the body to reach a temperature of 50° F, and (*b*) the temperature of the body after 20 minutes.

9.22. A body of unknown temperature is placed in a refrigerator at a constant temperature of 0° F. If after 20 minutes the temperature of the body is 40° F and after 40 minutes the temperature of the body is 20° F, find the initial temperature of the body.

9.23. A body at a temperature of 50° F is placed in an oven whose temperature is kept at 150° F. If after 10 minutes the temperature of the body is 75° F, find the time required for the body to reach a temperature of 100° F.

9.24. A certain radioactive material is known to decay at a rate proportional to the amount present. If initially there are 100 milligrams of the material present and if after two years it is observed that 5% of the original mass has decayed, find (*a*) an expression for the mass at any time t and (*b*) the time necessary for 10% of the original mass to have decayed.

9.25. A certain radioactive material is known to decay at a rate proportional to the amount present. If after one hour it is observed that 10% of the material has decayed, find the half-life of the material. (*Hint.* Designate the initial mass of the material by N_0. It is not necessary to know N_0 explicitly.)

9.26. The population of a certain state is known to grow at a rate proportional to the number of people presently living in the state. If after 10 years the population has trebled and if after 20 years the population is 150,000, find the number of people initially living in the state.

9.27. A body of mass 10 slugs is dropped from a height of 1000 feet with no initial velocity. The body encounters an air resistance proportional to its velocity. If the limiting velocity is known to be 320 ft/sec, find (*a*) an expression for the velocity of the body at any time t, (*b*) an expression for the position of the body at any time t, and (*c*) the time required for the body to attain a velocity of 160 ft/sec.

9.28. A body of mass m is thrown vertically into the air with an initial velocity v_0. The body encounters no air resistance. Find (*a*) the equation of motion in the coordinate system of Fig. 9-6, (*b*) an expression for the velocity of the body at any time t, (*c*) the time t_m at which the body reaches its maximum height, (*d*) an expression for the position of the body at any time t, and (*e*) the maximum height attained by the body.

9.29. A body of mass 1 slug is dropped with an initial velocity of 1 ft/sec and encounters a force due to air resistance given exactly by $-8v^2$. Find the velocity at any time t.

9.30. A tank initially holds 10 gal of fresh water. At $t = 0$, a brine solution containing $\frac{1}{2}$ lb of salt per gallon is poured into the tank at a rate of 2 gal/min, while the well-stirred mixture leaves the tank at the same rate. Find (*a*) the amount and (*b*) the concentration of salt in the tank at any time t.

9.31. A tank initially holds 80 gallons of a brine solution containing $\frac{1}{8}$ lb of salt per gallon. At $t = 0$, another brine solution containing 1 lb of salt per gallon is poured into the tank at the rate of 4 gal/min, while the well-stirred mixture leaves the tank at the rate of 8 gal/min. Find the amount of salt in the tank when the tank contains exactly 40 gal of solution.

9.32. An RC circuit has an emf of 5 volts, a resistance of 10 ohms, a capacitance of 10^{-2} farad, and initially a charge of 5 coulombs on the capacitor. Find (*a*) the transient current and (*b*) the steady-state current.

9.33. An RL circuit with no emf source has an initial current given by I_0. Find the current at any time t.

9.34. An RL circuit has an emf given (in volts) by $4 \sin t$, a resistance of 100 ohms, an inductance of 4 henries, and no initial current. Find the current at any time t.

9.35. The steady-state current in a circuit is known to be $\frac{5}{17} \sin t - \frac{3}{17} \cos t$. Rewrite this current in the form $A \sin (t - \phi)$.

9.36. Rewrite the steady-state current of Problem 9.17 in the form $A \cos (2t + \phi)$. [*Hint*: Use the identity $\cos (x + y) \equiv \cos x \cos y - \sin x \sin y$.]

9.37. Find the orthogonal trajectories of the family of curves $x^2 - y^2 = c^2$.

9.38. Find the orthogonal trajectories of the family of curves $y = ce^x$.

9.39. Find the orthogonal trajectories of the family of curves $x^2 - y^2 = cx$.

Answers to Supplementary Problems

9.21. $T = -100e^{-0.029t} + 100$; (*a*) 23.9 min, (*b*) 44° F

9.22. $T = 80e^{-0.035t}$; $T_0 = 80°$ F

9.23. $T = -100e^{-0.029t} + 150$; $t_{100} = 23.9$ min

9.24. (*a*) $N = 100e^{-0.026t}$ (*b*) 4.05 yr

9.25. $N = N_0 e^{-0.105t}$; $t_{1/2} = 6.6$ hr

9.26. $N = 16{,}620e^{0.11t}$; $N_0 = 16{,}620$

9.27. (a) $v = -320e^{-0.1t} + 320$ (b) $x = 3200e^{-0.1t} + 320t - 3200$ (c) 6.9 sec

9.28. (a) $\frac{dv}{dt} = -g$ (c) $t_m = \frac{v_0}{g}$ (e) $x_m = \frac{v_0^2}{2g}$

(b) $v = -gt + v_0$ (d) $x = -\frac{1}{2}gt^2 + v_0 t$

9.29. $\frac{2+v}{2-v} = 3e^{32t}$ or $v = 2(3e^{32t} - 1)/(3e^{32t} + 1)$

9.30. (a) $Q = -5e^{-0.2t} + 5$ (b) $\frac{Q}{V} = \frac{1}{2}(-e^{-0.2t} + 1)$

9.31. $Q = -\frac{7}{40}(20 - t)^2 + 4(20 - t)$; at $t = 10$, $Q = 22.5$ lb

(Note that $a = 80(1/8) = 10$ lb.)

9.32. (a) $-\frac{99}{2}e^{-10t}$ (b) 0 amp

9.33. $I = I_0 e^{-(R/L)t}$

9.34. $I = \frac{1}{626}(e^{-25t} + 25 \sin t - \cos t)$

9.35. $A = \frac{2}{\sqrt{34}}$ $\phi = \arctan \frac{3}{5}$

9.36. $A = \frac{8}{\sqrt{5}}$ $\phi = \arctan \frac{1}{2}$

9.37. $xy = k$

9.38. $y^2 = -2x + k$

9.39. $x^2 y + \frac{1}{3}y^3 = k$

Chapter 10

Linear Differential Equations: General Remarks

10.1 DEFINITIONS. UNIQUENESS THEOREM

An nth-order linear differential equation has the form

$$b_n(x)y^{(n)} + b_{n-1}(x)y^{(n-1)} + \cdots + b_2(x)y'' + b_1(x)y' + b_0(x)y = g(x) \tag{10.1}$$

where $g(x)$ and the coefficients $b_j(x)$ $(j = 0, 1, 2, \ldots, n)$ depend solely on the variable x. In other words, they do *not* depend on y or on any derivative of y.

Theorem 10.1. Consider the initial-value problem given by the linear differential equation (*10.1*) and the n initial conditions

$$y(x_0) = c_0, \;\; y'(x_0) = c_1, \;\; y''(x_0) = c_2, \;\; \ldots, \;\; y^{(n-1)}(x_0) = c_{n-1} \tag{10.2}$$

If $g(x)$ and $b_j(x)$ $(j = 0, 1, 2, \ldots, n)$ are continuous in some interval $\mathcal{I}$ containing x_0 and if $b_n(x) \neq 0$ in $\mathcal{I}$, then the initial-value problem given by (*10.1*) and (*10.2*) has a unique (only one) solution defined throughout $\mathcal{I}$.

(See Problems 10.4–10.6.)

When the conditions on $b_n(x)$ in Theorem 10.1 hold, we can divide (*10.1*) by $b_n(x)$ to get

$$y^{(n)} + a_{n-1}y^{(n-1)} + \cdots + a_2(x)y'' + a_1(x)y' + a_0(x)y = \phi(x) \tag{10.3}$$

where $a_j(x) = b_j(x)/b_n(x)$ $(j = 0, 1, \ldots, n-1)$ and $\phi(x) = g(x)/b_n(x)$.

Example 10.1. A second-order linear differential equation has the form

$$b_2(x)y'' + b_1(x)y' + b_0(x)y = g(x) \tag{10.4}$$

or, if $b_2(x) \neq 0$,

$$y'' + a_1(x)y' + a_0(x)y = \phi(x) \tag{10.5}$$

If $\phi(x) \equiv 0$ [or $g(x) \equiv 0$], then (*10.3*) and (*10.5*) [or (*10.1*) and (*10.4*)] are said to be *homogeneous*. If not, they are *nonhomogeneous*. (See Problem 10.2.)

If all the coefficients $a_j(x)$ [or $b_j(x)$] in (*10.3*) and (*10.5*) [or (*10.1*) and (*10.4*)] are constants, the differential equations are said to have *constant coefficients*. If any one coefficient is not constant, the differential equations have *variable coefficients*. (See Problem 10.3.)

10.2 THE LINEAR DIFFERENTIAL OPERATOR

Let us define the differential operator $\mathbf{L}(y)$ by

$$\mathbf{L}(y) \equiv y^{(n)} + a_{n-1}(x)y^{(n-1)} + \cdots + a_2(x)y'' + a_1(x)y' + a_0(x)y \tag{10.6}$$

where $a_i(x)$ $(i = 0, 1, 2, \ldots, n-1)$ is continuous on some interval of interest. Then (*10.3*) can be rewritten as

$$\mathbf{L}(y) = \phi(x) \tag{10.7}$$

and, in particular, a linear *homogeneous* differential equation can be expressed as

$$\mathbf{L}(y) = 0 \tag{10.8}$$

Theorem 10.2. The operator $\mathbf{L}(y)$ is linear; that is,

$$\mathbf{L}(c_1y_1 + c_2y_2) = c_1\mathbf{L}(y_1) + c_2\mathbf{L}(y_2) \tag{10.9}$$

where c_1 and c_2 are arbitrary constants and y_1 and y_2 are arbitrary n-times differentiable functions.

(See Problem 10.7.)

Theorem 10.3 (Principle of superposition). If y_1 and y_2 are two solutions of $\mathbf{L}(y) = 0$, then $c_1y_1 + c_2y_2$ is also a solution of $\mathbf{L}(y) = 0$ for any two constants c_1 and c_2.

(See Problem 10.8.)

Solved Problems

10.1. State the order and whether the equation is linear:

(a) $2xy'' + x^2y' - (\sin x)y = 2$ (c) $y'' - y = 0$

(b) $yy''' + xy' + y = x^2$ (d) $3y' + xy = e^{-x^2}$

(a) *Second-order.* Here $b_2(x) = 2x$, $b_1(x) = x^2$, $b_0(x) = -\sin x$, and $g(x) = 2$. Since none of these terms depends on y or any derivative of y, the differential equation is *linear.*

(b) *Third-order.* Since $b_3 = y$, which does depend on y, the differential equation is *nonlinear.*

(c) *Second-order.* Here $b_2(x) = 1$, $b_1(x) = 0$, $b_0(x) = 1$, and $g(x) = 0$. None of these terms depends on y or any derivative of y; hence the differential equation is *linear.*

(d) *First-order.* Here $b_1(x) = 3$, $b_0(x) = x$, and $g(x) = e^{-x^2}$; hence the differential equation is *linear.* (See also Chapter 8.)

10.2. Which of the linear differential equations given in Problem 10.1 are homogeneous?

(a) is nonhomogeneous, since $\phi(x) = 1/x \not\equiv 0$; (c) is homogeneous, since $\phi(x) \equiv 0$; (d) is nonhomogeneous, since $\phi(x) = \frac{1}{3}e^{-x^2} \not\equiv 0$.

10.3. Which of the linear differential equations given in Problem 10.1 have constant coefficients?

Only (c) has constant coefficients, for only in this equation are *all* the coefficients constants.

10.4. Specialize Theorem 10.1 for second-order initial-value problems.

Theorem: Consider the initial-value problem

$$b_2(x)y'' + b_1(x)y' + b_0(x)y = g(x); \quad y(x_0) = c_0, \quad y'(x_0) = c_1 \tag{1}$$

If $g(x)$, $b_2(x)$, $b_1(x)$, and $b_0(x)$ are continuous on some interval $\mathcal{J}$ containing x_0, and if $b_2(x) \neq 0$ on $\mathcal{J}$, then the initial-value problem (1) has a unique solution defined throughout $\mathcal{J}$.

10.5. The initial-value problem $y' = 2\sqrt{|y|}$; $y(0) = 0$ has the two solutions $y = x|x|$ and $y \equiv 0$. Does this result violate Theorem 10.1?

No. Here $\phi = 2\sqrt{|y|}$, which depends on y; therefore, the differential equation is not linear and Theorem 10.1 does not apply. (See also Problem 3.6.)

10.6. Determine all solutions of the initial-value problem $y'' + e^x y' + (x+1)y = 0$; $y(1) = 0$, $y'(1) = 0$.

Here, $b_2(x) = 1$, $b_1(x) = e^x$, $b_0(x) = x + 1$, and $g(x) \equiv 0$ satisfy the hypotheses of Theorem 10.1; thus, the solution to the initial-value problem is unique. By inspection, $y \equiv 0$ is a solution. It follows that $y \equiv 0$ is the only solution.

10.7. Prove Theorem 10.2 for a second-order linear differential equation.

From *(10.5)* and *(10.6)*, $\mathbf{L}(y) = y'' + a_1(x)y' + a_0(x)y$. Thus,

$$\begin{aligned}
\mathbf{L}(c_1y_1 + c_2y_2) &= (c_1y_1 + c_2y_2)'' + a_1(x)(c_1y_1 + c_2y_2)' + a_0(x)(c_1y_1 + c_2y_2) \\
&= c_1y_1'' + c_2y_2'' + a_1(x)c_1y_1' + a_1(x)c_2y_2' + a_0(x)c_1y_1 + a_0(x)c_2y_2 \\
&= c_1[y_1'' + a_1(x)y_1' + a_0(x)y_1] + c_2[y_2'' + a_1(x)y_2' + a_0(x)y_2] \\
&= c_1\mathbf{L}(y_1) + c_2\mathbf{L}(y_2)
\end{aligned}$$

10.8. Prove Theorem 10.3.

Let y_1 and y_2 be two solutions of $\mathbf{L}(y) = 0$; that is, $\mathbf{L}(y_1) = 0$ and $\mathbf{L}(y_2) = 0$. Then *(10.9)* gives

$$\mathbf{L}(c_1y_1 + c_2y_2) = c_1\mathbf{L}(y_1) + c_2\mathbf{L}(y_2) = c_1(0) + c_2(0) = 0$$

Thus, $c_1y_1 + c_2y_2$ is also a solution of $\mathbf{L}(y) = 0$.

Supplementary Problems

In Problems 10.9 through 10.16, find the order of the given differential equations and determine if the equations are linear. For those equations that are linear, also determine if they are homogeneous and/or have constant coefficients.

10.9. $y'' + xy' + 2y = 0$.

10.10. $y''' - y = x$.

10.11. $y' + 5y = 0$.

10.12. $y^{(4)} + x^2y''' + xy'' - e^xy' + 2y = x^2 + x + 1$.

10.13. $y'' + 2xy' + y = 4xy^2$.

10.14. $y' - 2y = xy$.

10.15. $y'' + yy' = x^2$.

10.16. $y''' + (x^2 - 1)y'' - 2y' + y = 5 \sin x$.

10.17. The initial-value problem $y' - \frac{2}{x}y = 0;\ y(0) = 0$ has two solutions $y \equiv 0$ and $y = x^2$. Why doesn't this result violate Theorem 10.1?

10.18. Does Theorem 10.1 apply to the initial-value problem $y' - \frac{2}{x}y = 0;\ y(1) = 3$?

10.19. The initial-value problem $xy' - 2y = 0;\ y(0) = 0$ has two solutions $y \equiv 0$ and $y = x^2$. Why doesn't this result violate Theorem 10.1?

Answers to Supplementary Problems

10.9. second-order, linear, homogeneous, variable coefficients

10.10. third-order, linear, nonhomogeneous, constant coefficients

10.11. first-order, linear, homogeneous, constant coefficients

10.12. fourth-order, linear, nonhomogeneous, variable coefficients

10.13. second-order, nonlinear

10.14. (Rewrite as $y' - (2 + x)y = 0$.) first-order, linear, homogeneous, variable coefficients

10.15. second-order, nonlinear

10.16. third-order, linear, nonhomogeneous, variable coefficients

10.17. Theorem 10.1 does not apply, since $a_0(x) = -\frac{2}{x}$ is not continuous about $x_0 = 0$.

10.18. Yes; $a_0(x)$ is continuous about $x_0 = 1$.

10.19. Theorem 10.1 does not apply, since $b_1(x)$ is zero at the origin.

Chapter 11

Linear Differential Equations: Theory of Solutions

11.1 LINEAR DEPENDENCE

A set of functions $\{y_1(x), y_2(x), \ldots, y_n(x)\}$ is *linearly dependent* on $a \leq x \leq b$ if there exist constants $c_1, c_2, \ldots, c_n$, *not all zero,* such that

$$c_1y_1(x) + c_2y_2(x) + \cdots + c_ny_n(x) \equiv 0 \tag{11.1}$$

on $a \leq x \leq b$.

Example 11.1. The set $\{x, 5x, 1, \sin x\}$ is linearly dependent on $[-1, 1]$ since there exist constants $c_1 = -5$, $c_2 = 1$, $c_3 = 0$, and $c_4 = 0$, *not all zero,* such that (*11.1*) is satisfied. In particular,

$$-5 \cdot x + 1 \cdot 5x + 0 \cdot 1 + 0 \cdot \sin x \equiv 0$$

Note that $c_1 = c_2 = \cdots = c_n = 0$ is a set of constants that always satisfies (*11.1*). A set of functions is linearly dependent if there exists *another* set of constants, *not all zero,* that also satisfies (*11.1*).

11.2 LINEAR INDEPENDENCE

A set of functions $\{y_1(x), y_2(x), \ldots, y_n(x)\}$ is *linearly independent* on $a \leq x \leq b$ if it is not linearly dependent there; that is, if the only constants that satisfy (*11.1*) for $a \leq x \leq b$ are $c_1 = c_2 = \cdots = c_n = 0$. (See Problems 11.1–11.4.)

11.3 LINEARLY INDEPENDENT SOLUTIONS. THE WRONSKIAN

Theorem 11.1. The nth-order linear *homogeneous* differential equation $\mathbf{L}(y) = 0$ always has n linearly independent solutions. If $y_1(x), y_2(x), \ldots, y_n(x)$ represent these solutions, then the general solution of $\mathbf{L}(y) = 0$ is

$$y(x) = c_1y_1(x) + c_2y_2(x) + \cdots + c_ny_n(x) \tag{11.2}$$

where $c_1, c_2, \cdots, c_n$ denote arbitrary constants.

Example 11.2. Two solutions of $y'' + 4y = 0$ are $y_1(x) = \sin 2x$ and $y_2(x) = \cos 2x$. Since these solutions are linearly independent (see Example 11.4), the general solution is

$$y(x) = c_1 \sin 2x + c_2 \cos 2x$$

Theorem 11.1 makes clear the importance of being able to establish whether or not a set of solutions of $\mathbf{L}(y) = 0$ is linearly independent. The question usually cannot be settled directly from (*11.1*); one cannot very well try out all possible values of the c's. Fortunately, an alternative method exists.

Definition: Let $\{z_1(x), z_2(x), \cdots, z_n(x)\}$ be a set of functions on $a \leq x \leq b$, each of which possesses $n-1$ derivatives. The determinant

$$W(z_1, z_2, \ldots, z_n) = \begin{vmatrix} z_1 & z_2 & \cdots & z_n \\ z_1' & z_2' & \cdots & z_n' \\ z_1'' & z_2'' & \cdots & z_n'' \\ \vdots & \vdots & & \vdots \\ z_1^{(n-1)} & z_2^{(n-1)} & \cdots & z_n^{(n-1)} \end{vmatrix} \tag{11.3}$$

is called the *Wronskian* of the given set of functions.

Example 11.3. The Wronskian of the set $\{x, x^2, x^3\}$ is

$$W(x, x^2, x^3) = \begin{vmatrix} x & x^2 & x^3 \\ \frac{d(x)}{dx} & \frac{d(x^2)}{dx} & \frac{d(x^3)}{dx} \\ \frac{d^2(x)}{dx^2} & \frac{d^2(x^2)}{dx^2} & \frac{d^2(x^3)}{dx^2} \end{vmatrix}$$

$$= \begin{vmatrix} x & x^2 & x^3 \\ 1 & 2x & 3x^2 \\ 0 & 2 & 6x \end{vmatrix} = 2x^3$$

This example shows that the Wronskian is in general a nonconstant function.

Theorem 11.2. Let $\{y_1(x), y_2(x), \ldots, y_n(x)\}$ be a set of n solutions of the nth-order linear homogeneous differential equation $\mathbf{L}(y) = 0$. This set is linearly independent on $a \leq x \leq b$ if and only if the Wronskian of the set is not identically zero there.

Example 11.4. The two solutions $y_1(x) = \sin 2x$ and $y_2(x) = \cos 2x$ of $y'' + 4y = 0$ are linearly independent for all x, since

$$W(\sin 2x, \cos 2x) = \begin{vmatrix} \sin 2x & \cos 2x \\ \frac{d(\sin 2x)}{dx} & \frac{d(\cos 2x)}{dx} \end{vmatrix}$$

$$= \begin{vmatrix} \sin 2x & \cos 2x \\ 2\cos 2x & -2\sin 2x \end{vmatrix} = -2 \neq 0$$

Caution: (1) Recall from Section 10.2 that the coefficients of $\mathbf{L}(y)$ must be continuous on the interval of interest, which for Theorem 11.2 is $[a, b]$. If these continuity conditions are not satisfied, the theorem is false (see Problem 11.7). (2) Theorem 11.2 does *not* apply for an *arbitrary* set of functions. It can only be used to test for linear independence when the functions under consideration are *all* solutions of the same equation $\mathbf{L}(y) = 0$. (See Problems 11.4–11.6.) For arbitrary functions, one must test for linear dependence or independence directly in terms of satisfying (*11.1*). (See Problems 11.1–11.4.)

Now consider the general linear differential equation (nonhomogeneous) $\mathbf{L}(y) = \phi(x)$. Let y_p represent any *particular* solution (Section 2.2) of this equation and let y_h (henceforth called the *homogeneous* or *complementary solution*) represent the *general* solution of the associated homogeneous equation $\mathbf{L}(y) = 0$ (see Theorem 11.1). Then,

Theorem 11.3. The general solution to $\mathbf{L}(y) = \phi(x)$ is

$$y = y_h + y_p \tag{11.4}$$

Example 11.5. Consider the differential equation $y'' + 4y = x$. A particular solution is $y_p = \frac{1}{4}x$ (see Chapter 14). The general solution to the associated homogeneous equation, $y'' + 4y = 0$, is $y_h = c_1 \sin 2x + c_2 \cos 2x$ (see Example 11.2). The general solution to the given differential equation is, therefore,

$$y = y_h + y_p = c_1 \sin 2x + c_2 \cos 2x + \frac{1}{4}x$$

Solved Problems

11.1. Determine whether the set $\{e^x, e^{-x}\}$ is linearly dependent on $(-\infty, \infty)$.

Consider the equation

$$c_1 e^x + c_2 e^{-x} \equiv 0 \tag{1}$$

We must determine whether there exist values of c_1 and c_2, *not both zero*, that will satisfy (*1*). Rewriting (*1*), we have $c_2 e^{-x} \equiv -c_1 e^x$ or

$$c_2 \equiv -c_1 e^{2x} \tag{2}$$

For any nonzero value of c_1, the left side of (*2*) is a constant whereas the right side is not; hence the equality in (*2*) is not valid. It follows that the *only* solution to (*2*), and therefore to (*1*), is $c_1 = c_2 = 0$. Thus, the set is not linearly dependent; rather it is linearly independent.

11.2. Is the set $\{x^2, x, 1\}$ linearly dependent on $(-\infty, \infty)$?

Consider the equation

$$c_1 x^2 + c_2 x + c_3 \equiv 0 \tag{1}$$

Since this equation is valid for all x only if $c_1 = c_2 = c_3 = 0$, the given set is linearly independent. Note that if any of the c's were not zero, then the quadratic equation (*1*) could hold for at most two values of x, the roots of the equation, *and not for all x*.

11.3. Determine whether the set $\{1 - x, 1 + x, 1 - 3x\}$ is linearly dependent on $(-\infty, \infty)$.

Consider the equation

$$c_1(1 - x) + c_2(1 + x) + c_3(1 - 3x) \equiv 0 \tag{1}$$

which can be rewritten as

$$(-c_1 + c_2 - 3c_3)x + (c_1 + c_2 + c_3) \equiv 0$$

This linear equation can be satisfied for all x only if both coefficients are zero. Thus,

$$-c_1 + c_2 - 3c_3 = 0 \quad \text{and} \quad c_1 + c_2 + c_3 = 0$$

Solving these equations simultaneously, we find that $c_1 = -2c_3$, $c_2 = c_3$, with c_3 arbitrary. Choosing $c_3 = 1$ (any other nonzero number would do), we obtain $c_1 = -2$, $c_2 = 1$, and $c_3 = 1$ as a set of constants, not all zero, that satisfy (*1*). Thus, the given set of functions is linearly dependent.

11.4. Determine whether the set $\{x^3, |x^3|\}$ is linearly dependent on $[-1, 1]$.

Consider the equation

$$c_1 x^3 + c_2 |x^3| \equiv 0 \tag{1}$$

Recall that $|x^3| = x^3$ if $x \geqq 0$ and $|x^3| = -x^3$ if $x < 0$. Thus, when $x \geqq 0$, (1) becomes

$$c_1 x^3 + c_2 x^3 \equiv 0 \tag{2}$$

whereas when $x < 0$, (1) becomes

$$c_1 x^3 - c_2 x^3 \equiv 0 \tag{3}$$

Solving (2) and (3) simultaneously for c_1 and c_2, we find that the *only* solution is $c_1 = c_2 = 0$. The given set is, therefore, linearly independent.

11.5. Find $W(x^3, |x^3|)$ on $[-1, 1]$.

We have

$$|x^3| = \begin{cases} x^3 & \text{if } x \geqq 0 \\ -x^3 & \text{if } x < 0 \end{cases} \qquad \frac{d(|x^3|)}{dx} = \begin{cases} 3x^2 & \text{if } x > 0 \\ 0 & \text{if } x = 0 \\ -3x^2 & \text{if } x < 0 \end{cases}$$

Then, for $x > 0$,
$$W(x^3, |x^3|) = \begin{vmatrix} x^3 & x^3 \\ 3x^2 & 3x^2 \end{vmatrix} \equiv 0$$

For $x < 0$,
$$W(x^3, |x^3|) = \begin{vmatrix} x^3 & -x^3 \\ 3x^2 & -3x^2 \end{vmatrix} \equiv 0$$

For $x = 0$,
$$W(x^3, |x^3|) = \begin{vmatrix} 0 & 0 \\ 0 & 0 \end{vmatrix} = 0$$

Thus, $W(x^3, |x^3|) \equiv 0$ on $[-1, 1]$.

11.6. Do the results of Problems 11.4 and 11.5 contradict Theorem 11.2?

No. Since the Wronskian of two linearly independent functions is identically zero, it follows from Theorem 11.2 that these two functions, x^3 and $|x^3|$, are *not* both solutions of the *same* linear homogeneous differential equation of the form $\mathbf{L}(y) = 0$. (See Remark (2) following Example 11.4.)

11.7. Two solutions of $y'' - \dfrac{2}{x}y' = 0$ on $[-1, 1]$ are $y = x^3$ and $y = |x^3|$. Does this result contradict the solution to Problem 11.6?

No. Although $W(x^3, |x^3|) \equiv 0$ and both $y = x^3$ and $y = |x^3|$ are linearly independent solutions of the same linear homogeneous differential equation $y'' - \dfrac{2}{x}y' = 0$, this differential equation is not of the form $\mathbf{L}(y) = 0$. (The coefficient $-2/x$ is discontinuous at $x = 0$. See Remark (1) following Example 11.4.)

11.8. It can be shown by direct substitution that e^x and e^{-x} are both solutions of the differential equation $y'' - y = 0$. Prove that these functions are linearly independent.

We have

$$W(e^x, e^{-x}) = \begin{vmatrix} e^x & e^{-x} \\ e^x & -e^{-x} \end{vmatrix} = e^x(-e^{-x}) - e^{-x}(e^x) = -2 \not\equiv 0$$

Since the Wronskian is not identically zero, and the two functions are both solutions of the same differential equation of the form $\mathbf{L}(y) = 0$, it follows from Theorem 11.2 that the functions are linearly independent.

For an alternate method of solution, see Problem 11.1.

11.9. Find the general solution of $y'' - y = 0$.

From Problem 11.8, e^x and e^{-x} are two linearly independent solutions of $y'' - y = 0$. Using Theorem 11.1, we obtain the general solution as $y = c_1e^x + c_2e^{-x}$.

11.10. Find the general solution of $y'' - y = 2\sin x$, if it is known that a particular solution is $y_p = -\sin x$.

From Problem 11.9, the general solution to the associated homogeneous equation $y'' - y = 0$ is $y_h = c_1e^x + c_2e^{-x}$. It now follows from Theorem 11.3 that the general solution to the given differential equation is

$$y = y_h + y_p = c_1e^x + c_2e^{-x} - \sin x$$

11.11. Two solutions of $y'' - 2y' + y = 0$ are (see Chapter 12) e^{-x} and $5e^{-x}$. Is the general solution $y = c_1e^{-x} + c_2 5e^{-x}$?

We calculate:
$$W(e^{-x}, 5e^{-x}) = \begin{vmatrix} e^{-x} & 5e^{-x} \\ -e^{-x} & -5e^{-x} \end{vmatrix} \equiv 0$$

Therefore the functions e^{-x} and $5e^{-x}$ are linearly dependent (see Theorem 11.2), and we conclude from Theorem 11.1 that $y = c_1e^{-x} + c_2 5e^{-x}$ is *not* the general solution.

11.12. Find the general solution of $y'' - 2y' + y = x^2$, if one solution is (see Chapter 14) $y = x^2 + 4x + 6$, and if two solutions of $y'' - 2y' + y = 0$ are (see Chapter 12) e^x and xe^x.

Because
$$W(e^x, xe^x) = \begin{vmatrix} e^x & xe^x \\ e^x & e^x + xe^x \end{vmatrix} = e^{2x} \not\equiv 0$$

it follows from Theorem 11.2 that e^x and xe^x are linearly independent on $(-\infty, \infty)$. We have, therefore, from Theorem 11.1 that the general solution to $y'' - 2y' + y = 0$ is

$$y_h = c_1e^x + c_2xe^x$$

Since we are given that $y_p = x^2 + 4x + 6$, it follows from Theorem 11.3 that

$$y = c_1e^x + c_2xe^x + x^2 + 4x + 6$$

11.13. Prove Theorem 11.3.

Since $\mathbf{L}(y_h) = 0$ and $\mathbf{L}(y_p) = \phi(x)$, it follows from the linearity of $\mathbf{L}$ that

$$\mathbf{L}(y) = \mathbf{L}(y_h + y_p) = \mathbf{L}(y_h) + \mathbf{L}(y_p) = 0 + \phi(x) = \phi(x)$$

Thus, y is a solution.

To prove that it is the general solution, we must show that every solution of $\mathbf{L}(y) = \phi(x)$ is of the form *(11.4)*. Let y be any solution of $\mathbf{L}(y) = \phi(x)$ and set $z = y - y_p$. Then

$$\mathbf{L}(z) = \mathbf{L}(y - y_p) = \mathbf{L}(y) - \mathbf{L}(y_p) = \phi(x) - \phi(x) = 0$$

so that z is a solution to the homogeneous equation $\mathbf{L}(y) = 0$. Since $z = y - y_p$, it follows that $y = z + y_p$, where z is a solution of $\mathbf{L}(y) = 0$.

Supplementary Problems

In Problems 11.14 through 11.18, determine whether the given sets of functions are linearly dependent on $(-\infty, \infty)$.

11.14. $\{e^{2x}, e^{-2x}\}$.

11.15. $\{e^{\lambda_1 x}, e^{\lambda_2 x}\}$, $\lambda_1 \neq \lambda_2$.

11.16. $\{x, 1, 2x-7\}$.

11.17. $\{x+1, x-1\}$.

11.18. $\{x+1, x^2+x, 2x^2-x-3\}$.

11.19. Find the Wronskian of (a) $\{x^2, x\}$; (b) $\{\sin x,\ 2\cos x,\ 3\sin x + \cos x\}$; (c) $\{e^x, e^{-x}, e^{2x}\}$.

11.20. Find the general solution of $y''+y=x^2$, if one solution is $y=x^2-2$, and if two solutions of $y''+y=0$ are $\sin x$ and $\cos x$.

11.21. Find the general solution of $y''-y=x^2$, if one solution is $y=-x^2-2$, and if two solutions of $y''-y=0$ are e^x and $3e^x$.

11.22. Find the general solution of $y'''-y''-y+1=5$, if one solution is $y=-4$, and if three solutions of $y'''-y''-y+1=0$ are e^x, e^{-x}, and xe^x.

Answers to Supplementary Problems

11.14. independent

11.15. independent

11.16. dependent; $c_1=-2,\ c_2=7,\ c_3=1$

11.17. independent

11.18. dependent; $c_1=3,\ c_2=-2,\ c_3=1$

11.19. (a) $-x^2$; (b) 0; (c) $-6e^{2x}$

11.20. $y = c_1 \sin x + c_2 \cos x + x^2 - 2$

11.21. Since e^x and $3e^x$ are linearly dependent, there is not enough information given to find the general solution.

11.22. $y = c_1 e^x + c_2 e^{-x} + c_3 x e^x - 4$

Chapter 12

Second-Order Linear Homogeneous Differential Equations with Constant Coefficients

12.1 THE CHARACTERISTIC EQUATION

Corresponding to the differential equation

$$y'' + a_1 y' + a_0 y = 0 \tag{12.1}$$

in which a_1 and a_0 are constants, is the algebraic equation

$$\lambda^2 + a_1\lambda + a_0 = 0 \tag{12.2}$$

which is obtained from (*12.1*) by replacing y'', y', and y by λ^2, λ^1, and $\lambda^0 = 1$, respectively. Equation (*12.2*) is called the *characteristic equation* of (*12.1*).

Example 12.1. The characteristic equation of $y'' + 3y' - 4y = 0$ is $\lambda^2 + 3\lambda - 4 = 0$; the characteristic equation of $y'' - 2y' + y = 0$ is $\lambda^2 - 2\lambda + 1 = 0$.

The characteristic equation can be factored into

$$(\lambda - \lambda_1)(\lambda - \lambda_2) = 0 \tag{12.3}$$

12.2 SOLUTION IN TERMS OF THE CHARACTERISTIC ROOTS

The solution of (*12.1*) is obtained directly from the roots of (*12.3*). There are three cases to consider.

Case 1. λ_1 and λ_2 both real and distinct. Two linearly independent solutions are $e^{\lambda_1 x}$ and $e^{\lambda_2 x}$, and the general solution is (Theorem 11.1)

$$y = c_1 e^{\lambda_1 x} + c_2 e^{\lambda_2 x} \tag{12.4}$$

(See Problems 12.1–12.3.) In the special case $\lambda_2 = -\lambda_1$, the solution (*12.4*) can be rewritten as

$$y = k_1 \cosh \lambda_1 x + k_2 \sinh \lambda_1 x \tag{12.5}$$

Here $k_1 = c_1 + c_2$ and $k_2 = c_1 - c_2$, and we have used the identities

$$\cosh \lambda_1 x \equiv \tfrac{1}{2}(e^{\lambda_1 x} + e^{-\lambda_1 x}) \qquad \sinh \lambda_1 x \equiv \tfrac{1}{2}(e^{\lambda_1 x} - e^{-\lambda_1 x})$$

(See Problem 12.3.)

Case 2. $\lambda_1 = a + ib$, **a complex number.** Since a_1 and a_0 in (*12.1*) and (*12.2*) are assumed real, the roots of (*12.2*) must appear in conjugate pairs; thus, the other root is $\lambda_2 = a - ib$. Two linearly independent solutions are $e^{(a+ib)x}$ and $e^{(a-ib)x}$, and the general complex solution is

$$y = d_1 e^{(a+ib)x} + d_2 e^{(a-ib)x}$$

Using Euler's relations

$$e^{ibx} = \cos bx + i \sin bx \qquad e^{-ibx} = \cos bx - i \sin bx$$

we can rewrite the solution as

$$\begin{aligned} y &= d_1 e^{ax} e^{ibx} + d_2 e^{ax} e^{-ibx} = e^{ax}(d_1 e^{ibx} + d_2 e^{-ibx}) \\ &= e^{ax}[d_1(\cos bx + i \sin bx) + d_2(\cos bx - i \sin bx)] \\ &= e^{ax}[(d_1 + d_2) \cos bx + i(d_1 - d_2) \sin bx] \end{aligned}$$

If we define $c_1 = d_1 + d_2$ and $c_2 = i(d_1 - d_2)$ as two new arbitrary constants, we can give the general solution by

$$y = c_1 e^{ax} \cos bx + c_2 e^{ax} \sin bx \tag{12.6}$$

Equation (*12.6*) is real if and only if c_1 and c_2 are both real, which occurs if and only if d_1 and d_2 are complex conjugates. Since we are interested in the general *real* solution to (*12.1*), we restrict d_1 and d_2 to be a conjugate pair. (See Problems 12.4–12.6.)

Case 3. $\lambda_1 = \lambda_2$. Two linearly independent solutions are $e^{\lambda_1 x}$ and $xe^{\lambda_1 x}$, and the general solution is

$$y = c_1 e^{\lambda_1 x} + c_2 x e^{\lambda_1 x} \tag{12.7}$$

(See Problems 12.7 and 12.8.)

Warning: The above solutions *are not valid* if the differential equation is not linear or does not have constant coefficients. Consider, for example, the equation $y'' - x^2 y = 0$. The roots of the characteristic equation are $\lambda_1 = x$ and $\lambda_2 = -x$, but the solution is *not*

$$y = c_1 e^{(x)x} + c_2 e^{(-x)x} = c_1 e^{x^2} + c_2 e^{-x^2}$$

Linear equations with variable coefficients are considered in Chapters 18 through 21.

Solved Problems

12.1. Solve $y'' - y' - 2y = 0$.

The characteristic equation is $\lambda^2 - \lambda - 2 = 0$, which can be factored into $(\lambda + 1)(\lambda - 2) = 0$. Since the roots $\lambda_1 = -1$ and $\lambda_2 = 2$ are real and distinct, the solution is given by (*12.4*) as

$$y = c_1 e^{-x} + c_2 e^{2x}$$

12.2. Solve $y'' - 7y' = 0$.

The characteristic equation is $\lambda^2 - 7\lambda = 0$, which can be factored into $(\lambda - 0)(\lambda - 7) = 0$. Since the roots $\lambda_1 = 0$ and $\lambda_2 = 7$ are real and distinct, the solution is given by *(12.4)* as

$$y = c_1 e^{0x} + c_2 e^{7x} = c_1 + c_2 e^{7x}$$

12.3. Solve $y'' - 5y = 0$.

The characteristic equation is $\lambda^2 - 5 = 0$, which can be factored into $(\lambda - \sqrt{5})(\lambda + \sqrt{5}) = 0$. Since the roots $\lambda_1 = \sqrt{5}$ and $\lambda_2 = -\sqrt{5}$ are real and distinct, the solution is given by *(12.4)* as

$$y = c_1 e^{\sqrt{5}x} + c_2 e^{-\sqrt{5}x}$$

or by *(12.5)* as

$$y = k_1 \cosh \sqrt{5}\, x + k_2 \sinh \sqrt{5}\, x$$

12.4. Solve $y'' + 4y' + 5y = 0$.

The characteristic equation is $\lambda^2 + 4\lambda + 5 = 0$, which has roots $\lambda_1 = -2 + i$ and $\lambda_2 = -2 - i$. The solution is given by *(12.6)* as

$$y = c_1 e^{-2x} \cos x + c_2 e^{-2x} \sin x$$

12.5. Solve $y'' + 4y = 0$.

The characteristic equation is $\lambda^2 + 4 = 0$, which has roots $\lambda_1 = 2i$ and $\lambda_2 = -2i$. The solution is given by *(12.6)* as

$$y = c_1 \cos 2x + c_2 \sin 2x$$

Note that these complex roots have real parts equal to zero.

12.6. Solve $y'' - 3y' + 4y = 0$.

The characteristic equation is $\lambda^2 - 3\lambda + 4 = 0$, which has roots

$$\lambda_1 = \frac{3}{2} + i\frac{\sqrt{7}}{2} \quad \text{and} \quad \lambda_2 = \frac{3}{2} - i\frac{\sqrt{7}}{2}$$

The solution is given by *(12.6)* as

$$y = c_1 e^{(3/2)x} \cos \frac{\sqrt{7}}{2} x + c_2 e^{(3/2)x} \sin \frac{\sqrt{7}}{2} x$$

12.7. Solve $y'' + 4y' + 4y = 0$.

The characteristic equation is $\lambda^2 + 4\lambda + 4 = 0$, which has roots $\lambda_1 = \lambda_2 = -2$. The solution is given by *(12.7)* as

$$y = c_1 e^{-2x} + c_2 x e^{-2x}$$

12.8. Solve $y'' = 0$.

The characteristic equation is $\lambda^2 = 0$, which has roots $\lambda_1 = \lambda_2 = 0$. The solution is given by *(12.7)* as

$$y = c_1 e^{0x} + c_2 x e^{0x} = c_1 + c_2 x$$

Supplementary Problems

Solve the following differential equations.

12.9. $y'' - y = 0.$

12.10. $y'' - y' - 30y = 0.$

12.11. $y'' - 2y' + y = 0.$

12.12. $y'' + y = 0.$

12.13. $y'' + 2y' + 2y = 0.$

12.14. $y'' - 7y = 0.$

12.15. $y'' + 6y' + 9y = 0.$

12.16. $y'' + 2y' + 3y = 0.$

12.17. $y'' - 3y' - 5y = 0.$

12.18. $y'' + y' + \frac{1}{4}y = 0.$

Answers to Supplementary Problems

12.9. $y = c_1e^{x} + c_2e^{-x}$

12.10. $y = c_1e^{-5x} + c_2e^{6x}$

12.11. $y = c_1e^{x} + c_2xe^{x}$

12.12. $y = c_1 \cos x + c_2 \sin x$

12.13. $y = c_1e^{-x} \cos x + c_2e^{-x} \sin x$

12.14. $y = c_1e^{\sqrt{7}x} + c_2e^{-\sqrt{7}x}$

12.15. $y = c_1e^{-3x} + c_2xe^{-3x}$

12.16. $y = c_1e^{-x} \cos \sqrt{2}\, x + c_2e^{-x} \sin \sqrt{2}\, x$

12.17. $y = c_1e^{[(3+\sqrt{29})/2]x} + c_2e^{[(3-\sqrt{29})/2]x}$

$$= e^{(3/2)x}\left(k_1 \cosh \frac{\sqrt{29}}{2}x + k_2 \sinh \frac{\sqrt{29}}{2}x\right)$$

12.18. $y = c_1e^{-(1/2)x} + c_2xe^{-(1/2)x}$

Chapter 13

nth-Order Linear Homogeneous Differential Equations with Constant Coefficients

13.1 THE CHARACTERISTIC EQUATION

One method for solving an nth-order equation is a direct extension of the method given in the previous chapter for solving a second-order equation.

The differential equation under consideration is

$$y^{(n)} + a_{n-1}y^{(n-1)} + \cdots + a_1y' + a_0y = 0 \tag{13.1}$$

where a_j $(j = 0, 1, \ldots, n-1)$ is a constant. The characteristic equation associated with (*13.1*) is

$$\lambda^n + a_{n-1}\lambda^{n-1} + \cdots + a_1\lambda + a_0 = 0 \tag{13.2}$$

It is obtained from (*13.1*) by replacing $y^{(j)}$ by λ^j $(j = 0, 1, \ldots, n)$.

Example 13.1. The characteristic equation of $y^{(4)} - 3y''' + 2y'' - y = 0$ is $\lambda^4 - 3\lambda^3 + 2\lambda^2 - 1 = 0$.

13.2 SOLUTION IN TERMS OF THE CHARACTERISTIC ROOTS

The roots of the characteristic equation again determine the solution of (*13.1*). If the roots $\lambda_1, \lambda_2, \ldots, \lambda_n$ are all real and distinct, the solution is

$$y = c_1e^{\lambda_1 x} + c_2e^{\lambda_2 x} + \cdots + c_ne^{\lambda_n x} \tag{13.3}$$

If the roots are distinct, but some are complex, then the solution is again given by (*13.3*). Once again, those terms involving complex exponentials can be combined to yield terms involving sines and cosines. If λ_k is a root of multiplicity p (that is, if $(\lambda - \lambda_k)^p$ is a factor of the characteristic equation, but $(\lambda - \lambda_k)^{p+1}$ is not) then there will be p linearly independent solutions associated with λ_k given by $e^{\lambda_k x}, xe^{\lambda_k x}, x^2e^{\lambda_k x}, \ldots, x^{p-1}e^{\lambda_k x}$. These solutions are combined in the usual way with the solutions associated with the other roots to obtain the complete solution. (See Problems 13.6–13.9.)

In theory it is always possible to factor the characteristic equation, but in practice this can be extremely difficult, especially for differential equations of high order. In such cases, one must often use numerical techniques to approximate the roots or develop other methods of solution. See Chapters 32–35.

Solved Problems

13.1. Solve $y''' - 6y'' + 11y' - 6y = 0$.

The characteristic equation is $\lambda^3 - 6\lambda^2 + 11\lambda - 6 = 0$, which can be factored into

$$(\lambda - 1)(\lambda - 2)(\lambda - 3) = 0$$

The roots are $\lambda_1 = 1$, $\lambda_2 = 2$, and $\lambda_3 = 3$; hence the solution is

$$y = c_1e^{x} + c_2e^{2x} + c_3e^{3x}$$

13.2. Solve $y^{(4)} - 9y'' + 20y = 0$.

The characteristic equation is $\lambda^4 - 9\lambda^2 + 20 = 0$, which can be factored into

$$(\lambda - 2)(\lambda + 2)(\lambda - \sqrt{5})(\lambda + \sqrt{5}) = 0$$

The roots are $\lambda_1 = 2$, $\lambda_2 = -2$, $\lambda_3 = \sqrt{5}$, and $\lambda_4 = -\sqrt{5}$; hence the solution is

$$\begin{aligned} y &= c_1e^{2x} + c_2e^{-2x} + c_3e^{\sqrt{5}x} + c_4e^{-\sqrt{5}x} \\ &= k_1 \cosh 2x + k_2 \sinh 2x + k_3 \cosh \sqrt{5}\,x + k_4 \sinh \sqrt{5}\,x \end{aligned}$$

13.3. Solve $y' - 5y = 0$.

The characteristic equation is $\lambda - 5 = 0$, which has the single root $\lambda_1 = 5$. The solution is then $y = c_1e^{5x}$. (Compare this result with Problem 8.5. The two methods are equivalent for first-order homogeneous linear differential equations with constant coefficients.)

13.4. Solve $y''' - 6y'' + 2y' + 36y = 0$.

The characteristic equation, $\lambda^3 - 6\lambda^2 + 2\lambda + 36 = 0$, has roots $\lambda_1 = -2$, $\lambda_2 = 4 + i\sqrt{2}$, and $\lambda_3 = 4 - i\sqrt{2}$. The solution is

$$y = c_1e^{-2x} + d_2e^{(4+i\sqrt{2})x} + d_3e^{(4-i\sqrt{2})x}$$

which can be rewritten, using Euler's relations (see Section 12.2), as

$$y = c_1e^{-2x} + c_2e^{4x} \cos \sqrt{2}\,x + c_3e^{4x} \sin \sqrt{2}\,x$$

13.5. Solve $y^{(4)} - 4y''' + 7y'' - 4y' + 6y = 0$.

The characteristic equation, $\lambda^4 - 4\lambda^3 + 7\lambda^2 - 4\lambda + 6 = 0$, has roots $\lambda_1 = 2 + i\sqrt{2}$, $\lambda_2 = 2 - i\sqrt{2}$, $\lambda_3 = i$, and $\lambda_4 = -i$. The solution is

$$y = d_1e^{(2+i\sqrt{2})x} + d_2e^{(2-i\sqrt{2})x} + d_3e^{ix} + d_4e^{-ix}$$

If, using Euler's relations, we combine the first two terms and then similarly combine the last two terms, we can rewrite the solution as

$$y = c_1e^{2x} \cos \sqrt{2}\,x + c_2e^{2x} \sin \sqrt{2}\,x + c_3 \cos x + c_4 \sin x$$

13.6. Solve $y^{(4)} + 8y''' + 24y'' + 32y' + 16y = 0$.

The characteristic equation, $\lambda^4 + 8\lambda^3 + 24\lambda^2 + 32\lambda + 16 = 0$, can be factored into $(\lambda + 2)^4 = 0$. Here $\lambda_1 = -2$ is a root of multiplicity four; hence the solution is

$$y = c_1e^{-2x} + c_2xe^{-2x} + c_3x^2e^{-2x} + c_4x^3e^{-2x}$$

13.7. Solve $y^{(5)} - y^{(4)} - 2y''' + 2y'' + y' - y = 0$.

The characteristic equation can be factored into $(\lambda - 1)^3(\lambda + 1)^2 = 0$; hence, $\lambda_1 = 1$ is a root of multiplicity three and $\lambda_2 = -1$ is a root of multiplicity two. The solution is

$$y = c_1e^{x} + c_2xe^{x} + c_3x^2e^{x} + c_4e^{-x} + c_5xe^{-x}$$

13.8. Solve $y^{(4)} - 8y''' + 32y'' - 64y' + 64y = 0.$

The characteristic equation has roots $2 \pm i2$ and $2 \pm i2$; hence $\lambda_1 = 2 + i2$ and $\lambda_2 = 2 - i2$ are both roots of multiplicity two. The solution is

$$\begin{aligned} y &= d_1 e^{(2+i2)x} + d_2 x e^{(2+i2)x} + d_3 e^{(2-i2)x} + d_4 x e^{(2-i2)x} \\ &= e^{2x}(d_1 e^{i2x} + d_3 e^{-i2x}) + x e^{2x}(d_2 e^{i2x} + d_4 e^{-i2x}) \\ &= e^{2x}(c_1 \cos 2x + c_3 \sin 2x) + x e^{2x}(c_2 \cos 2x + c_4 \sin 2x) \\ &= (c_1 + c_2 x)e^{2x} \cos 2x + (c_3 + c_4 x)e^{2x} \sin 2x \end{aligned}$$

13.9. Solve $y^{(6)} - 5y^{(4)} + 16y''' + 36y'' - 16y' - 32y = 0.$

The roots of the characteristic equation are $2 \pm i2$, -2, -2, 1, and -1. All roots are distinct, except for the root $\lambda_3 = -2$, which is of multiplicity two. The solution is

$$y = c_1 e^{2x} \cos 2x + c_2 e^{2x} \sin 2x + c_3 e^{-2x} + c_4 x e^{-2x} + c_5 e^{x} + c_6 e^{-x}$$

Supplementary Problems

Solve the following differential equations.

13.10. $y''' - 2y'' - y' + 2y = 0.$

13.11. $y''' - y'' - y' + y = 0.$

13.12. $y''' - 3y'' + 3y' - y = 0.$

13.13. $y''' - y'' + y' - y = 0.$

13.14. $y^{(4)} + 2y'' + y = 0.$

13.15. $y^{(4)} - y = 0.$

13.16. $y^{(4)} + 2y''' - 2y' - y = 0.$

13.17. $y^{(4)} - 4y'' + 16y' + 32y = 0.$

13.18. $y^{(4)} + 5y''' = 0.$

13.19. $y^{(4)} + 2y''' + 3y'' + 2y' + y = 0.$

Answers to Supplementary Problems

13.10. $y = c_1 e^{-x} + c_2 e^{x} + c_3 e^{2x}$

13.11. $y = c_1 e^{x} + c_2 x e^{x} + c_3 e^{-x}$

13.12. $y = c_1 e^{x} + c_2 x e^{x} + c_3 x^2 e^{x}$

13.13. $y = c_1 e^{x} + c_2 \cos x + c_3 \sin x$

13.14. $y = (c_1 + c_2 x) \cos x + (c_3 + c_4 x) \sin x$

13.15. $y = c_1 e^{x} + c_2 e^{-x} + c_3 \cos x + c_4 \sin x$

13.16. $y = c_1 e^{x} + c_2 e^{-x} + c_3 x e^{-x} + c_4 x^2 e^{-x}$

13.17. $y = c_1 e^{-2x} + c_2 x e^{-2x} + c_3 e^{2x} \cos 2x + c_4 e^{2x} \sin 2x$

13.18. $y = c_1 + c_2 x + c_3 x^2 + c_4 e^{-5x}$

13.19. $y = (c_1 + c_3 x)e^{-(1/2)x} \cos \frac{\sqrt{3}}{2} x + (c_2 + c_4 x)e^{-(1/2)x} \sin \frac{\sqrt{3}}{2} x$

Chapter 14

The Method of Undetermined Coefficients

The general solution to the linear differential equation $\mathbf{L}(y) = \phi(x)$ (See Section 10.2) is given by Theorem 11.3 as $y = y_h + y_p$. Methods for obtaining y_h when the differential equation has constant coefficients are given in Chapters 12 and 13. In this chapter and in the next, we give methods for obtaining y_p *once* y_h *is known.*

The *method of undetermined coefficients* is initiated by guessing the form of y_p up to arbitrary multiplicative constants. These arbitrary constants are then evaluated by substituting the proposed solution into the given differential equation and equating the coefficients of like terms.

14.1 SIMPLE FORM OF THE METHOD

We assume that the desired y_p can be expressed as a sum of those terms that form $\phi(x)$ and all the derivatives of $\phi(x)$ (disregarding multiplicative constants).

Case 1: $\phi(x) = p_n(x)$**, an** n**th-degree polynomial in** x**.** Assume a solution of the form

$$y_p = A_n x^n + A_{n-1}x^{n-1} + \cdots + A_1 x + A_0 \tag{14.1}$$

where A_j $(j = 0, 1, 2, \ldots, n)$ is a constant to be determined. (See Problem 14.1.)

Case 2: $\phi(x) = e^{\alpha x}p_n(x)$**, where** α **is a known constant and** $p_n(x)$ **is as in Case 1.** Assume a solution of the form

$$y_p = e^{\alpha x}(A_n x^n + A_{n-1}x^{n-1} + \cdots + A_1 x + A_0) \tag{14.2}$$

where A_j is as in Case 1. (See Problems 14.2 and 14.4.)

Case 3: $\phi(x) = e^{\alpha x}p_n(x)\sin\beta x$**, where** α **and** β **are known constants and** $p_n(x)$ **is as in Case 1.** Assume a solution of the form

$$y_p = e^{\alpha x}\sin\beta x\,(A_n x^n + \cdots + A_1 x + A_0) + e^{\alpha x}\cos\beta x\,(B_n x^n + \cdots + B_1 x + B_0) \tag{14.3}$$

where A_j and B_j $(j = 0, 1, \ldots, n)$ are constants which still must be determined. (See Problem 14.3.)

Case 4: $\phi(x) = e^{\alpha x}p_n(x)\cos\beta x$**, where** α**,** β**, and** $p_n(x)$ **are as in Case 3.** Assume a solution of the same form as *(14.3)*.

14.2 MODIFICATIONS

If any term of the assumed solution, disregarding multiplicative constants, is also a term of y_h (the homogeneous solution), then the assumed solution must be modified by multiplying it by x^m, where m is the smallest positive integer such that the product of x^m with the assumed solution has no terms in common with y_h. (See Problems 14.5 and 14.6.)

Example 14.1. Consider a differential equation with $y_h = c_1x + c_0$ and $\phi(x) = 9x^2 + 2x - 1$. Since $\phi(x)$ is a second-degree polynomial (Case 1 above), we first try $y_p = A_2x^2 + A_1x + A_0$. Note, however, that this assumed solution has terms, disregarding multiplicative constants, in common with y_h: in particular, the first-power term and the constant term. Hence, we must determine the smallest positive integer m such that $x^m(A_2x^2 + A_1x + A_0)$ has no terms in common with y_h.

For $m = 1$, we obtain

$$x(A_2x^2 + A_1x + A_0) = A_2x^3 + A_1x^2 + A_0x$$

which still has a first-power term in common with y_h. For $m = 2$, we obtain

$$x^2(A_2x^2 + A_1x + A_0) = A_2x^4 + A_1x^3 + A_0x^2$$

which has no terms in common with y_h; therefore, we assume an expression of this form for y_p.

14.3 GENERALIZATIONS

If $\phi(x)$ is the sum (or difference) of terms already considered, then we take y_p to be the sum (or difference) of the corresponding assumed solutions and algebraically combine arbitrary constants where possible. (See Problems 14.7–14.9.)

Example 14.2. Consider $\phi(x) = (x-1)\sin x + (x+1)\cos x$. An assumed solution for $(x-1)\sin x$ is given by *(14.3)* (with $\alpha = 0$) as

$$(A_1x + A_0)\sin x + (B_1x + B_0)\cos x$$

and an assumed solution for $(x+1)\cos x$ is given also by *(14.3)* as

$$(C_1x + C_0)\sin x + (D_1x + D_0)\cos x$$

(Note that we have used C and D in the last expression, since the constants A and B already have been used.) We therefore take

$$y_p = (A_1x + A_0)\sin x + (B_1x + B_0)\cos x + (C_1x + C_0)\sin x + (D_1x + D_0)\cos x$$

Combining like terms, we arrive at

$$y_p = (E_1x + E_0)\sin x + (F_1x + F_0)\cos x$$

as the assumed solution, where $E_j = A_j + C_j$ and $F_j = B_j + D_j$ $(j = 0, 1)$.

14.4 LIMITATIONS OF THE METHOD

In general, if $\phi(x)$ is not one of the types of functions considered above, or if the differential equation *does not have constant coefficients,* then the method given in Chapter 15 should be preferred.

Solved Problems

14.1. Solve $y'' - y' - 2y = 4x^2$.

From Problem 12.1, $y_h = c_1e^{-x} + c_2e^{2x}$. Here $\phi(x) = 4x^2$, a second-degree polynomial. Using *(14.1)*, we assume that

$$y_p = A_2x^2 + A_1x + A_0 \tag{1}$$

Thus, $y_p' = 2A_2x + A_1$ and $y_p'' = 2A_2$. Substituting these results into the differential equation, we have

$$2A_2 - (2A_2x + A_1) - 2(A_2x^2 + A_1x + A_0) = 4x^2$$

or, equivalently,

$$-2A_2x^2 + (-2A_2 - 2A_1)x + (2A_2 - A_1 - 2A_0) = 4x^2 + (0)x + 0$$

Equating the coefficients of like powers of x, we obtain

$$-2A_2 = 4 \qquad -2A_2 - 2A_1 = 0 \qquad 2A_2 - A_1 - 2A_0 = 0$$

Solving this system, we find that $A_2 = -2$, $A_1 = 2$, and $A_0 = -3$. Hence (1) becomes

$$y_p = -2x^2 + 2x - 3$$

and the general solution is

$$y = y_h + y_p = c_1e^{-x} + c_2e^{2x} - 2x^2 + 2x - 3$$

14.2. Solve $y'' - y' - 2y = e^{3x}$.

From Problem 12.1, $y_h = c_1e^{-x} + c_2e^{2x}$. Here $\phi(x) = e^{\alpha x}p_n(x)$, where $\alpha = 3$ and $p_n(x) = 1$, a zero-degree polynomial. Using (14.2), we assume that

$$y_p = A_0e^{3x} \tag{1}$$

Thus, $y_p' = 3A_0e^{3x}$ and $y_p'' = 9A_0e^{3x}$. Substituting these results into the differential equation, we have

$$9A_0e^{3x} - 3A_0e^{3x} - 2A_0e^{3x} = e^{3x} \qquad \text{or} \qquad 4A_0e^{3x} = e^{3x}$$

It follows that $4A_0 = 1$, or $A_0 = \frac{1}{4}$, so that (1) becomes $y_p = \frac{1}{4}e^{3x}$. The general solution then is

$$y = c_1e^{-x} + c_2e^{2x} + \frac{1}{4}e^{3x}$$

14.3. Solve $y'' - y' - 2y = \sin 2x$.

Again by Problem 12.1, $y_h = c_1e^{-x} + c_2e^{2x}$. Here $\phi(x) = e^{\alpha x}p_n(x) \sin \beta x$, where $\alpha = 0$, $\beta = 2$, and $p_n(x) = 1$. Using (14.3), we assume that

$$y_p = A_0 \sin 2x + B_0 \cos 2x \tag{1}$$

Thus, $y_p' = 2A_0 \cos 2x - 2B_0 \sin 2x$ and $y_p'' = -4A_0 \sin 2x - 4B_0 \cos 2x$. Substituting these results into the differential equation, we have

$$(-4A_0 \sin 2x - 4B_0 \cos 2x) - (2A_0 \cos 2x - 2B_0 \sin 2x) - 2(A_0 \sin 2x + B_0 \cos 2x) = \sin 2x$$

or, equivalently,

$$(-6A_0 + 2B_0) \sin 2x + (-6B_0 - 2A_0) \cos 2x = (1) \sin 2x + (0) \cos 2x$$

Equating coefficients of like terms, we obtain

$$-6A_0 + 2B_0 = 1 \qquad -2A_0 - 6B_0 = 0$$

Solving this system, we find that $A_0 = -3/20$ and $B_0 = 1/20$. Then from (1),

$$y_p = -\frac{3}{20} \sin 2x + \frac{1}{20} \cos 2x$$

and the general solution is

$$y = c_1e^{-x} + c_2e^{2x} - \frac{3}{20} \sin 2x + \frac{1}{20} \cos 2x$$

14.4. Solve $y''' - 6y'' + 11y' - 6y = 2xe^{-x}$.

From Problem 13.1, $y_h = c_1e^{x} + c_2e^{2x} + c_3e^{3x}$. Here $\phi(x) = e^{\alpha x}p_n(x)$, where $\alpha = -1$ and $p_n(x) = 2x$, a first-degree polynomial. Using (14.2), we assume that $y_p = e^{-x}(A_1x + A_0)$, or

$$y_p = A_1xe^{-x} + A_0e^{-x} \tag{1}$$

Thus,

$$y_p' = -A_1xe^{-x} + A_1e^{-x} - A_0e^{-x}$$

$$y_p'' = A_1xe^{-x} - 2A_1e^{-x} + A_0e^{-x}$$

$$y_p''' = -A_1xe^{-x} + 3A_1e^{-x} - A_0e^{-x}$$

Substituting these results into the differential equation and simplifying, we obtain

$$-24A_1xe^{-x} + (26A_1 - 24A_0)e^{-x} = 2xe^{-x} + (0)e^{-x}$$

Equating coefficients of like terms, we have

$$-24A_1 = 2 \qquad 26A_1 - 24A_0 = 0$$

from which $A_1 = -1/12$ and $A_0 = -13/144$.

Equation (*1*) becomes

$$y_p = -\frac{1}{12}xe^{-x} - \frac{13}{144}e^{-x}$$

and the general solution is

$$y = c_1e^x + c_2e^{2x} + c_3e^{3x} - \frac{1}{12}xe^{-x} - \frac{13}{144}e^{-x}$$

14.5. Solve $y'' = 9x^2 + 2x - 1$.

The solution of $y'' = 0$ is $y_h = c_1x + c_0$, which has terms in common with $\phi(x)$. Hence, we assume that (see Example 14.1)

$$y_p = A_2x^4 + A_1x^3 + A_0x^2 \tag{1}$$

Substituting (*1*) into the differential equation, we obtain

$$12A_2x^2 + 6A_1x + 2A_0 = 9x^2 + 2x - 1$$

from which $A_2 = 3/4$, $A_1 = 1/3$ and $A_0 = -1/2$. Then (*1*) becomes

$$y_p = \frac{3}{4}x^4 + \frac{1}{3}x^3 - \frac{1}{2}x^2$$

and the general solution is

$$y = c_1x + c_0 + \frac{3}{4}x^4 + \frac{1}{3}x^3 - \frac{1}{2}x^2$$

The solution also can be obtained simply by twice integrating both sides of the differential equation with respect to x.

14.6. Solve $y' - 5y = 2e^{5x}$.

From Problem 13.3, $y_h = c_1e^{5x}$. Since $\phi(x) = 2e^{5x}$, it would follow from (*14.2*) that the guess for y_p should be $y_p = A_0e^{5x}$. Note, however, that this y_p has exactly the same form as y_h; therefore (see Section 14.2), we must modify y_p. Multiplying y_p by x ($m = 1$), we obtain

$$y_p = A_0xe^{5x} \tag{1}$$

As this expression has no terms in common with y_h, it is a candidate for the particular solution. Substituting (*1*) and $y_p' = A_0e^{5x} + 5A_0xe^{5x}$ into the differential equation and simplifying, we obtain $A_0e^{5x} = 2e^{5x}$, from which $A_0 = 2$. Equation (*1*) becomes $y_p = 2xe^{5x}$, and the general solution is $y = (c_1 + 2x)e^{5x}$.

14.7. Solve $y' - 5y = (x-1)\sin x + (x+1)\cos x$.

From Problem 13.3, $y_h = c_1e^{5x}$. Using Example 14.2, we assume that

$$y_p = (E_1x + E_0)\sin x + (F_1x + F_0)\cos x \tag{1}$$

Thus,

$$y_p' = (E_1 - F_1x - F_0)\sin x + (E_1x + E_0 + F_1)\cos x$$

Substituting these values into the differential equation and simplifying, we obtain

$$(-5E_1 - F_1)x \sin x + (-5E_0 + E_1 - F_0) \sin x$$
$$+ (-5F_1 + E_1)x \cos x + (-5F_0 + E_0 + F_1) \cos x$$
$$= (1)x \sin x - (1) \sin x + (1)x \cos x + (1) \cos x$$

Equating coefficients of like terms, we have

$$\begin{aligned} -5E_1 \quad\quad - F_1 &= 1 \\ -5E_0 + E_1 - F_0 \quad\quad &= -1 \\ E_1 \quad\quad - 5F_1 &= 1 \\ E_0 \quad - 5F_0 + F_1 &= 1 \end{aligned}$$

Solving, we obtain $E_1 = -2/13$, $E_0 = 71/338$, $F_1 = -3/13$, and $F_0 = -69/338$. Then, from (1),

$$y_p = \left(-\frac{2}{13}x + \frac{71}{338}\right) \sin x + \left(-\frac{3}{13}x - \frac{69}{338}\right) \cos x$$

and the general solution is

$$y = c_1 e^{5x} + \left(\frac{-2}{13}x + \frac{71}{338}\right) \sin x - \left(\frac{3}{13}x + \frac{69}{338}\right) \cos x$$

14.8. Solve $y' - 5y = 3e^x - 2x + 1$.

From Problem 13.3, $y_h = c_1 e^{5x}$. Here, we can write $\phi(x)$ as the sum of two manageable functions: $\phi(x) = (3e^x) + (-2x + 1)$. For the term $3e^x$ we would assume a solution of the form $A_0 e^x$ (by (14.2) with $\alpha = 1$ and $p_n(x) = 3$); for the term $-2x + 1$ we would assume a solution of the form $B_1 x + B_0$. Thus, we try

$$y_p = A_0 e^x + B_1 x + B_0 \tag{1}$$

Substituting (1) into the differential equation and simplifying, we obtain

$$(-4A_0)e^x + (-5B_1)x + (B_1 - 5B_0) = (3)e^x + (-2)x + (1)$$

Equating coefficients of like terms, we find that $A_0 = -3/4$, $B_1 = 2/5$, and $B_0 = -3/25$. Hence, (1) becomes

$$y_p = -\frac{3}{4}e^x + \frac{2}{5}x - \frac{3}{25}$$

and the general solution is

$$y = c_1 e^{5x} - \frac{3}{4}e^x + \frac{2}{5}x - \frac{3}{25}$$

14.9. Solve $y' - 5y = x^2 e^x - x e^{5x}$.

From Problem 13.3, $y_h = c_1 e^{5x}$. Here $\phi(x) = x^2 e^x - x e^{5x}$, which is the difference of two terms, each in manageable form. For $x^2 e^x$ we would assume a solution of the form

$$e^x(A_2 x^2 + A_1 x + A_0) \tag{1}$$

For $x e^{5x}$ we would try initially a solution of the form

$$e^{5x}(B_1 x + B_0) = B_1 x e^{5x} + B_0 e^{5x}$$

But this supposed solution would have, disregarding multiplicative constants, the term e^{5x} in common with y_h. We are led, therefore, to the modified expression

$$x e^{5x}(B_1 x + B_0) = e^{5x}(B_1 x^2 + B_0 x) \tag{2}$$

We now take y_p to be the difference of (1) and (2):

$$y_p = e^x(A_2 x^2 + A_1 x + A_0) - e^{5x}(B_1 x^2 + B_0 x) \tag{3}$$

Substituting (3) into the differential equation and simplifying, we obtain

$$e^x[(-4A_2)x^2 + (2A_2 - 4A_1)x + (A_1 - 4A_0)] + e^{5x}[(-2B_1)x - B_0]$$
$$= e^x[(1)x^2 + (0)x + (0)] + e^{5x}[(-1)x + (0)]$$

Equating coefficients of like terms, we have

$$-4A_2 = 1 \qquad 2A_2 - 4A_1 = 0 \qquad A_1 - 4A_0 = 0$$
$$-2B_1 = -1 \qquad -B_0 = 0$$

from which

$$A_2 = -\frac{1}{4} \qquad A_1 = -\frac{1}{8} \qquad A_0 = -\frac{1}{32}$$

$$B_1 = \frac{1}{2} \qquad B_0 = 0$$

Equation (*3*) then gives

$$y_p = e^x\left(-\frac{1}{4}x^2 - \frac{1}{8}x - \frac{1}{32}\right) - \frac{1}{2}x^2e^{5x}$$

and the general solution is

$$y = c_1e^{5x} + e^x\left(-\frac{1}{4}x^2 - \frac{1}{8}x - \frac{1}{32}\right) - \frac{1}{2}x^2e^{5x}$$

Supplementary Problems

Find the general solutions of the following differential equations.

14.10. $y'' - 2y' + y = x^2 - 1.$

14.11. $y'' - 2y' + y = 3e^{2x}.$

14.12. $y'' - 2y' + y = 4\cos x.$

14.13. $y'' - 2y' + y = 3e^x.$

14.14. $y'' - 2y' + y = xe^x.$

14.15. $y' - y = e^x.$

14.16. $y' - y = xe^{2x} + 1.$

14.17. $y' - y = \sin x + \cos 2x.$

14.18. $y''' - 3y'' + 3y' - y = e^x + 1.$

Answers to Supplementary Problems

14.10. $y = c_1e^x + c_2xe^x + x^2 + 4x + 5$

14.11. $y = c_1e^x + c_2xe^x + 3e^{2x}$

14.12. $y = c_1e^x + c_2xe^x - 2\sin x$

14.13. $y = c_1e^x + c_2xe^x + \frac{3}{2}x^2e^x$

14.14. $y = c_1e^x + c_2xe^x + \frac{1}{6}x^3e^x$

14.15. $y = c_1e^x + xe^x$

14.16. $y = c_1e^x + xe^{2x} - e^{2x} - 1$

14.17. $y = c_1e^x - \frac{1}{2}\sin x - \frac{1}{2}\cos x + \frac{2}{5}\sin 2x - \frac{1}{5}\cos 2x$

14.18. $y = c_1e^x + c_2xe^x + c_3x^2e^x + \frac{1}{6}x^3e^x - 1$

Chapter 15

Variation of Parameters

Variation of parameters is an alternate method, more general than that of Chapter 14, for finding a particular solution of the nth-order linear differential equation

$$\mathbf{L}(y) = \phi(x) \tag{15.1}$$

once the solution of the associated homogeneous equation $\mathbf{L}(y) = 0$ is known. Recall from Theorem 11.1 that if $y_1(x), y_2(x), \ldots, y_n(x)$ are n linearly independent solutions of $\mathbf{L}(y) = 0$, then the general solution of $\mathbf{L}(y) = 0$ is

$$y_h = c_1y_1(x) + c_2y_2(x) + \cdots + c_ny_n(x) \tag{15.2}$$

15.1 VARIATION OF PARAMETERS

A particular solution of $\mathbf{L}(y) = \phi(x)$ has the form

$$y_p = v_1y_1 + v_2y_2 + \cdots + v_ny_n \tag{15.3}$$

where $y_i = y_i(x)$ $(i = 1, 2, \ldots, n)$ is given in (15.2) and v_i $(i = 1, 2, \ldots, n)$ is an unknown function of x which still must be determined.

To find v_i, first solve the following linear equations simultaneously for v_i':

$$\begin{aligned} v_1'y_1 + v_2'y_2 + \cdots + v_n'y_n &= 0 \\ v_1'y_1' + v_2'y_2' + \cdots + v_n'y_n' &= 0 \\ &\vdots \\ v_1'y_1^{(n-2)} + v_2'y_2^{(n-2)} + \cdots + v_n'y_n^{(n-2)} &= 0 \\ v_1'y_1^{(n-1)} + v_2'y_2^{(n-1)} + \cdots + v_n'y_n^{(n-1)} &= \phi(x) \end{aligned} \tag{15.4}$$

Then integrate each v_i' to obtain v_i, disregarding all constants of integration. This is permissible because we are seeking only *one* particular solution.

Example 15.1. For the special case $n = 3$, equations (15.4) reduce to

$$\begin{aligned} v_1'y_1 + v_2'y_2 + v_3'y_3 &= 0 \\ v_1'y_1' + v_2'y_2' + v_3'y_3' &= 0 \\ v_1'y_1'' + v_2'y_2'' + v_3'y_3'' &= \phi(x) \end{aligned} \tag{15.5}$$

For the case $n = 2$, equations (15.4) become

$$\begin{aligned} v_1'y_1 + v_2'y_2 &= 0 \\ v_1'y_1' + v_2'y_2' &= \phi(x) \end{aligned} \tag{15.6}$$

and for the case $n = 1$, we obtain the single equation

$$v_1'y_1 = \phi(x) \tag{15.7}$$

The first $n-1$ equations of (*15.4*) come from requiring that the first $n-1$ derivatives of y_p, (*15.3*), and y_h, (*15.2*), be formally the same (with the v's replacing the c's.) Then the last equation of (*15.4*) is a direct result of satisfying $\mathbf{L}(y_p) = \phi(x)$.

Since $y_1(x), y_2(x), \ldots, y_n(x)$ are n linearly independent solutions of the same equation $\mathbf{L}(y) = 0$, their Wronskian is not zero (Theorem 11.2). This means that the system (*15.4*) has a nonzero determinant and can be solved uniquely for $v_1'(x), v_2'(x), \ldots, v_n'(x)$.

15.2 SCOPE OF THE METHOD

The method of variation of parameters can be applied to *all* linear differential equations. It is therefore more powerful than the method of undetermined coefficients, which generally is restricted to linear differential equations with constant coefficients and particular forms of $\phi(x)$. Nonetheless, in those cases where both methods are applicable, the method of undetermined coefficients is usually the more efficient and, hence, the preferable method. (See Problem 15.3.)

As a practical matter, the integration of $v_i'(x)$ may be impossible to perform. In such event, other methods (in particular, numerical techniques) must be employed.

Solved Problems

15.1. Solve $y''' + y' = \sec x$.

Here $n = 3$ and $y_h = c_1 + c_2 \cos x + c_3 \sin x$; hence,

$$y_p = v_1 + v_2 \cos x + v_3 \sin x \tag{1}$$

Since $y_1 = 1$, $y_2 = \cos x$, $y_3 = \sin x$, and $\phi(x) = \sec x$, it follows from (*15.5*) that

$$v_1'(1) + v_2'(\cos x) + v_3'(\sin x) = 0$$

$$v_1'(0) + v_2'(-\sin x) + v_3'(\cos x) = 0$$

$$v_1'(0) + v_2'(-\cos x) + v_3'(-\sin x) = \sec x$$

Solving this set of equations simultaneously, we obtain $v_1' = \sec x$, $v_2' = -1$, and $v_3' = -\tan x$. Thus,

$$v_1 = \int v_1'\,dx = \int \sec x\,dx = \ln|\sec x + \tan x|$$

$$v_2 = \int v_2'\,dx = -\int dx = -x$$

$$v_3 = \int v_3'\,dx = \int -\tan x\,dx = -\int \frac{\sin x}{\cos x}\,dx = \ln|\cos x|$$

Substituting these values into (*1*), we obtain

$$y_p = \ln|\sec x + \tan x| - x\cos x + (\sin x)\ln|\cos x|$$

The general solution is therefore

$$y = y_h + y_p = c_1 + c_2\cos x + c_3 \sin x + \ln|\sec x + \tan x| - x\cos x + (\sin x)\ln|\cos x|$$

15.2. Solve $y'' - 2y' + y = \dfrac{e^x}{x}$.

Here $n = 2$ and $y_h = c_1e^x + c_2xe^x$; hence,

$$y_p = v_1e^x + v_2xe^x \tag{1}$$

Since $y_1 = e^x$, $y_2 = xe^x$, and $\phi(x) = e^x/x$, it follows from *(15.6)* that

$$v_1'(e^x) + v_2'(xe^x) = 0$$

$$v_1'(e^x) + v_2'(e^x + xe^x) = \frac{e^x}{x}$$

Solving this set of equations simultaneously, we obtain $v_1' = -1$ and $v_2' = 1/x$. Thus,

$$v_1 = \int v_1'\,dx = \int -1\,dx = -x$$

$$v_2 = \int v_2'\,dx = \int \frac{1}{x}\,dx = \ln|x|$$

Substituting these values into *(1)*, we obtain

$$y_p = -xe^x + xe^x \ln|x|$$

The general solution is therefore,

$$y = y_h + y_p = c_1e^x + c_2xe^x - xe^x + xe^x\ln|x|$$

$$= c_1e^x + c_3xe^x + xe^x\ln|x| \qquad (c_3 = c_2 - 1)$$

15.3. Solve $y'' - y' - 2y = e^{3x}$.

Here $n = 2$ and $y_h = c_1e^{-x} + c_2e^{2x}$; hence,

$$y_p = v_1e^{-x} + v_2e^{2x} \tag{1}$$

Since $y_1 = e^{-x}$, $y_2 = e^{2x}$, and $\phi(x) = e^{3x}$, it follows from *(15.6)* that

$$v_1'(e^{-x}) + v_2'(e^{2x}) = 0$$

$$v_1'(-e^{-x}) + v_2'(2e^{2x}) = e^{3x}$$

Solving this set of equations simultaneously, we obtain $v_1' = -e^{4x}/3$ and $v_2' = e^x/3$, from which $v_1 = -e^{4x}/12$ and $v_2 = e^x/3$. Substituting these results into *(1)*, we obtain

$$y_p = -\frac{1}{12}e^{4x}e^{-x} + \frac{1}{3}e^xe^{2x} = -\frac{1}{12}e^{3x} + \frac{1}{3}e^{3x} = \frac{1}{4}e^{3x}$$

The general solution is, therefore,

$$y = c_1e^{-x} + c_2e^{2x} + \frac{1}{4}e^{3x}$$

(Compare with Problem 14.2.)

15.4. Solve $y' + \dfrac{4}{x}y = x^4$.

Here $n = 1$ and (from Chapter 8) $y_h = c_1x^{-4}$; hence,

$$y_p = v_1x^{-4} \tag{1}$$

Since $y_1 = x^{-4}$ and $\phi(x) = x^4$, *(15.7)* becomes $v_1'x^{-4} = x^4$, from which we obtain $v_1' = x^8$ and $v_1 = x^9/9$. Equation *(1)* now becomes $y_p = x^5/9$, and the general solution is therefore

$$y = c_1x^{-4} + \frac{1}{9}x^5$$

(Compare with Problem 8.3.)

15.5. Solve $y^{(4)} = 5x$ by variation of parameters.

Here $n = 4$ and $y_h = c_1 + c_2x + c_3x^2 + c_4x^3$; hence,

$$y_p = v_1 + v_2x + v_3x^2 + v_4x^3 \tag{1}$$

Since $y_1 = 1$, $y_2 = x$, $y_3 = x^2$, $y_4 = x^3$, and $\phi(x) = 5x$, it follows from *(15.4)*, with $n = 4$, that

$$v_1'(1) + v_2'(x) + v_3'(x^2) + v_4'(x^3) = 0$$

$$v_1'(0) + v_2'(1) + v_3'(2x) + v_4'(3x^2) = 0$$

$$v_1'(0) + v_2'(0) + v_3'(2) + v_4'(6x) = 0$$

$$v_1'(0) + v_2'(0) + v_3'(0) + v_4'(6) = 5x$$

Solving this set of equations simultaneously, we obtain

$$v_1' = -\frac{5}{6}x^4 \qquad v_2' = \frac{5}{2}x^3 \qquad v_3' = -\frac{5}{2}x^2 \qquad v_4' = \frac{5}{6}x$$

whence

$$v_1 = -\frac{1}{6}x^5 \qquad v_2 = \frac{5}{8}x^4 \qquad v_3 = -\frac{5}{6}x^3 \qquad v_4 = \frac{5}{12}x^2$$

Then, from *(1)*,

$$y_p = -\frac{1}{6}x^5 + \frac{5}{8}x^4(x) - \frac{5}{6}x^3(x^2) + \frac{5}{12}x^2(x^3) = \frac{1}{24}x^5$$

and the general solution is

$$y_h = c_1 + c_2x + c_3x^2 + c_4x^3 + \frac{1}{24}x^5$$

The solution also can be obtained simply by integrating both sides of the differential equation four times with respect to x.

Supplementary Problems

Use variation of parameters to find the general solutions of the following differential equations.

15.6. $y'' - 2y' + y = \dfrac{e^x}{x^5}$.

15.7. $y'' + y = \sec x$.

15.8. $y'' + 4y = \sin^2 2x$.

15.9. $y' - \dfrac{1}{x}y = x^2$.

15.10. $y' + 2xy = x$.

15.11. $y''' = 12$.

Answers to Supplementary Problems

15.6. $y = c_1 e^x + c_2 x e^x + \frac{1}{12} x^{-3} e^x$

15.7. $y = c_1 \cos x + c_2 \sin x + (\cos x) \ln |\cos x| + x \sin x$

15.8. $y = c_1 \cos 2x + c_2 \sin 2x + \frac{1}{4} \cos^2 2x - \frac{1}{12} \cos^4 2x + \frac{1}{12} \sin^4 2x$

$= c_1 \cos 2x + c_2 \sin 2x + \frac{1}{6} \cos^2 2x + \frac{1}{12} \sin^2 2x$

where we have used the identity

$$\frac{1}{12}(\sin^4 2x - \cos^4 2x) \equiv \frac{1}{12}(\sin^2 2x - \cos^2 2x)(\sin^2 2x + \cos^2 2x) \equiv \frac{1}{12}(\sin^2 2x - \cos^2 2x)$$

15.9. $y = c_1 x + \frac{1}{2} x^3$

15.10. $y = c_1 e^{-x^2} + \frac{1}{2}$

15.11. $y = c_1 + c_2 x + c_3 x^2 + 2x^3$

Chapter 16

Initial-Value Problems

We saw in Chapter 2 (Problems 2.5 and 2.6) that an initial-value problem may be solved by applying the initial conditions to the general solution of the differential equation. It must be emphasized that the initial conditions are applied *only* to the general solution and *not* to the homogeneous solution y_h, even though it is y_h that possesses all the arbitrary constants that must be evaluated. The one exception is when the general solution is the homogeneous solution; that is, when the differential equation under consideration is itself homogeneous.

Solved Problems

16.1. Solve $y'' - y' - 2y = 4x^2;\ y(0) = 1,\ y'(0) = 4.$

The general solution of the differential equation is given in Problem 14.1 as

$$y = c_1e^{-x} + c_2e^{2x} - 2x^2 + 2x - 3 \tag{1}$$

Therefore,

$$y' = -c_1e^{-x} + 2c_2e^{2x} - 4x + 2 \tag{2}$$

Applying the first initial condition to (*1*), we obtain

$$c_1e^{-(0)} + c_2e^{2(0)} - 2(0)^2 + 2(0) - 3 = 1 \quad \text{or} \quad c_1 + c_2 = 4 \tag{3}$$

Applying the second initial condition to (*2*), we obtain

$$-c_1e^{-(0)} + 2c_2e^{2(0)} - 4(0) + 2 = 4 \quad \text{or} \quad -c_1 + 2c_2 = 2 \tag{4}$$

Solving (*3*) and (*4*) simultaneously, we find that $c_1 = 2$ and $c_2 = 2$. Substituting these values into (*1*), we obtain the solution of the initial-value problem as

$$y = 2e^{-x} + 2e^{2x} - 2x^2 + 2x - 3$$

16.2. Solve $y'' - 2y' + y = \dfrac{e^x}{x};\ y(1) = 0,\ y'(1) = 1.$

The general solution of the differential equation is given in Problem 15.2 as

$$y = c_1e^x + c_3xe^x + xe^x \ln|x| \tag{1}$$

Therefore,

$$y' = c_1e^x + c_3e^x + c_3xe^x + e^x\ln|x| + xe^x\ln|x| + e^x \tag{2}$$

Applying the first initial condition to (*1*), we obtain

$$c_1e^1 + c_3(1)e^1 + (1)e^1\ln 1 = 0$$

or (noting that $\ln 1 = 0$),

$$c_1e + c_3e = 0 \tag{3}$$

Applying the second initial condition to (2), we obtain

$$c_1e^1 + c_3e^1 + c_3(1)e^1 + e^1 \ln 1 + (1)e^1 \ln 1 + e^1 = 1$$

or

$$c_1e + 2c_3e = 1 - e \tag{4}$$

Solving (3) and (4) simultaneously, we find that $c_1 = -c_3 = (e-1)/e$. Substituting these values into (1), we obtain the solution of the initial-value problem as

$$y = e^{x-1}(e-1)(1-x) + xe^x \ln|x|$$

16.3. Solve $y'' + 4y' + 8y = \sin x;\ y(0) = 1,\ y'(0) = 0.$

Here $y_h = e^{-2x}(c_1 \cos 2x + c_2 \sin 2x)$, and, by the method of undetermined coefficients,

$$y_p = \frac{7}{65}\sin x - \frac{4}{65}\cos x$$

Thus, the general solution to the differential equation is

$$y = e^{-2x}(c_1 \cos 2x + c_2 \sin 2x) + \frac{7}{65}\sin x - \frac{4}{65}\cos x \tag{1}$$

Therefore,

$$y' = -2e^{-2x}(c_1 \cos 2x + c_2 \sin 2x) + e^{-2x}(-2c_1 \sin 2x + 2c_2 \cos 2x) + \frac{7}{65}\cos x + \frac{4}{65}\sin x$$

Applying the first initial condition to (1), we obtain

$$c_1 = \frac{69}{65} \tag{3}$$

Applying the second initial condition to (2), we obtain

$$-2c_1 + 2c_2 = -\frac{7}{65} \tag{4}$$

Solving (3) and (4) simultaneously, we find that $c_1 = 69/65$ and $c_2 = 131/130$. Substituting these values into (1), we obtain the solution of the initial-value problem as

$$y = e^{-2x}\left(\frac{69}{65}\cos 2x + \frac{131}{130}\sin 2x\right) + \frac{7}{65}\sin x - \frac{4}{65}\cos x$$

16.4. Solve $y''' - 6y'' + 11y' - 6y = 0;\ y(\pi) = 0,\ y'(\pi) = 0,\ y''(\pi) = 1.$

From Problem 13.1, we have

$$\begin{aligned} y_h &= c_1e^x + c_2e^{2x} + c_3e^{3x} \\ y_h' &= c_1e^x + 2c_2e^{2x} + 3c_3e^{3x} \\ y_h'' &= c_1e^x + 4c_2e^{2x} + 9c_3e^{3x} \end{aligned} \tag{1}$$

Since the given differential equation is homogeneous, y_h is also the general solution. Applying each initial condition separately, we obtain

$$\begin{aligned} c_1e^\pi + c_2e^{2\pi} + c_3e^{3\pi} &= 0 \\ c_1e^\pi + 2c_2e^{2\pi} + 3c_3e^{3\pi} &= 0 \\ c_1e^\pi + 4c_2e^{2\pi} + 9c_3e^{3\pi} &= 1 \end{aligned}$$

Solving these equations simultaneously, we find

$$c_1 = \frac{1}{2}e^{-\pi} \qquad c_2 = -e^{-2\pi} \qquad c_3 = \frac{1}{2}e^{-3\pi}$$

Substituting these values into the first equation (1), we obtain

$$y = \frac{1}{2}e^{(x-\pi)} - e^{2(x-\pi)} + \frac{1}{2}e^{3(x-\pi)}$$

Supplementary Problems

Solve the following initial-value problems.

16.5. $y'' - y' - 2y = e^{3x}; \quad y(0) = 1, \ y'(0) = 2.$

16.6. $y'' - y' - 2y = e^{3x}; \quad y(0) = 2, \ y'(0) = 1.$

16.7. $y'' - y' - 2y = 0; \quad y(0) = 2, \ y'(0) = 1.$

16.8. $y'' - y' - 2y = e^{3x}; \quad y(1) = 2, \ y'(1) = 1.$

16.9. $y'' + y = x; \quad y(1) = 0, \ y'(1) = 1.$

16.10. $y'' + 4y = \sin^2 2x; \quad y(\pi) = 0, \ y'(\pi) = 0.$

16.11. $y'' + y = 0; \quad y(2) = 0, \ y'(2) = 0.$

16.12. $y''' = 12; \quad y(1) = 0, \ y'(1) = 0, \ y''(1) = 0.$

16.13. $\ddot{y} + 2\dot{y} + 2y = \sin 2t + \cos 2t; \quad y(0) = 0, \ \dot{y}(0) = 1.$

Answers to Supplementary Problems

16.5. $y = \frac{1}{12}e^{-x} + \frac{2}{3}e^{2x} + \frac{1}{4}e^{3x}$

16.6. $y = \frac{13}{12}e^{-x} + \frac{2}{3}e^{2x} + \frac{1}{4}e^{3x}$

16.7. $y = e^{-x} + e^{2x}$

16.8. $y = \left(1 + \frac{1}{12}e^3\right)e^{-(x-1)} + \left(1 - \frac{1}{3}e^3\right)e^{2(x-1)} + \frac{1}{4}e^{3x}$

16.9. $y = -\cos 1 \cos x - \sin 1 \sin x + x = -\cos(x-1) + x$

16.10. $y = -\frac{1}{6}\cos 2x + \frac{1}{4}\cos^2 2x - \frac{1}{12}\cos^4 2x + \frac{1}{12}\sin^4 2x$

16.11. $y \equiv 0$

16.12. $y = -2 + 6x - 6x^2 + 2x^3$

16.13. $y = e^{-t}\left(\frac{3}{10}\cos t + \frac{11}{10}\sin t\right) + \frac{1}{10}\sin 2t - \frac{3}{10}\cos 2t$

Chapter 17

Applications of Second-Order Linear Differential Equations with Constant Coefficients

Consider a physical system whose small oscillations are governed by the linear differential equation

$$\ddot{x} + a_1\dot{x} + a_0 x = f(t) \tag{17.1}$$

Here, the function $f(t)$ and the constants a_1 and a_0 are known, and $\ddot{x}$ and $\dot{x}$ represent the quantities d^2x/dt^2 and dx/dt, respectively. Two such systems are a simple weighted spring (see Problem 17.1) and a simple electrical circuit (see Problem 17.10).

If $f(t) \equiv 0$ and $a_1 = 0$, the motion is *free* and *undamped.* If $f(t)$ is identically zero but a_1 is not zero, the motion is *free* and *damped.* For damped motion, there are three separate cases to consider, according as the roots of the associated characteristic equation (see Chapter 12) are (1) real and distinct, (2) equal, or (3) complex conjugate. These cases are respectively classified as (1) *overdamped,* (2) *critically damped,* and (3) *oscillatory damped* (or, in electrical problems, *underdamped*). (See Problems 17.4–17.7.) If $f(t)$ is not identically zero, the motion is *forced.* (See Problems 17.8 and 17.9.)

A motion or current is *transient* if it "dies out" (that is, goes to zero) as $t \to \infty$. A *steady-state* motion or current is one that is not transient and does not become unbounded. Free damped systems always yield transient motions, while forced damped systems (assuming the external force to be sinusoidal) yield both transient and steady-state motions.

The following three laws of physics will be needed in the problems for this chapter:

Newton's second law (see page 37).

Kirchhoff's loop law: The algebraic sum of the voltage drops in a simple closed electrical circuit is zero.

Hooke's law: $F = -kl$, where F represents the restoring force in a spring, which is equal and opposite to the force applied to the spring; l denotes the extension of the spring as a result of the applied force; and k is the spring constant.

Example 17.1. A steel ball weighing 128 lb is suspended from a spring, whereupon the spring is stretched 2 ft from its natural length. What is the value of the spring constant?

The applied force responsible for the 2-ft displacement is the weight of the ball, 128 lb. Thus, $F = -128$ lb. Hooke's law then gives $-128 = -k(2)$, or $k = 64$ lb/ft.

Solved Problems

VIBRATING SPRINGS

In the following problems the mass of the spring itself is to be neglected.

17.1. A spring with a mass m attached to its lower end is suspended vertically from a mounting and allowed to come to rest in an equilibrium position. The system is then set in motion by releasing the mass with an initial velocity v_0 at a distance x_0 below its equilibrium position and by simultaneously applying to the mass an external force $F(t)$ in the downward direction. Show that the motion of this system is governed by (*17.1*).

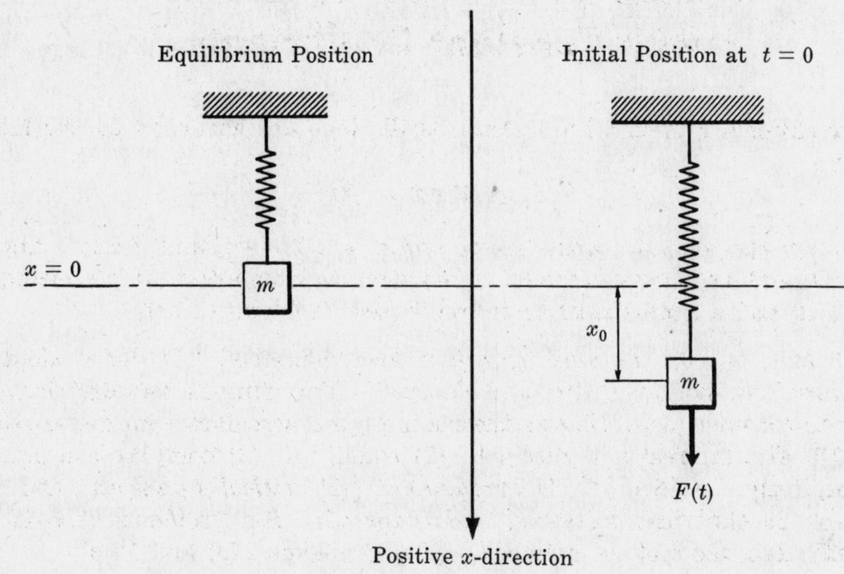

Fig. 17-1

For convenience, we choose the downward direction as the positive direction and take the origin to be the center of gravity of the mass in the equilibrium position (see Fig. 17-1). Furthermore, we assume that air resistance is present and is proportional to the velocity of the mass. Thus, at any time t, there are three forces acting on the system: (1) $F(t)$, measured in the positive direction; (2) a restoring force given by Hooke's law as $F_s = -kx$, $k > 0$; and (3) a force due to air resistance given by $F_a = -a\dot{x}$, $a > 0$, where a is the constant of proportionality. Note that the restoring force F_s always acts in a direction that will tend to return the system to the equilibrium position: if the mass is below the equilibrium position, then x is positive and $-kx$ is negative; whereas if the mass is above the equilibrium position, then x is negative and $-kx$ is positive. Also note that because $a > 0$ the force F_a due to air resistance acts in the opposite direction of the velocity and thus tends to retard, or damp, the motion of the mass.

It now follows from Newton's second law that $m\ddot{x} = -kx - a\dot{x} + F(t)$, or

$$\ddot{x} + \frac{a}{m}\dot{x} + \frac{k}{m}x = \frac{F(t)}{m} \tag{1}$$

If we define $a_1 = a/m$, $a_0 = k/m$, and $f(t) = F(t)/m$, then (*1*) is exactly (*17.1*). Since the system starts at $t = 0$ with an initial velocity v_0 and from an initial position x_0, we have along with (*1*) the initial conditions

$$x(0) = x_0 \qquad \dot{x}(0) = v_0 \tag{2}$$

The force of gravity does not explicitly appear in (*1*), but it is present nonetheless. We automatically compensated for this force by measuring distance from the equilibrium position of the spring. If one wishes to exhibit gravity explicitly, then distance must be measured from the bottom end of the *natural length* of the spring. That is, the motion of a vibrating spring can be given by

$$\ddot{x} + \frac{a}{m}\dot{x} + \frac{k}{m}x = g + \frac{F(t)}{m}$$

if the origin, $x = 0$, is the terminal point of the unstretched spring before the mass m is attached.

17.2. Find a solution to Problem 17.1 if the vibrations are free and undamped.

With $F(t) \equiv 0$ and $a = 0$, equation (*1*) *of* Problem 17.1 becomes

$$\ddot{x} + \frac{k}{m}x = 0 \tag{1}$$

The roots of the characteristic equation for (*1*) are $\lambda_1 = \sqrt{-k/m}$ and $\lambda_2 = -\sqrt{-k/m}$, or, since both k and m are positive, $\lambda_1 = i\sqrt{k/m}$ and $\lambda_2 = -i\sqrt{k/m}$. The solution to (*1*) is (see Section 12.2)

$$x = c_1 \cos\sqrt{k/m}\,t + c_2 \sin\sqrt{k/m}\,t \tag{2}$$

Applying the initial conditions, (*2*) of Problem 17.1, we obtain $c_1 = x_0$ and $c_2 = v_0\sqrt{m/k}$. Thus (*2*) becomes

$$x = x_0 \cos\sqrt{k/m}\,t + v_0\sqrt{m/k}\sin\sqrt{k/m}\,t \tag{3}$$

Using the techniques given in Problem 9.36, (*3*) can be further simplified to

$$x = A\cos(\sqrt{k/m}\,t - \phi) \tag{4}$$

where $A = \sqrt{x_0^2 + v_0^2(m/k)}$ and where the phase angle ϕ is given implicitly by

$$\cos\phi = \frac{x_0}{A} \quad \text{and} \quad \sin\phi = \frac{v_0\sqrt{m/k}}{A}$$

Any motion described by (*2*) is called *simple harmonic motion.* The *circular frequency* of such motion is given by $\omega_n = \sqrt{k/m}$, while the *natural frequency*, or number of complete oscillations per second, is

$$f_n = \frac{\omega_n}{2\pi} = \frac{1}{2\pi}\sqrt{\frac{k}{m}} \tag{5}$$

The *period* of the system, or the time required to complete one oscillation, is

$$T = \frac{1}{f_n} = 2\pi\sqrt{\frac{m}{k}} \tag{6}$$

17.3. A steel ball weighing 128 lb is suspended from a spring, whereupon the spring is stretched 2 ft from its natural length. The ball is started in motion with no initial velocity by displacing it 6 in. above the equilibrium position. Assuming no air resistance, find (*a*) the position of the ball at $t = \pi/12$ sec, (*b*) the natural frequency, and (*c*) the period.

(*a*) This is a free undamped motion, for which $m = 128/32 = 4$ slugs, $k = 64$ lb/ft (see Example 17.1), $x_0 = -\frac{1}{2}$ ft (the minus sign is required since the ball is initially displaced *above* the equilibrium position, which is in the *negative* direction), and $v_0 = 0$ ft/sec. Substituting these values into (*3*) of Problem 17.2, we obtain $x(t) = -\frac{1}{2}\cos 4t$. Therefore, at $t = \pi/12$,

$$x\left(\frac{\pi}{12}\right) = -\frac{1}{2}\cos\frac{4\pi}{12} = -\frac{1}{4}\text{ ft}$$

(*b*) From (*5*) of Problem 17.2,

$$f_n = \frac{1}{2\pi}\sqrt{\frac{k}{m}} = \frac{2}{\pi}\text{ cycles/sec}$$

(c) From (6) of Problem 17.2,

$$T = \frac{1}{f_n} = \frac{\pi}{2} \simeq 1.57 \text{ sec}$$

17.4. A 10-kilogram mass is attached to a spring, stretching it 0.7 meter from its natural length. The mass is started in motion from the equilibrium position with an initial velocity of 1 m/sec in the upward direction. Find the subsequent motion, if the force due to air resistance is $-90\dot{x}$ newton.

Taking $g = 9.8$ m/sec^2, we have $w = mg = 98$ newtons and $k = w/l = 140$ N/m. Furthermore, $a = 90$ and $F(t) \equiv 0$ (there is no external force). Hence, (1) of Problem 17.1 becomes

$$\ddot{x} + 9\dot{x} + 14x = 0 \tag{1}$$

The roots of the associated characteristic equation are $\lambda_1 = -2$ and $\lambda_2 = -7$, which are real and distinct; hence this problem is an example of overdamped motion. The solution of (1) is

$$x = c_1 e^{-2t} + c_2 e^{-7t}$$

The initial conditions are $x(0) = 0$ (the mass starts at the equilibrium position) and $\dot{x}(0) = -1$ (the initial velocity is in the negative direction). Applying these conditions, we find that $c_1 = -c_2 = -\frac{1}{5}$, so that $x = \frac{1}{5}(e^{-7t} - e^{-2t})$. Note that $x \to 0$ as $t \to \infty$; thus, the motion is transient.

17.5. A mass of 1/4 slug is attached to a spring, whereupon the spring is stretched 1.28 ft from its natural length. The mass is started in motion from the equilibrium position with an initial velocity of 4 ft/sec in the downward direction. Find the subsequent motion of the mass if the force due to air resistance is $-2\dot{x}$ lb.

Here $m = 1/4$, $a = 2$, $F(t) \equiv 0$ (there is no external force), and, from Hooke's law, $k = mg/l = (1/4)(32)/1.28 = 6.25$. Hence, (1) of Problem 17.1 becomes

$$\ddot{x} + 8\dot{x} + 25x = 0 \tag{1}$$

The roots of the associated characteristic equation are $\lambda_1 = -4 + i3$ and $\lambda_2 = -4 - i3$, which are complex conjugate; hence this problem is an example of oscillatory damped motion. The solution of (1) is

$$x = e^{-4t}(c_1 \cos 3t + c_2 \sin 3t)$$

The initial conditions are $x(0) = 0$ and $\dot{x}(0) = 4$. Applying these conditions, we find that $c_1 = 0$ and $c_2 = \frac{4}{3}$; thus, $x = \frac{4}{3}e^{-4t} \sin 3t$. Since $x \to 0$ as $t \to \infty$, the motion is transient.

17.6. A mass of 1/4 slug is attached to a spring having a spring constant of 1 lb/ft. The mass is started in motion by initially displacing it 2 ft in the downward direction and giving it an initial velocity of 2 ft/sec in the upward direction. Find the subsequent motion of the mass, if the force due to air resistance is $-1\dot{x}$ lb.

Here $m = 1/4$, $a = 1$, $k = 1$, and $F(t) \equiv 0$. Equation (1) of Problem 17.1 becomes

$$\ddot{x} + 4\dot{x} + 4x = 0 \tag{1}$$

The roots of the associated characteristic equation are $\lambda_1 = \lambda_2 = -2$, which are equal; hence this problem is an example of critically damped motion. The solution of (1) is

$$x = c_1 e^{-2t} + c_2 t e^{-2t}$$

The initial conditions are $x(0) = 2$ and $\dot{x}(0) = -2$ (the initial velocity is in the negative direction). Applying these conditions, we find that $c_1 = c_2 = 2$. Thus,

$$x = 2e^{-2t} + 2te^{-2t}$$

Since $x \to 0$ as $t \to \infty$, the motion is transient.

17.7. Show that the types of motions that result from free damped problems are completely determined by the quantity $a^2 - 4km$.

For free damped motions $F(t) \equiv 0$ and (*1*) of Problem 17.1 becomes

$$\ddot{x} + \frac{a}{m}\dot{x} + \frac{k}{m}x = 0$$

The roots of the associated charactertistic equation are

$$\lambda_1 = \frac{-a + \sqrt{a^2 - 4km}}{2m} \qquad \lambda_2 = \frac{-a - \sqrt{a^2 - 4km}}{2m}$$

If $a^2 - 4km > 0$, the roots are real and distinct; if $a^2 - 4km = 0$, the roots are equal; if $a^2 - 4km < 0$, the roots are complex conjugate. The corresponding motions are, respectively, overdamped, critically damped, and oscillatory damped. Since the real parts of both roots are always negative, the resulting motion in all three cases is transient. (For overdamped motion, we need only note that $\sqrt{a^2 - 4km} < a$, whereas for the other two cases the real parts are both $-a/2m$.)

17.8. A 10-kg mass is attached to a spring having a spring constant of 140 N/m. The mass is started in motion from the equilibrium position with an initial velocity of 1 m/sec in the upward direction and with an applied external force $F(t) = 5 \sin t$. Find the subsequent motion of the mass if the force due to air resistance is $-90\dot{x}$ N.

Here $m = 10$, $k = 140$, $a = 90$, and $F(t) = 5 \sin t$. The equation of motion, (*1*) of Problem 17.1, becomes

$$\ddot{x} + 9\dot{x} + 14x = \frac{1}{2}\sin t \tag{1}$$

The general solution to the associated homogeneous equation $\ddot{x} + 9\dot{x} + 14x = 0$ is (see Problem 17.4)

$$x_h = c_1 e^{-2t} + c_2 e^{-7t}$$

Using the method of undetermined coefficients (Section 14.1, Case 3), we find

$$x_p = \frac{13}{500}\sin t - \frac{9}{500}\cos t$$

The general solution of (*1*) is therefore

$$x = x_h + x_p = c_1 e^{-2t} + c_2 e^{-7t} + \frac{13}{500}\sin t - \frac{9}{500}\cos t$$

Applying the initial conditions, $x(0) = 0$ and $\dot{x}(0) = -1$, we obtain

$$x = \frac{1}{500}(-90e^{-2t} + 99e^{-7t} + 13\sin t - 9\cos t)$$

Note that the exponential terms, which come from x_h and hence represent an associated free overdamped motion (see Problem 17.4), quickly die out. These terms are the transient part of the solution. The terms coming from x_p, however, do not die out as $t \to \infty$; they are the steady-state part of the solution.

17.9. A 128-lb weight is attached to a spring having a spring constant of 64 lb/ft. The weight is started in motion with no initial velocity by displacing it 6 in. above the equilibrium position and by simultaneously applying to the weight an external force $F(t) = 8 \sin 4t$. Assuming no air resistance, find the subsequent motion of the weight.

Here $m = 4$, $k = 64$, $a = 0$, and $F(t) = 8 \sin 4t$; hence, (*1*) of Problem 17.1 becomes

$$\ddot{x} + 16x = 2\sin 4t \tag{1}$$

This problem is, therefore, an example of forced undamped motion. The solution to the associated homogeneous equation is

$$x_h = c_1 \cos 4t + c_2 \sin 4t$$

A particular solution is found by the method of undetermined coefficients (the modification described in Section 14.2 is necessary here): $x_p = -\frac{1}{4}t\cos 4t$. The solution to (1) is then

$$x = c_1 \cos 4t + c_2 \sin 4t - \frac{1}{4}t\cos 4t$$

Applying the initial conditions, $x(0) = -\frac{1}{2}$ and $\dot{x}(0) = 0$, we obtain

$$x = -\frac{1}{2}\cos 4t + \frac{1}{16}\sin 4t - \frac{1}{4}t\cos 4t$$

Note that $|x| \to \infty$ as $t \to \infty$. This phenomenon is called *pure resonance*. It is due to the forcing function $F(t)$ having the same circular frequency as the circular frequency (see Problem 17.2) of the associated free undamped system.

ELECTRICAL CIRCUITS

17.10. Show that a simple electrical circuit consisting of a resistor, a capacitor, an inductor, and an electromotive force (usually a battery or a generator) connected in series is governed by (*17.1*). Also, assuming an initial charge on the capacitor of q_0 coulombs and an initial current in the circuit of I_0 amperes, find initial conditions for the system.

The circuit is shown in Fig. 17-2, where R is the resistance in ohms, C is the capacitance in farads, L is the inductance in henries, $E(t)$ is the electromotive force (emf) in volts, and I is the current in amperes. It is known that the voltage drops across a resistor, a capacitor, and an inductor are respectively RI, $\frac{1}{C}q$, and $L\frac{dI}{dt}$ where q is the charge on the capacitor. The voltage drop across an emf is $-E(t)$. Thus, from Kirchhoff's loop law, we have

$$RI + L\frac{dI}{dt} + \frac{1}{C}q - E(t) = 0 \tag{1}$$

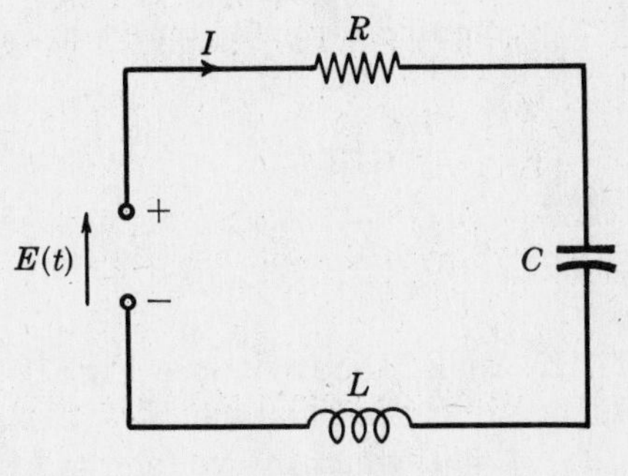

Fig. 17-2

Recall, from Chapter 9, that

$$I = \frac{dq}{dt} \qquad \frac{dI}{dt} = \frac{d^2q}{dt^2} \tag{2}$$

Substituting these values into (1), we obtain

$$\frac{d^2q}{dt^2} + \frac{R}{L}\frac{dq}{dt} + \frac{1}{LC}q = \frac{1}{L}E(t) \tag{3}$$

If we define $a_1 = R/L$, $a_0 = 1/LC$, and $f(t) = E(t)/L$, then (3) is exactly (*17.1*) with x replaced by q. The initial conditions for q are

$$q(0) = q_0 \qquad \left.\frac{dq}{dt}\right|_{t=0} = I(0) = I_0 \tag{4}$$

To obtain the differential equation governing the current, we first differentiate (1) with respect to t and then substitute (2) directly into the resulting equation. The new equation is

$$\frac{d^2I}{dt^2} + \frac{R}{L}\frac{dI}{dt} + \frac{1}{LC}I = \frac{1}{L}\frac{dE(t)}{dt} \tag{5}$$

If we now define $a_1 = R/L$, $a_0 = 1/LC$, and $f(t) = (1/L)(dE/dt)$, then (5) is exactly (*17.1*) with x now replaced by I. The first initial condition is $I(0) = I_0$. The second initial condition is obtained from (1) by solving for dI/dt and then setting $t = 0$. Thus,

$$\left.\frac{dI}{dt}\right|_{t=0} = \frac{1}{L}E(0) - \frac{R}{L}I_0 - \frac{1}{LC}q_0 \tag{6}$$

We see that the current in the circuit can be obtained either by solving (5) directly, or by solving (3) for the charge and then differentiating the charge to obtain the current.

Because (*3*) and (*5*) are identical in form to (*1*) of Problem 17.1, *the solutions of* (*3*) *and* (*5*) *must be identical in form to the solutions for vibrating springs.* That is, for free undamped systems ($R = 0$ and $E(t) \equiv 0$), the solutions are simple harmonic motions (see Problem 17.2); whereas for free damped systems ($E(t) \equiv 0$), the motions are either overdamped, critically damped, or oscillatory damped (underdamped), depending upon the sign of $R^2 - 4(L/C)$. (See Problem 17.7.)

17.11. An RCL circuit (see Fig. 17-2) has $R = 180$ ohms, $C = 1/280$ farad, $L = 20$ henries, and an applied voltage $E(t) = 10 \sin t$. Assuming no initial charge on the capacitor, but an initial current of 1 ampere at $t = 0$ when the voltage is first applied, find the subsequent charge on the capacitor.

Substituting the given quantities into (*3*) of Problem 17.10, we obtain

$$\ddot{q} + 9\dot{q} + 14q = \frac{1}{2}\sin t$$

This equation is identical in form to (*1*) of Problem 17.8; hence, the solution must be identical in form to the solution of that equation. Thus,

$$q = c_1 e^{-2t} + c_2 e^{-7t} + \frac{13}{500}\sin t - \frac{9}{500}\cos t$$

Applying the initial conditions $q(0) = 0$ and $\dot{q}(0) = 1$, we obtain $c_1 = 110/500$ and $c_2 = -101/500$. Hence,

$$q = \frac{1}{500}(110e^{-2t} - 101e^{-7t} + 13\sin t - 9\cos t)$$

As in Problem 17.8, the solution is the sum of transient and steady-state terms.

17.12. An RCL circuit (see Fig. 17-2) has $R = 10$ ohms, $C = 10^{-2}$ farad, $L = \frac{1}{2}$ henry, and an applied voltage $E = 12$ volts. Assuming no initial current and no initial charge at $t = 0$ when the voltage is first applied, find the subsequent current in the system.

Substituting the given values into (*5*) of Problem 17.10, we obtain the homogeneous equation (since $E(t) = 12$, $dE/dt = 0$)

$$\frac{d^2I}{dt^2} + 20\frac{dI}{dt} + 200I = 0$$

The roots of the associated characteristic equation are $\lambda_1 = -10 + 10i$ and $\lambda_2 = -10 - 10i$; hence, this is an example of a free underdamped system for the current. The solution is

$$I = e^{-10t}(c_1 \cos 10t + c_2 \sin 10t) \qquad (1)$$

The initial conditions are $I(0) = 0$ and, from (*6*) of Problem 17.10,

$$\left.\frac{dI}{dt}\right|_{t=0} = \frac{12}{1/2} - \left(\frac{10}{1/2}\right)(0) - \frac{1}{(1/2)(10^{-2})}(0) = 24$$

Applying these conditions to (*1*), we obtain $c_1 = 0$ and $c_2 = \frac{12}{5}$; thus, $I = \frac{12}{5}e^{-10t}\sin 10t$, which is completely transient.

17.13. Solve Problem 17.12 by first finding the charge on the capacitor.

We first solve for the charge q and then use $I = dq/dt$ to obtain the current. Substituting the values given in Problem 17.12 into (*3*) of Problem 17.10, we have $\ddot{q} + 20\dot{q} + 200q = 24$, which represents a forced system for the charge, in contrast to the free damped system obtained in Problem 17.12 for the current. Using the method of undetermined coefficients to find a particular solution, we obtain the general solution

$$q = e^{-10t}(c_1 \cos 10t + c_2 \sin 10t) + \frac{3}{25}$$

Initial conditions for the charge are $q(0) = 0$ and $\dot{q}(0) = 0$; applying them, we obtain $c_1 = c_2 = -3/25$. Therefore,

$$q = -e^{-10t}\left(\frac{3}{25}\cos 10t + \frac{3}{25}\sin 10t\right) + \frac{3}{25}$$

and

$$I = \frac{dq}{dt} = \frac{12}{5}e^{-10t}\sin 10t$$

as before.

Note that although the current is completely transient, the charge on the capacitor is the sum of both transient and steady-state terms.

Supplementary Problems

17.14. A 10-lb weight is suspended from a spring and stretches it 2 inches from its natural length. Find the spring constant.

17.15. A $\frac{1}{4}$-slug mass is hung onto a spring, whereupon the spring is stretched 6 inches from its natural length. The mass is then started in motion from the equilibrium position with an initial velocity of 4 ft/sec in the upward direction. Find the subsequent motion of the mass, if the force due to air resistance is $-2\dot{x}$ lb.

17.16. A $\frac{1}{2}$-slug mass is attached to a spring so that the spring is stretched 2 ft from its natural length. The mass is started in motion with no initial velocity by displacing it $\frac{1}{2}$ ft in the upward direction. Find the subsequent motion of the mass, if the medium offers a resistance of $-4\dot{x}$ lb.

17.17. A 32-lb weight is attached to a spring, stretching it 8 ft from its natural length. The weight is started in motion by displacing it 1 ft in the upward direction and by giving it an initial velocity of 2 ft/sec in the downward direction. Find the subsequent motion of the weight, if the medium offers negligible resistance.

17.18. A $\frac{1}{2}$-slug mass is attached to a spring having a spring constant of 6 lb/ft. The mass is set into motion by displacing it 6 in. below its equilibrium position with no initial velocity. Find the subsequent motion of the mass, if the force due to the medium is $-4\dot{x}$ lb.

17.19. A 1-slug mass is attached to a spring having a spring constant of 8 lb/ft. The mass is initially set into motion from the equilibrium position with no initial velocity by applying an external force $F(t) = 16\cos 4t$. Find the subsequent motion of the mass, if the force due to air resistance is $-4\dot{x}$ lb.

17.20. An RCL circuit, as shown in Fig. 17-2, with $R = 6$ ohms, $C = 0.02$ farad, and $L = 0.1$ henry, has an applied voltage $E(t) = 6$ volts. Assuming no initial current and no initial charge at $t = 0$ when the voltage is first applied, find the subsequent charge on the capacitor and the current in the circuit.

17.21. An RCL circuit, as shown in Fig. 17-2, with $R = 6$ ohms, $C = 0.02$ farad, and $L = 0.1$ henry, has no applied voltage . Find the subsequent current in the circuit if the initial charge on the capacitor is $\frac{1}{10}$ coulomb and the initial current is zero.

17.22. An RCL circuit, as shown in Fig. 17-2, has $R = 5$ ohms, $C = 10^{-2}$ farad, $L = \frac{1}{8}$ henry, and no applied voltage. Find the subsequent steady-state current in the circuit. (*Hint.* Initial conditions are not needed.)

17.23. An RCL circuit, as shown in Fig. 17-2, with $R = 5$ ohms, $C = 10^{-2}$ farad, and $L = \frac{1}{8}$ henry, has applied voltage $E(t) = \sin t$. Find the steady-state current in the circuit. (*Hint.* Initial conditions are not needed.)

Answers to Supplementary Problems

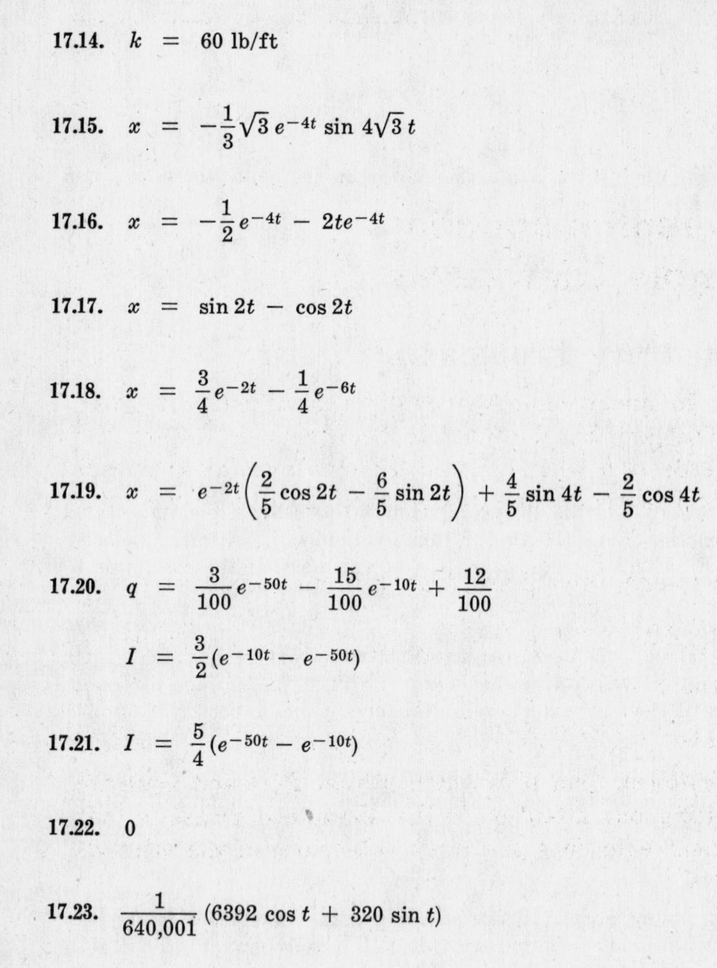

17.14. $k = 60$ lb/ft

17.15. $x = -\frac{1}{3}\sqrt{3}\,e^{-4t}\sin 4\sqrt{3}\,t$

17.16. $x = -\frac{1}{2}e^{-4t} - 2te^{-4t}$

17.17. $x = \sin 2t - \cos 2t$

17.18. $x = \frac{3}{4}e^{-2t} - \frac{1}{4}e^{-6t}$

17.19. $x = e^{-2t}\left(\frac{2}{5}\cos 2t - \frac{6}{5}\sin 2t\right) + \frac{4}{5}\sin 4t - \frac{2}{5}\cos 4t$

17.20. $q = \frac{3}{100}e^{-50t} - \frac{15}{100}e^{-10t} + \frac{12}{100}$

$I = \frac{3}{2}(e^{-10t} - e^{-50t})$

17.21. $I = \frac{5}{4}(e^{-50t} - e^{-10t})$

17.22. 0

17.23. $\frac{1}{640{,}001}(6392\cos t + 320\sin t)$

Chapter 18

Linear Differential Equations with Variable Coefficients

18.1 INTRODUCTION

In Chapters 12 and 13, the solutions to linear homogeneous differential equations with *constant coefficients* were discussed. We now consider linear homogeneous differential equations with *variable coefficients* (see Section 10.1). For these equations, all theorems given in Chapters 10 and 11, in particular Theorem 11.3, remain valid, since those theorems apply to all linear differential equations.

Specifically, we are interested in the *second-order* linear homogeneous equation

$$b_2(x)y'' + b_1(x)y' + b_0(x)y = 0 \tag{18.1}$$

Dividing by $b_2(x)$, we can rewrite (*18.1*) as

$$y'' + P(x)y' + Q(x)y = 0 \tag{18.2}$$

where $P(x) = b_1(x)/b_2(x)$ and $Q(x) = b_0(x)/b_2(x)$, and it is assumed that $P(x)$ and $Q(x)$ are not both constants. The reason for limiting our attention to the second-order case is that the methods for solving higher-order linear equations are direct extensions of the methods to be developed for second-order equations.

18.2 ANALYTIC FUNCTIONS

A function $f(x)$ is *analytic* at x_0 if its Taylor series about x_0,

$$\sum_{n=0}^{\infty} \frac{f^{(n)}(x_0)(x - x_0)^n}{n!} \tag{18.3}$$

converges to $f(x)$ in some neighborhood of x_0. (See Problems 18.1–18.3.)

Polynomials, $\sin x$, $\cos x$, and e^x are analytic everywhere; so too are sums, differences, and products of these functions. Quotients of any two of these functions are analytic at all points where the denominator is not zero.

18.3 ORDINARY POINTS AND SINGULAR POINTS

The point x_0 is an *ordinary point* of the differential equation (*18.2*) if both $P(x)$ and $Q(x)$ are analytic at x_0. If either of these functions is not analytic at x_0, then x_0 is a *singular point* of (*18.2*).

The point x_0 is a *regular singular point* of (*18.2*) if (1) x_0 is a singular point of (*18.2*) and (2) both functions $(x - x_0)P(x)$ and $(x - x_0)^2 Q(x)$ are analytic at x_0. Singular points which are not regular are called *irregular*.

Solved Problems

18.1. Determine whether $\ln x$ possesses a Taylor series about $x = 1$.

Here $x_0 = 1$ and $f(x) = \ln x$. Thus,

$$f'(x) = \frac{1}{x}, \quad f''(x) = -\frac{1}{x^2}, \quad \ldots, \quad f^{(n)}(x) = \frac{(-1)^{n-1}(n-1)!}{x^n} \qquad (n \geqq 1)$$

Therefore we have $f(1) = \ln 1 = 0$ and

$$f'(1) = 1, \quad f''(1) = -1, \quad \ldots, \quad f^{(n)}(1) = (-1)^{n-1}(n-1)! \qquad (n \geqq 1)$$

Substituting these values into *(18.3)*, where we recall that $0! = 1$ and $n! = n(n-1)!$, we find

$$\begin{aligned}
\sum_{n=0}^{\infty} \frac{f^{(n)}(1)(x-1)^n}{n!} &= \frac{f(1)(x-1)^0}{0!} + \sum_{n=1}^{\infty} \frac{f^{(n)}(1)(x-1)^n}{n!} \\
&= 0 + \sum_{n=1}^{\infty} \frac{(-1)^{n-1}(n-1)!\,(x-1)^n}{n!} \\
&= \sum_{n=1}^{\infty} \frac{(-1)^{n-1}}{n}(x-1)^n \\
&= (x-1) - \frac{1}{2}(x-1)^2 + \frac{1}{3}(x-1)^3 - \cdots
\end{aligned}$$

18.2. Does $\ln x$ possess a Taylor series about $x = 0$?

No. Neither $\ln x$ nor any of its derivatives exists at $x = 0$; therefore, *(18.3)* cannot be evaluated at $x = 0$.

18.3. Use the ratio test to determine those values of x for which the Taylor series found in Problem 18.1 converges.

By the ratio test, the series $\sum_{n=0}^{\infty} a_n$ converges if $\lim_{n\to\infty} \left|\frac{a_{n+1}}{a_n}\right| < 1$. If the limit $L > 1$ or $L = +\infty$, the series diverges; whereas if $L = 1$, no conclusion can be inferred. For the series of Problem 18.1,

$$\sum_{n=1}^{\infty} \frac{(-1)^{n-1}}{n}(x-1)^n \tag{1}$$

we have $a_0 = 0$ and $a_n = \dfrac{(-1)^{n-1}(x-1)^n}{n}$ $(n \geqq 1)$; hence,

$$\lim_{n\to\infty}\left|\frac{a_{n+1}}{a_n}\right| = \lim_{n\to\infty}\left|\frac{\dfrac{(-1)^n(x-1)^{n+1}}{n+1}}{\dfrac{(-1)^{n-1}(x-1)^n}{n}}\right| = \lim_{n\to\infty}\frac{n}{n+1}|x-1| = |x-1|$$

We conclude from the ratio test that the series *(1)* converges when $|x-1| < 1$ or, equivalently, when $0 < x < 2$.

The points $x = 0$ and $x = 2$ must be checked separately, since the ratio test is inconclusive at these points. Substituting $x = 0$ into *(1)*, we obtain

$$\sum_{n=1}^{\infty} \frac{(-1)^{n-1}(-1)^n}{n} = -\sum_{n=1}^{\infty} \frac{1}{n}$$

This is the harmonic series, which is known to diverge. Substituting $x = 2$ into *(1)*, we obtain

$$\sum_{n=1}^{\infty} \frac{(-1)^{n-1}(1)^n}{n} = -\sum_{n=1}^{\infty} \frac{(-1)^n}{n}$$

which converges by the alternating series test. Thus, the series of Problem 18.1 converges for $0 < x \leqq 2$.

18.4. Determine whether $x = 0$ is an ordinary point of the differential equation

$$y'' - xy' + 2y = 0$$

Here $P(x) = -x$ and $Q(x) = 2$ are both polynomials, hence they are analytic everywhere. Therefore, every value of x, in particular $x = 0$, is an ordinary point.

18.5. Determine whether $x = 1$ or $x = 2$ is an ordinary point of the differential equation

$$(x^2 - 4)y'' + y = 0$$

We first put the differential equation into form (*18.2*) by dividing by $x^2 - 4$. Then $P(x) \equiv 0$ and $Q(x) = 1/(x^2 - 4)$. Since both $P(x)$ and $Q(x)$ are analytic at $x = 1$, this point is an ordinary point. At $x = 2$, however, the denominator of $Q(x)$ is zero; hence $Q(x)$ is not analytic there. Thus, $x = 2$ is not an ordinary point but a singular point.

Note that $(x-2)P(x) \equiv 0$ and $(x-2)^2 Q(x) = (x-2)/(x+2)$ are analytic at $x = 2$, so that $x = 2$ is a regular singular point (see Section 18.3).

18.6. Determine whether $x = 0$ is an ordinary point of the differential equation

$$2x^2y'' + 7x(x+1)y' - 3y = 0$$

Dividing by $2x^2$, we have

$$P(x) = \frac{7(x+1)}{2x} \qquad Q(x) = \frac{-3}{2x^2}$$

As neither function is analytic at $x = 0$ (both denominators are zero there), $x = 0$ is not an ordinary point but, rather, a singular point.

Note that $(x-0)P(x) = \frac{7}{2}(x+1)$ and $(x-0)^2 Q(x) = -\frac{3}{2}$ are both analytic at $x = 0$; thus, $x = 0$ is a regular singular point.

18.7. Determine whether $x = 0$ is an ordinary point of the differential equation

$$x^2y'' + 2y' + xy = 0$$

Here $P(x) = 2/x^2$ and $Q(x) = 1/x$. Neither of these functions is analytic at $x = 0$, so $x = 0$ is not an ordinary point but a singular point. Furthermore, since $(x-0)P(x) = 2/x$ is not analytic at $x = 0$, $x = 0$ is not a regular singular point either; it is an irregular singular point.

Supplementary Problems

18.8. Find a Taylor series for $\sin x$ about $x = 0$ and then use the ratio test to prove that this Taylor series converges everywhere.

18.9. Find a Taylor series for e^x about $x = 1$ and then use the ratio test to prove that this Taylor series converges everywhere.

18.10. Determine those values of x for which the Taylor series for $1/x^2$ about $x = 1$ converges.

In Problems 18.11 through 18.19, determine whether the given value of x is an ordinary point, a regular singular point, or an irregular singular point of the given differential equation.

18.11. $x = 1$; $y'' + 3y' + 2xy = 0$.

18.12. $x = 2$; $(x-2)y'' + 3(x^2 - 3x + 2)y' + (x-2)^2 y = 0$.

18.13. $x = 0$; $(x+1)y'' + \frac{1}{x}y' + xy = 0$.

18.14. $x = -1$; $(x+1)y'' + \frac{1}{x}y' + xy = 0$.

18.15. $x = 0$; $x^3 y'' + y = 0$.

18.16. $x = 0$; $x^3 y'' + xy = 0$.

18.17. $x = 0$; $e^x y'' + (\sin x)y' + xy = 0$.

18.18. $x = -1$; $(x+1)^3 y'' + (x^2 - 1)(x+1)y' + (x-1)y = 0$.

18.19. $x = 2$; $x^4(x^2-4)y'' + (x+1)y' + (x^2 - 3x + 2)y = 0$.

Answers to Supplementary Problems

18.8. $\sum_{n=0}^{\infty} \frac{(-1)^n x^{2n+1}}{(2n+1)!}$

18.9. $e \sum_{n=0}^{\infty} \frac{(x-1)^n}{n!}$

18.10. $\frac{1}{x^2} = \sum_{n=0}^{\infty} (-1)^n (n+1)(x-1)^n$; converges for $0 < x < 2$

18.11. ordinary point

18.12. ordinary point

18.13. regular singular point

18.14. regular singular point

18.15. irregular singular point

18.16. regular singular point

18.17. ordinary point

18.18. irregular singular point

18.19. regular singular point

Chapter 19

Power-Series Solutions About an Ordinary Point

19.1 METHOD FOR HOMOGENEOUS EQUATIONS

Consider the linear second-order homogeneous differential equation

$$y'' + P(x)y' + Q(x)y = 0 \tag{19.1}$$

Theorem 19.1. If x_0 is an ordinary point of (*19.1*), then the general solution in an interval containing x_0 is

$$y = \sum_{n=0}^{\infty} a_n(x - x_0)^n \equiv a_0y_1(x) + a_1y_2(x) \tag{19.2}$$

where a_0 and a_1 are arbitrary constants and $y_1(x)$ and $y_2(x)$ are linearly independent functions analytic at x_0.

To evaluate the coefficients a_n in the solution furnished by Theorem 19.1, proceed as follows. *First,* substitute in the left side of (*19.1*) the power series

$$y = \sum_{n=0}^{\infty} a_n(x - x_0)^n \tag{19.3}$$

together with the power series for y' and y'' obtained from (*19.3*) by termwise differentiation, as well as the power-series expansions of $P(x)$ and $Q(x)$ about $x = x_0$. *Second,* collect powers of $x - x_0$ and set each collected coefficient equal to zero. *Third,* solve the resulting equations for a_n $(n = 2, 3, 4, \ldots)$ in terms of a_0 and a_1. The functions $y_1(x)$ and $y_2(x)$ can now be obtained by combining all terms in (*19.3*) that contain a_0 [the result being $a_0y_1(x)$] and all terms that contain a_1 [the result being $a_1y_2(x)$]. (See Problems 19.1–19.4.)

Remark 1. If the ordinary point $x_0 \neq 0$, it generally simplifies the algebra if x_0 is translated to the origin by the change of variables $t = x - x_0$. The solution of the new differential equation that results can be obtained by the power-series method about $t = 0$. Then the solution of the original equation is easily gotten by back-substitution. (See Problems 19.3 and 19.4.)

Remark 2. If, in (*19.1*), $P(x)$ and $Q(x)$ are both constants, then every point is an ordinary point, and the power-series method is applicable about any point. For equations of this type, however, the methods given in Chapter 12 are clearly more efficient (see Problem 19.2). If $P(x)$ and $Q(x)$ are quotients of polynomials, it is usually easier first to multiply (*19.1*) by their lowest common denominator. To the resulting equation, which is of the form (*18.1*) with polynomial coefficients, the power-series method applies without change. (See Problems 19.9 and 19.11.)

Remark 3. Initial-value problems can be solved in the usual way: first the general solution is obtained, which for *(19.1)* is given by *(19.2)*, and then the initial conditions are applied. An alternate method is given in Problems 19.6 and 19.7.

19.2 METHOD FOR NONHOMOGENEOUS EQUATIONS

The general solution near $x = x_0$ to the differential equation

$$y'' + P(x)y' + Q(x)y = \phi(x) \tag{19.4}$$

where $P(x)$ and $Q(x)$ are analytic about x_0 and where $\phi(x)$ is a known function that is not identically zero, is $y = y_h + y_p$ (see Theorem 11.3). Here, y_h can be found by the power-series method of Section 19.1, and y_p can be obtained by variation of parameters. In general, the method of undetermined coefficients is not applicable (see Problem 19.8). Alternatively, if $\phi(x)$ also is analytic at x_0, the general solution can be obtained completely by applying the power-series method directly to *(19.4)*, in which case we equate coefficients of $(x - x_0)^n$ on the two sides of the equation. This last procedure is often the most efficient. (See Problems 19.9 and 19.10.)

Solved Problems

19.1. Find the general solution near $x = 0$ of $y'' - xy' + 2y = 0$.

Note that $x = 0$ is an ordinary point of the given differential equation. Using the power-series method, we assume

$$y = a_0 + a_1x + a_2x^2 + a_3x^3 + a_4x^4 + \cdots + a_nx^n + a_{n+1}x^{n+1} + a_{n+2}x^{n+2} + \cdots \tag{1}$$

Therefore, differentiating termwise, we have

$$y' = a_1 + 2a_2x + 3a_3x^2 + 4a_4x^3 + \cdots + na_nx^{n-1} + (n+1)a_{n+1}x^n + (n+2)a_{n+2}x^{n+1} + \cdots \tag{2}$$

$$\begin{aligned} y'' = {} & 2a_2 + 6a_3x + 12a_4x^2 + \cdots + n(n-1)a_nx^{n-2} + (n+1)(n)a_{n+1}x^{n-1} \\ & + (n+2)(n+1)a_{n+2}x^n + \cdots \end{aligned} \tag{3}$$

Substituting *(1)*, *(2)*, and *(3)* into the differential equation, we find

$$\begin{aligned} & [2a_2 + 6a_3x + 12a_4x^2 + \cdots + n(n-1)a_nx^{n-2} \\ & \qquad + (n+1)(n)a_{n+1}x^{n-1} + (n+2)(n+1)a_{n+2}x^n + \cdots] \\ & - x[a_1 + 2a_2x + 3a_3x^2 + 4a_4x^3 + \cdots + na_nx^{n-1} \\ & \qquad + (n+1)a_{n+1}x^n + (n+2)a_{n+2}x^{n+1} + \cdots] \\ & + 2[a_0 + a_1x + a_2x^2 + a_3x^3 + a_4x^4 + \cdots + a_nx^n + a_{n+1}x^{n+1} \\ & \qquad + a_{n+2}x^{n+2} + \cdots] \\ & = 0 \end{aligned}$$

Combining terms that contain like powers of x, we have

$$\begin{aligned} & (2a_2 + 2a_0) + x(6a_3 + a_1) + x^2(12a_4) + x^3(20a_5 - a_3) \\ & \qquad + \cdots + x^n[(n+2)(n+1)a_{n+2} - na_n + 2a_n] + \cdots \\ & = 0 + 0x + 0x^2 + 0x^3 + \cdots + 0x^n + \cdots \end{aligned}$$

This last equation holds if and only if each coefficient in the left-hand side is zero. Thus,

$$2a_2 + 2a_0 = 0, \quad 6a_3 + a_1 = 0, \quad 12a_4 = 0, \quad 20a_5 - a_3 = 0, \quad \ldots \tag{4}$$

In general, $(n+2)(n+1)a_{n+2} - (n-2)a_n = 0$, or,

$$a_{n+2} = \frac{(n-2)}{(n+2)(n+1)} a_n \tag{5}$$

Equation (*5*) is the *recurrence formula* for this problem. In general, a recurrence formula is an equation that relates a_n (or else a_{n+1}, a_{n+2}, etc.) to preceding a's. In (*5*), a_{n+2} is related to a_n.

Solving equations (*4*), or, alternatively, substituting successive values of n into (*5*), we obtain

$$\begin{aligned}
a_2 &= -a_0 \\
a_3 &= -\frac{1}{6}a_1 \\
a_4 &= 0 \\
a_5 &= \frac{1}{20}a_3 = \frac{1}{20}\left(-\frac{1}{6}a_1\right) = -\frac{1}{120}a_1 \\
a_6 &= \frac{2}{30}a_4 = \frac{1}{15}(0) = 0 \\
a_7 &= \frac{3}{42}a_5 = \frac{1}{14}\left(-\frac{1}{120}\right)a_1 = -\frac{1}{1680}a_1 \\
a_8 &= \frac{4}{56}a_6 = \frac{1}{14}(0) = 0 \\
&\cdots\cdots\cdots\cdots
\end{aligned} \tag{6}$$

Note that since $a_4 = 0$, it follows from (*5*) that all the even coefficients beyond a_4 are also zero. Substituting (*6*) into (*1*), we have

$$\begin{aligned}
y &= a_0 + a_1x - a_0x^2 - \frac{1}{6}a_1x^3 + 0x^4 - \frac{1}{120}a_1x^5 + 0x^6 - \frac{1}{1680}a_1x^7 - \cdots \\
&= a_0(1-x^2) + a_1\left(x - \frac{1}{6}x^3 - \frac{1}{120}x^5 - \frac{1}{1680}x^7 - \cdots\right)
\end{aligned} \tag{7}$$

If we define $y_1(x) \equiv 1 - x^2$ and $y_2(x) \equiv x - \frac{1}{6}x^3 - \frac{1}{120}x^5 - \frac{1}{1680}x^7 - \cdots$, then the general solution (*7*) can be rewritten as $y = a_0y_1(x) + a_1y_2(x)$.

19.2. Use the power-series method to find the general solution near $x = 0$ of $y'' + y = 0$.

Since this equation has constant coefficients, the solution is $y = c_1 \cos x + c_2 \sin x$ (see Chapter 12). To obtain this solution by the power-series method, we first substitute (*1*) and (*3*) of Problem 19.1 into the differential equation, obtaining

$$\begin{aligned}
&[2a_2 + 6a_3x + 12a_4x^2 + \cdots + n(n-1)a_nx^{n-2} \\
&\qquad + (n+1)na_{n+1}x^{n-1} + (n+2)(n+1)a_{n+2}x^n + \cdots] \\
&\quad + (a_0 + a_1x + a_2x^2 + a_3x^3 + a_4x^4 + \cdots + a_nx^n \\
&\qquad + a_{n+1}x^{n+1} + a_{n+2}x^{n+2} + \cdots) = 0
\end{aligned}$$

or

$$\begin{aligned}
&(2a_2 + a_0) + x(6a_3 + a_1) + x^2(12a_4 + a_2) + x^3(20a_5 + a_3) \\
&\qquad + \cdots + x^n[(n+2)(n+1)a_{n+2} + a_n] + \cdots \\
&\quad = 0 + 0x + 0x^2 + \cdots + 0x^n + \cdots
\end{aligned}$$

Equating each coefficient to zero, we have

$$2a_2 + a_0 = 0, \quad 6a_3 + a_1 = 0, \quad 12a_4 + a_2 = 0, \quad 20a_5 + a_3 = 0, \quad \ldots \tag{1}$$

In general $(n+2)(n+1)a_{n+2} + a_n = 0$, which is equivalent to

$$a_{n+2} = \frac{-1}{(n+2)(n+1)} a_n \tag{2}$$

the recurrence formula for this problem. Solving equations (*1*), or, alternatively, substituting successive values of n into (*2*), we obtain

$$a_2 = -\frac{1}{2} a_0 = -\frac{1}{2!} a_0$$

$$a_3 = -\frac{1}{6} a_1 = -\frac{1}{3!} a_1$$

$$a_4 = -\frac{1}{(4)(3)} a_2 = -\frac{1}{(4)(3)}\left(-\frac{1}{2!} a_0\right) = \frac{1}{4!} a_0$$

$$a_5 = -\frac{1}{(5)(4)} a_3 = -\frac{1}{(5)(4)}\left(-\frac{1}{3!} a_1\right) = \frac{1}{5!} a_1$$

$$a_6 = -\frac{1}{(6)(5)} a_4 = -\frac{1}{(6)(5)}\left(\frac{1}{4!} a_0\right) = -\frac{1}{6!} a_0$$

$$a_7 = -\frac{1}{(7)(6)} a_5 = -\frac{1}{(7)(6)}\left(\frac{1}{5!} a_1\right) = -\frac{1}{7!} a_1$$

$$\cdots\cdots\cdots\cdots\cdots\cdots\cdots\cdots$$

Recall that for a positive integer n, n factorial, which is denoted by $n!$, is defined by

$$n! = n(n-1)(n-2)\cdots(3)(2)(1)$$

and $0!$ is defined as one. Thus, $4! = (4)(3)(2)(1) = 24$ and $5! = (5)(4)(3)(2)(1) = 5(4!) = 120$. In general, $n! = n(n-1)!$.

Now substituting the above values for $a_2, a_3, a_4, \ldots$ into (*1*) of Problem 19.1, we have

$$\begin{aligned} y &= a_0 + a_1 x - \frac{1}{2!} a_0 x^2 - \frac{1}{3!} a_1 x^3 + \frac{1}{4!} a_0 x^4 \\ &\quad + \frac{1}{5!} a_1 x^5 - \frac{1}{6!} a_0 x^6 - \frac{1}{7!} a_1 x^7 + \cdots \\ &= a_0\left(1 - \frac{1}{2!} x^2 + \frac{1}{4!} x^4 - \frac{1}{6!} x^6 + \cdots\right) \\ &\quad + a_1\left(x - \frac{1}{3!} x^3 + \frac{1}{5!} x^5 - \frac{1}{7!} x^7 + \cdots\right) \end{aligned} \tag{3}$$

But

$$\cos x = \sum_{n=0}^{\infty} \frac{(-1)^n x^{2n}}{(2n)!} = 1 - \frac{1}{2!} x^2 + \frac{1}{4!} x^4 - \frac{1}{6!} x^6 + \cdots$$

$$\sin x = \sum_{n=0}^{\infty} \frac{(-1)^n x^{2n+1}}{(2n+1)!} = x - \frac{1}{3!} x^3 + \frac{1}{5!} x^5 - \frac{1}{7!} x^7 + \cdots$$

Substituting these two results into (*3*) and letting $c_1 = a_0$ and $c_2 = a_1$, we obtain, as before,

$$y = c_1 \cos x + c_2 \sin x$$

19.3. Find the general solution near $x = 2$ of $y'' - (x-2)y' + 2y = 0$.

For this problem, $x_0 = 2$, which is an ordinary point of the given differential equation. We first make the change of variables $t = x - 2$ (see Remark 1, page 97). From the chain rule we find the corresponding transformations of the derivatives of y:

$$\frac{dy}{dx} = \frac{dy}{dt}\frac{dt}{dx} = \frac{dy}{dt}(1) = \frac{dy}{dt}$$

$$\frac{d^2y}{dx^2} = \frac{d}{dx}\left(\frac{dy}{dx}\right) = \frac{d}{dx}\left(\frac{dy}{dt}\right) = \frac{d}{dt}\left(\frac{dy}{dt}\right)\frac{dt}{dx} = \frac{d^2y}{dt^2}(1) = \frac{d^2y}{dt^2}$$

Substituting these results into the differential equation, we obtain

$$\frac{d^2y}{dt^2} - t\frac{dy}{dt} + 2y = 0$$

and this equation is to be solved near $t = 0$. From Problem 19.1, with x replaced by t, we see that the solution is

$$y = a_0(1 - t^2) + a_1\left(t - \frac{1}{6}t^3 - \frac{1}{120}t^5 - \frac{1}{1680}t^7 - \cdots\right)$$

Substituting $t = x - 2$ into this last equation, we obtain the solution to the original problem as

$$y = a_0[1 - (x-2)^2] + a_1\left[(x-2) - \frac{1}{6}(x-2)^3 - \frac{1}{120}(x-2)^5 - \frac{1}{1680}(x-2)^7 - \cdots\right]$$

19.4. Find the general solution near $x = -1$ of $y'' + xy' + (2x-1)y = 0$.

Here $x_0 = -1$ is an ordinary point. Applying Remark 1, page 97, we first make the substitution $t = x - (-1) = x + 1$. Then, as in Problem 19.3, $\frac{dy}{dx} = \frac{dy}{dt}$ and $\frac{d^2y}{dx^2} = \frac{d^2y}{dt^2}$. Substituting these results into the differential equation, we obtain

$$\frac{d^2y}{dt^2} + (t-1)\frac{dy}{dt} + (2t-3)y = 0 \tag{1}$$

We now seek a solution to (*1*) near $t = 0$. Substituting (*1*), (*2*), and (*3*) of Problem 19.1 into (*1*), with x replaced by t, we have

$$\begin{aligned}
&[2a_2 + 6a_3t + 12a_4t^2 + \cdots + n(n-1)a_nt^{n-2} \\
&\qquad + (n+1)na_{n+1}t^{n-1} + (n+2)(n+1)a_{n+2}t^n + \cdots] \\
&+ (t-1)[a_1 + 2a_2t + 3a_3t^2 + 4a_4t^3 + \cdots + na_nt^{n-1} \\
&\qquad + (n+1)a_{n+1}t^n + (n+2)a_{n+2}t^{n+1} + \cdots] \\
&+ (2t-3)[a_0 + a_1t + a_2t^2 + a_3t^3 + a_4t^4 + \cdots + a_nt^n \\
&\qquad + a_{n+1}t^{n+1} + a_{n+2}t^{n+2} + \cdots] = 0
\end{aligned}$$

or

$$\begin{aligned}
&(2a_2 - a_1 - 3a_0) + t(6a_3 + a_1 - 2a_2 + 2a_0 - 3a_1) \\
&\qquad + t^2(12a_4 + 2a_2 - 3a_3 + 2a_1 - 3a_2) + \cdots \\
&\qquad + t^n[(n+2)(n+1)a_{n+2} + na_n - (n+1)a_{n+1} + 2a_{n-1} - 3a_n] + \cdots \\
&= 0 + 0t + 0t^2 + \cdots + 0t^n + \cdots
\end{aligned}$$

Equating each coefficient to zero, we obtain

$$2a_2 - a_1 - 3a_0 = 0, \quad 6a_3 - 2a_2 - 2a_1 + 2a_0 = 0, \quad 12a_4 - 3a_3 - a_2 + 2a_1 = 0, \quad \ldots \tag{2}$$

In general, $$(n+2)(n+1)a_{n+2} - (n+1)a_{n+1} + (n-3)a_n + 2a_{n-1} = 0$$

which is equivalent to

$$a_{n+2} = \frac{1}{n+2}a_{n+1} - \frac{(n-3)}{(n+2)(n+1)}a_n - \frac{2}{(n+2)(n+1)}a_{n-1} \tag{3}$$

Equation (*3*) is the recurrence formula for this problem. Note, however, that it is not valid for $n = 0$, since a_{-1} is an undefined quantity. To obtain an equation for $n = 0$, we use the first equation (*2*), which gives $a_2 = \frac{1}{2}a_1 + \frac{3}{2}a_0$. Then, using in (*3*) successive values of n starting at $n = 1$, we find that

$$a_3 = \frac{1}{3}a_2 + \frac{1}{3}a_1 - \frac{1}{3}a_0 = \frac{1}{3}\left(\frac{1}{2}a_1 + \frac{3}{2}a_0\right) + \frac{1}{3}a_1 - \frac{1}{3}a_0 = \frac{1}{2}a_1 + \frac{1}{6}a_0$$

$$a_4 = \frac{1}{4}a_3 + \frac{1}{12}a_2 - \frac{1}{6}a_1 = \frac{1}{4}\left(\frac{1}{2}a_1 + \frac{1}{6}a_0\right) + \frac{1}{12}\left(\frac{1}{2}a_1 + \frac{3}{2}a_0\right) - \frac{1}{6}a_1 = \frac{1}{6}a_0$$

. . . .

Thus, the solution to (1) is

$$y = a_0 + a_1 t + \left(\frac{1}{2}a_1 + \frac{3}{2}a_0\right)t^2 + \left(\frac{1}{2}a_1 + \frac{1}{6}a_0\right)t^3 + \left(\frac{1}{6}a_0\right)t^4 + \cdots$$
$$= a_0\left(1 + \frac{3}{2}t^2 + \frac{1}{6}t^3 + \frac{1}{6}t^4 + \cdots\right) + a_1\left(t + \frac{1}{2}t^2 + \frac{1}{2}t^3 + 0t^4 + \cdots\right)$$

Recalling that $t = x + 1$, we finally obtain the solution to the original problem as

$$y = a_0\left[1 + \frac{3}{2}(x+1)^2 + \frac{1}{6}(x+1)^3 + \frac{1}{6}(x+1)^4 + \cdots\right]$$
$$+ a_1\left[(x+1) + \frac{1}{2}(x+1)^2 + \frac{1}{2}(x+1)^3 + 0(x+1)^4 + \cdots\right] \quad (4)$$

19.5. Solve $y'' + xy' + (2x-1)y = 0;\ y(-1) = 2,\ y'(-1) = -2.$

Since the initial conditions are prescribed at $x = -1$, it is advantageous to obtain the general solution to the differential equation near $x = -1$. This has already been done in (4) of Problem 19.4. Applying the initial conditions, we find that $a_0 = 2$ and $a_1 = -2$. Thus, the solution is

$$y = 2\left[1 + \frac{3}{2}(x+1)^2 + \frac{1}{6}(x+1)^3 + \frac{1}{6}(x+1)^4 + \cdots\right]$$
$$- 2\left[(x+1) + \frac{1}{2}(x+1)^2 + \frac{1}{2}(x+1)^3 + 0(x+1)^4 + \cdots\right]$$
$$= 2 - 2(x+1) + 2(x+1)^2 - \frac{2}{3}(x+1)^3 + \frac{1}{3}(x+1)^4 + \cdots$$

19.6. Solve Problem 19.5 by another method.

TAYLOR SERIES METHOD. An alternate method for solving initial-value problems rests on the assumption that the solution can be expanded in a Taylor series about the initial point x_0; i.e.

$$y = \sum_{n=0}^{\infty} \frac{y^{(n)}(x_0)}{n!}(x - x_0)^n$$
$$= \frac{y(x_0)}{0!} + \frac{y'(x_0)}{1!}(x - x_0) + \frac{y''(x_0)}{2!}(x - x_0)^2 + \cdots \quad (1)$$

The terms $y(x_0)$ and $y'(x_0)$ are given as initial conditions; the other terms $y^{(n)}(x_0)$ $(n = 2, 3, \cdots)$ can be obtained by successively differentiating the differential equation. For Problem 19.5, we have $x_0 = -1$, $y(x_0) = y(-1) = 2$, and $y'(x_0) = y'(-1) = -2$. Solving the differential equation of Problem 19.5 for y'', we find that

$$y'' = -xy' - (2x-1)y \quad (2)$$

We obtain $y''(x_0) = y''(-1)$ by substituting $x_0 = -1$ into (2) and using the given initial conditions. Thus,

$$y''(-1) = -(-1)y'(-1) - [2(-1) - 1]y(-1) = 1(-2) - (-3)(2) = 4 \quad (3)$$

To obtain $y'''(-1)$, we differentiate (2) and then substitute $x_0 = -1$ into the resulting equation. Thus,

$$y'''(x) = -y' - xy'' - 2y - (2x-1)y' \quad (4)$$

and
$$y'''(-1) = -y'(-1) - (-1)y''(-1) - 2y(-1) - [2(-1)-1]y'(-1)$$
$$= -(-2) + 4 - 2(2) - (-3)(-2) = -4 \tag{5}$$

To obtain $y^{(4)}(-1)$, we differentiate (4) and then substitute $x_0 = -1$ into the resulting equation. Thus,
$$y^{(4)}(x) = -xy''' - (2x+1)y'' - 4y' \tag{6}$$
and
$$y^{(4)}(-1) = -(-1)y'''(-1) - [2(-1)+1]y''(-1) - 4y'(-1)$$
$$= -4 - (-1)(4) - 4(-2) = 8 \tag{7}$$

This process can be kept up indefinitely. Substituting (3), (5), (7), and the initial conditions into (1), we obtain, as before,
$$y = 2 + \frac{-2}{1!}(x+1) + \frac{4}{2!}(x+1)^2 + \frac{-4}{3!}(x+1)^3 + \frac{8}{4!}(x+1)^4 + \cdots$$
$$= 2 - 2(x+1) + 2(x+1)^2 - \frac{2}{3}(x+1)^3 + \frac{1}{3}(x+1)^4 + \cdots$$

One advantage in using this alternate method, as compared to the usual method of first solving the differential equation and then applying the initial conditions, is that the Taylor series method is easier to apply when only the first few terms of the solution are required. One disadvantage is that the recurrence formula cannot be found by the Taylor series method, and, therefore, a general expression for the nth term of the solution cannot be obtained. Note that this alternate method is also useful in solving differential equations without initial conditions. In such cases, we set $y(x_0) = a_0$ and $y'(x_0) = a_1$, where a_0 and a_1 are unknown constants, and procede as before.

19.7. Use the method outlined in Problem 19.6 to solve $y'' - 2xy = 0$; $y(2) = 1$, $y'(2) = 0$.

Using (1) of Problem 19.6, we assume a solution of the form
$$y(x) = \frac{y(2)}{0!} + \frac{y'(2)}{1!}(x-2) + \frac{y''(2)}{2!}(x-2)^2 + \frac{y'''(2)}{3!}(x-2)^3 + \cdots \tag{1}$$

From the differential equation,
$$y''(x) = 2xy, \quad y'''(x) = 2y + 2xy', \quad y^{(4)}(x) = 4y' + 2xy'', \quad \ldots$$

Substituting $x = 2$ into these equations and using the initial conditions, we find that
$$y''(2) = 2(2)y(2) = 4(1) = 4$$
$$y'''(2) = 2y(2) + 2(2)y'(2) = 2(1) + 4(0) = 2$$
$$y^{(4)}(2) = 4y'(2) + 2(2)y''(2) = 4(0) + 4(4) = 16$$
$$\cdots\cdots\cdots\cdots\cdots\cdots\cdots\cdots\cdots$$

Substituting these results into (1), we obtain the solution as
$$y = 1 + 2(x-2)^2 + \frac{1}{3}(x-2)^3 + \frac{2}{3}(x-2)^4 + \cdots$$

19.8. Show that the method of undetermined coefficients cannot be used to obtain a particular solution of $y'' + xy = 2$.

By the method of undetermined coefficients, we assume a particular solution of the form $y_p = A_0x^m$, where m might be zero if the simple guess $y_p = A_0$ does not require modification (see Section 14.2). Substituting y_p into the differential equation, we find
$$m(m-1)A_0x^{m-2} + A_0x^{m+1} = 2 \tag{1}$$

Regardless of the value of m, it is impossible to assign A_0 any *constant* value that will satisfy (1). It follows that the method of undetermined coefficients is not applicable.

19.9. Use the power-series method to find the general solution near $x = 0$ of

$$(x^2+4)y'' + xy = x+2$$

Dividing the given equation by x^2+4, we see that $x=0$ is an ordinary point and that $\phi(x) = (x+2)/(x^2+4)$ is analytic there. Hence, the power-series method is applicable to the entire equation, which, furthermore, we may leave in the form originally given (compare Remark 2, page 97.) Substituting *(1)*, *(2)*, and *(3)* of Problem 19.1 in the original equation, we find

$$(x^2+4)[2a_2 + 6a_3x + 12a_4x^2 + \cdots + n(n-1)a_nx^{n-2} + (n+1)na_{n+1}x^{n-1} + (n+2)(n+1)a_{n+2}x^n + \cdots] + x[a_0 + a_1x + a_2x^2 + a_3x^3 + \cdots + a_{n-1}x^{n-1} + \cdots] = x + 2$$

or

$$(8a_2) + x(24a_3 + a_0) + x^2(2a_2 + 48a_4 + a_1) + x^3(6a_3 + 80a_5 + a_2) + \cdots + x^n[n(n-1)a_n + 4(n+2)(n+1)a_{n+2} + a_{n-1}] + \cdots = 2 + (1)x + (0)x^2 + (0)x^3 + \cdots \tag{1}$$

Equating coefficients of like powers of x, we have

$$8a_2 = 2, \quad 24a_3 + a_0 = 1, \quad 2a_2 + 48a_4 + a_1 = 0, \quad 6a_3 + 80a_5 + a_2 = 0, \quad \cdots \tag{2}$$

In general,

$$n(n-1)a_n + 4(n+2)(n+1)a_{n+2} + a_{n-1} = 0 \qquad (n = 2, 3, \ldots)$$

which is equivalent to

$$a_{n+2} = -\frac{n(n-1)}{4(n+2)(n+1)}a_n - \frac{1}{4(n+2)(n+1)}a_{n-1} \tag{3}$$

$(n = 2, 3, \ldots)$. Note that the recurrence formula *(3)* is not valid for $n=0$ or $n=1$, since the coefficients of x^0 and x^1 on the right side of *(1)* are not zero. Instead, we use the first two equations *(2)* to obtain

$$a_2 = \frac{1}{4} \qquad a_3 = \frac{1}{24} - \frac{1}{24}a_0$$

Then, from *(3)*,

$$a_4 = -\frac{1}{24}a_2 - \frac{1}{48}a_1 = -\frac{1}{24}\left(\frac{1}{4}\right) - \frac{1}{48}a_1 = -\frac{1}{96} - \frac{1}{48}a_1$$

$$a_5 = -\frac{3}{40}a_3 - \frac{1}{80}a_2 = -\frac{3}{40}\left(\frac{1}{24} - \frac{1}{24}a_0\right) - \frac{1}{80}\left(\frac{1}{4}\right) = \frac{-1}{160} + \frac{1}{320}a_0$$

. . . .

Thus,

$$y = a_0 + a_1x + \frac{1}{4}x^2 + \left(\frac{1}{24} - \frac{1}{24}a_0\right)x^3 + \left(-\frac{1}{96} - \frac{1}{48}a_1\right)x^4 + \left(\frac{-1}{160} + \frac{1}{320}a_0\right)x^5 + \cdots$$

$$= \left(\frac{1}{4}x^2 + \frac{1}{24}x^3 - \frac{1}{96}x^4 - \frac{1}{160}x^5 + \cdots\right) + a_0\left(1 - \frac{1}{24}x^3 + \frac{1}{320}x^5 + \cdots\right) + a_1\left(x - \frac{1}{48}x^4 + \cdots\right)$$

The first series is the particular solution. The second and third series together represent the general solution of the associated homogeneous equation $(x^2+4)y'' + xy = 0$.

19.10. Find the general solution near $x=1$ of $y'' + (x-1)y = e^x$.

Letting $t = x-1$, the differential equation becomes (see Problem 19.3)

$$\frac{d^2y}{dt^2} + ty = e^{t+1}$$

Substituting (1), (2), and (3) of Problem 19.1, with x replaced by t, into this new equation, we have

$$[2a_2 + 6a_3t + 12a_4t^2 + \cdots + (n+2)(n+1)a_{n+2}t^n + \cdots]$$
$$+ t(a_0 + a_1t + a_2t^2 + \cdots + a_{n-1}t^{n-1} + \cdots) = e^{t+1}$$

Recall that e^{t+1} has the Taylor expansion $e^{t+1} = e\sum_{n=0}^{\infty} t^n/n!$ about $t=0$. Thus, the last equation can be rewritten as

$$(2a_2) + t(6a_3 + a_0) + t^2(12a_4 + a_1) + \cdots$$
$$+ t^n[(n+2)(n+1)a_{n+2} + a_{n-1}] + \cdots$$
$$= \frac{e}{0!} + \frac{e}{1!}t + \frac{e}{2!}t^2 + \cdots + \frac{e}{n!}t^n + \cdots$$

Equating coefficients of like powers of t, we have

$$2a_2 = \frac{e}{0!}, \quad 6a_3 + a_0 = \frac{e}{1!}, \quad 12a_4 + a_1 = \frac{e}{2!}, \quad \cdots \tag{1}$$

In general, $(n+2)(n+1)a_{n+2} + a_{n-1} = e/n!$ for $n = 1, 2, \ldots$, or,

$$a_{n+2} = -\frac{1}{(n+2)(n+1)}a_{n-1} + \frac{e}{(n+2)(n+1)n!} \tag{2}$$

for $n = 1, 2, \ldots$. It follows from (1) that $a_2 = e/2$, and then from (2) that

$$a_3 = -\frac{1}{6}a_0 + \frac{e}{6}, \quad a_4 = -\frac{1}{12}a_1 + \frac{e}{24}, \quad \cdots$$

Thus,
$$y = a_0 + a_1t + \frac{e}{2}t^2 + \left(-\frac{1}{6}a_0 + \frac{e}{6}\right)t^3 + \left(-\frac{1}{12}a_1 + \frac{e}{24}\right)t^4 + \cdots$$
$$= e\left(\frac{1}{2}t^2 + \frac{1}{6}t^3 + \frac{1}{24}t^4 + \cdots\right) + a_0\left(1 - \frac{1}{6}t^3 + \cdots\right) + a_1\left(t - \frac{1}{12}t^4 + \cdots\right)$$

and the solution to the original differential equation is

$$y = e\left[\frac{1}{2}(x-1)^2 + \frac{1}{6}(x-1)^3 + \frac{1}{24}(x-1)^4 + \cdots\right]$$
$$+ a_0\left[1 - \frac{1}{6}(x-1)^3 + \cdots\right] + a_1\left[(x-1) - \frac{1}{12}(x-1)^4 + \cdots\right]$$

19.11. Show that whenever n is a positive integer, one solution near $x=0$ of *Legendre's equation*

$$(1-x^2)y'' - 2xy' + n(n+1)y = 0$$

is a polynomial of degree n.

Here $x=0$ is an ordinary point, and it is convenient to retain the form (18.1) for the equation, rather than to transform to (19.1) (see Remark 2, page 97). Substituting (1), (2), and (3) of Problem 19.1 into Legendre's equation (to avoid confusion, we replace the dummy index n in (1), (2), and (3) by the dummy index k) and simplifying, we have

$$[2a_2 + (n^2+n)a_0] + x[6a_3 + (n^2+n-2)a_1] + \cdots$$
$$+ x^k[(k+2)(k+1)a_{k+2} + (n^2+n-k^2-k)a_k] + \cdots = 0$$

Noting that $n^2+n-k^2-k = (n-k)(n+k+1)$, we obtain the recurrence formula

$$a_{k+2} = -\frac{(n-k)(n+k+1)}{(k+2)(k+1)}a_k \tag{1}$$

Because of the factor $n-k$ in (1), we find, upon letting $k=n$, that $a_{n+2}=0$. It follows at once that $0 = a_{n+4} = a_{n+6} = a_{n+8} = \cdots$. Thus, if n is odd, all odd coefficients a_k $(k>n)$ are zero; whereas if n is even, all even coefficients a_k $(k > n)$ are zero. Therefore, either $y_1(x)$ or $y_2(x)$ in *(19.2)* (depending on whether n is even or odd, respectively) will contain only a finite number of nonzero terms up to and including a term in x^n; hence, it is a polynomial of degree n.

Since a_0 and a_1 are arbitrary, it is customary to choose them so that $y_1(x)$ or $y_2(x)$, whichever is the polynomial, will satisfy the condition $y(1) = 1$. The resulting polynomial, denoted by $P_n(x)$, is known as the *Legendre polynomial of degree n*. The first few of these are

$$P_0(x) = 1 \qquad P_1(x) = x \qquad P_2(x) = \frac{1}{2}(3x^2 - 1)$$

$$P_3(x) = \frac{1}{2}(5x^3 - 3x) \qquad P_4(x) = \frac{1}{8}(35x^4 - 30x^2 + 3)$$

Supplementary Problems

19.12. Find the general solution near $x = 0$ of $y'' - y' = 0$. Check your answer by solving the equation by the method of Chapter 12 and then expanding the result in a power series about $x = 0$.

In Problems 19.13–19.19, find the general solution of the given differential equation near the indicated point.

19.13. $x = 0$; $(x^2 - 1)y'' + xy' - y = 0$.

19.14. $x = 0$; $y'' - xy = 0$.

19.15. $x = 1$; $y'' - xy = 0$.

19.16. $x = -2$; $y'' - x^2y' + (x+2)y = 0$.

19.17. $x = 0$; $(x^2 + 4)y'' + y = x$.

19.18. $x = 1$; $y'' - (x-1)y' = x^2 - 2x$.

19.19. $x = 0$; $y'' - xy' = e^{-x}$.

19.20. Use the Taylor series method outlined in Problem 19.6 to solve

(*a*) $y'' - 2xy' + x^2y = 0$; $y(0) = 1$, $y'(0) = -1$

(*b*) $y'' - 2xy = x^2$; $y(1) = 0$, $y'(1) = 2$

Answers to Supplementary Problems

19.12. $y = a_0 + a_1\left(x + \frac{x^2}{2} + \frac{x^3}{6} + \cdots\right) = c_1 + c_2e^x$, where $c_1 = a_0 - a_1$ and $c_2 = a_1$

19.13. RF (recurrence formula): $a_{n+2} = \dfrac{n-1}{n+2}a_n$

$$y = a_0\left(1 - \frac{1}{2}x^2 - \frac{1}{8}x^4 - \frac{1}{16}x^6 - \cdots\right) + a_1x$$

19.14. RF: $a_{n+2} = \frac{1}{(n+2)(n+1)}a_{n-1}$

$$y = a_0\left(1 + \frac{1}{6}x^3 + \frac{1}{180}x^6 + \cdots\right) + a_1\left(x + \frac{1}{12}x^4 + \frac{1}{504}x^7 + \cdots\right)$$

19.15. RF: $a_{n+2} = \frac{1}{(n+2)(n+1)}(a_n + a_{n-1})$

$$y = a_0\left[1 + \frac{1}{2}(x-1)^2 + \frac{1}{6}(x-1)^3 + \frac{1}{24}(x-1)^4 + \cdots\right]$$
$$+ a_1\left[(x-1) + \frac{1}{6}(x-1)^3 + \frac{1}{12}(x-1)^4 + \cdots\right]$$

19.16. RF: $a_{n+2} = \frac{n-2}{(n+2)(n+1)}a_{n-1} - \frac{4n}{(n+2)(n+1)}a_n + \frac{4}{n+2}a_{n+1}$

$$y = a_0\left[1 - \frac{1}{6}(x+2)^3 - \frac{1}{6}(x+2)^4 + \cdots\right]$$
$$+ a_1\left[(x+2) + 2(x+2)^2 + 2(x+2)^3 + \frac{2}{3}(x+2)^4 + \cdots\right]$$

19.17. RF: $a_{n+2} = -\frac{n^2-n+1}{4(n+2)(n+1)}a_n, \quad n > 1$

$$y = \left(\frac{1}{24}x^3 - \frac{7}{1920}x^5 + \cdots\right) + a_0\left(1 - \frac{1}{8}x^2 + \frac{1}{128}x^4 + \cdots\right)$$
$$+ a_1\left(x - \frac{1}{24}x^3 + \frac{7}{1920}x^5 + \cdots\right)$$

19.18. RF: $a_{n+2} = \frac{n}{(n+2)(n+1)}a_n, \quad n > 2$

$$y = -\frac{1}{2}(x-1)^2 + a_0 + a_1\left[(x-1) + \frac{1}{6}(x-1)^3 + \frac{1}{40}(x-1)^5 + \cdots\right]$$

19.19. RF: $a_{n+2} = \frac{n}{(n+2)(n+1)}a_n + \frac{(-1)^n}{n!\,(n+2)(n+1)}$

$$y = \left(\frac{1}{2}x^2 - \frac{1}{6}x^3 + \frac{1}{8}x^4 - \frac{1}{30}x^5 + \cdots\right)$$
$$+ a_0 + a_1\left(x + \frac{1}{6}x^3 + \frac{1}{40}x^5 + \cdots\right)$$

19.20. (*a*) $y = 1 - x - \frac{1}{3}x^3 - \frac{1}{12}x^4 - \cdots$

(*b*) $y = 2(x-1) + \frac{1}{2}(x-1)^2 + (x-1)^3 + \cdots$

Chapter 20

Regular Singular Points and the Method of Frobenius

20.1 EXISTENCE THEOREM

Consider the homogeneous differential equation

$$y'' + P(x)y' + Q(x)y = 0 \tag{20.1}$$

having a regular singular point (see Section 18.3) at x_0. We can assume that $x_0 = 0$; if this is not the case, then the change of variables $t = x - x_0$ (see Remark 1, page 97) will translate x_0 to the origin.

Theorem 20.1. If $x = 0$ is a regular singular point of (*20.1*), then the equation has at least one solution of the form

$$y = x^\lambda \sum_{n=0}^{\infty} a_n x^n \tag{20.2}$$

where λ and a_n $(n = 0, 1, 2, \ldots)$ are constants. This solution is valid in an interval $0 < x < R$ for some real number R.

20.2 METHOD OF FROBENIUS

In this method, one assumes a solution to (*20.1*) of the form (*20.2*). One then proceeds as in the power-series method of Chapter 19 to determine successively the constants λ and a_n $(n = 0, 1, 2, \ldots)$.

The constant λ will be determined by a quadratic equation, called the *indicial equation*. The two roots of the indicial equation can be real or complex. If complex they will occur in a conjugate pair and the complex solutions that they produce can be combined (by using Euler's relations and the identity $x^{a \pm ib} = x^a e^{\pm ib \ln x}$) to form real solutions. In this book we shall, for simplicity, suppose that both roots of the indicial equation are real. Then, if λ is taken as *the larger* indicial root, $\lambda = \lambda_1 \geqq \lambda_2$, the method of Frobenius always yields a solution

$$y_1(x) = x^{\lambda_1} \sum_{n=0}^{\infty} a_n(\lambda_1) x^n \tag{20.3}$$

to (*20.1*). (We have written $a_n(\lambda_1)$ to indicate the coefficients produced by the method when $\lambda = \lambda_1$.)

If $P(x)$ and $Q(x)$ are quotients of polynomials, it is usually easier first to multiply (*20.1*) by their lowest common denominator and then to apply the method of Frobenius to the resulting equation.

20.3 GENERAL SOLUTION

The method of Frobenius, modified if necessary as in Problem 20.4 or Problem 20.6, also produces a second solution $y_2(x)$ which is linearly independent of $y_1(x)$. The general solution of *(20.1)* is then obtained by superposition. We summarize the final results in the following theorem.

Theorem 20.2. Let $x = 0$ be a regular singular point of *(20.1)* and let λ_1 and $\lambda_2 \leq \lambda_1$ be the roots of the associated indicial equation. Then the general solution of *(20.1)* is $y = c_1 y_1(x) + c_2 y_2(x)$, where c_1 and c_2 are arbitrary constants, $y_1(x)$ is given by *(20.3)*, and:

Case 1. If $\lambda_1 - \lambda_2$ is not an integer, then

$$y_2(x) = x^{\lambda_2} \sum_{n=0}^{\infty} a_n(\lambda_2) x^n \tag{20.4}$$

Case 2. If $\lambda_1 = \lambda_2$, then

$$y_2(x) = y_1(x) \ln x + x^{\lambda_1} \sum_{n=0}^{\infty} b_n(\lambda_1) x^n \tag{20.5}$$

Case 3. If $\lambda_1 - \lambda_2 = N$, a positive integer, then

$$y_2(x) = d_{-1} y_1(x) \ln x + x^{\lambda_2} \sum_{n=0}^{\infty} d_n(\lambda_2) x^n \tag{20.6}$$

The coefficients $a_n(\lambda_1)$, $a_n(\lambda_2)$, $b_n(\lambda_1)$, $d_n(\lambda_2)$ and d_{-1} are all constants that may, on occasion, be zero. In all cases the solution is valid in an interval $0 < x < R$.

We see from Theorem 20.2 that the simple method of Frobenius can be used to obtain $y_2(x)$ in Case 1, as well as in Case 3 if the constant d_{-1} is zero, since the form of $y_2(x)$ is then identical to the form of $y_1(x)$. The remaining possibilities for $y_2(x)$ are covered by the modifications of the method which are described in Problems 20.4 and 20.6.

Occasionally, a particular equation can be solved more efficiently by another method. See Problems 20.11 and 20.12.

Solved Problems

THE INDICIAL ROOTS HAVE A NONINTEGRAL DIFFERENCE

20.1. Find the general solution near $x = 0$ of $8x^2y'' + 10xy' + (x-1)y = 0$.

Here $P(x) = \dfrac{5}{4x}$ and $Q(x) = \dfrac{1}{8x} - \dfrac{1}{8x^2}$, so $x = 0$ is a regular singular point. Using the method of Frobenius, we assume that

$$y = x^{\lambda} \sum_{n=0}^{\infty} a_n x^n = \sum_{n=0}^{\infty} a_n x^{\lambda+n}$$

$$= a_0 x^{\lambda} + a_1 x^{\lambda+1} + a_2 x^{\lambda+2} + \cdots + a_{n-1} x^{\lambda+n-1} + a_n x^{\lambda+n} + a_{n+1} x^{\lambda+n+1} + \cdots \tag{1}$$

and, correspondingly,

$$y' = \lambda a_0 x^{\lambda-1} + (\lambda+1) a_1 x^{\lambda} + (\lambda+2) a_2 x^{\lambda+1} + \cdots$$
$$+ (\lambda+n-1) a_{n-1} x^{\lambda+n-2} + (\lambda+n) a_n x^{\lambda+n-1} + (\lambda+n+1) a_{n+1} x^{\lambda+n} + \cdots \tag{2}$$

$$y'' = \lambda(\lambda-1)a_0x^{\lambda-2} + (\lambda+1)(\lambda)a_1x^{\lambda-1} + (\lambda+2)(\lambda+1)a_2x^{\lambda} + \cdots$$
$$+ (\lambda+n-1)(\lambda+n-2)a_{n-1}x^{\lambda+n-3} + (\lambda+n)(\lambda+n-1)a_nx^{\lambda+n-2}$$
$$+ (\lambda+n+1)(\lambda+n)a_{n+1}x^{\lambda+n-1} + \cdots \qquad (3)$$

Substituting (*1*), (*2*), and (*3*) into the differential equation and combining, we obtain

$$x^{\lambda}[8\lambda(\lambda-1)a_0 + 10\lambda a_0 - a_0]$$
$$+ x^{\lambda+1}[8(\lambda+1)\lambda a_1 + 10(\lambda+1)a_1 + a_0 - a_1] + \cdots$$
$$+ x^{\lambda+n}[8(\lambda+n)(\lambda+n-1)a_n + 10(\lambda+n)a_n + a_{n-1} - a_n] + \cdots$$
$$= 0$$

Dividing by x^{λ} and simplifying, we have

$$[8\lambda^2 + 2\lambda - 1]a_0 + x[(8\lambda^2 + 18\lambda + 9)a_1 + a_0] + \cdots$$
$$+ x^n\{[8(\lambda+n)^2 + 2(\lambda+n) - 1]a_n + a_{n-1}\} + \cdots = 0$$

Factoring the coefficient of a_n and equating the coefficient of each power of x to zero, we find

$$(8\lambda^2 + 2\lambda - 1)a_0 = 0 \qquad (4)$$

and, for $n \geqq 1$,

$$[4(\lambda+n) - 1][2(\lambda+n) + 1]a_n + a_{n-1} = 0$$

or,

$$a_n = \frac{-1}{[4(\lambda+n) - 1][2(\lambda+n) + 1]}a_{n-1} \qquad (5)$$

From (*4*), either $a_0 = 0$ or

$$8\lambda^2 + 2\lambda - 1 = 0 \qquad (6)$$

It is convenient to keep a_0 arbitrary; therefore, we must choose λ to satisfy (*6*), which is the indicial equation. The roots are $\lambda_1 = \frac{1}{4}$ and $\lambda_2 = -\frac{1}{2}$. Since $\lambda_1 - \lambda_2 = \frac{3}{4}$, the solution is given by (*20.3*) and (*20.4*). Substituting $\lambda = \frac{1}{4}$ into the recurrence formula (*5*) and simplifying, we obtain

$$a_n = \frac{-1}{2n(4n+3)}a_{n-1} \qquad (n \geqq 1)$$

Thus,

$$a_1 = \frac{-1}{14}a_0, \quad a_2 = \frac{-1}{44}a_1 = \frac{1}{616}a_0, \quad \ldots$$

and

$$y_1(x) = a_0x^{1/4}\left(1 - \frac{1}{14}x + \frac{1}{616}x^2 + \cdots\right)$$

Substituting $\lambda = -\frac{1}{2}$ into (*5*) and simplifying, we obtain

$$a_n = \frac{-1}{2n(4n-3)}a_{n-1}$$

Thus,

$$a_1 = -\frac{1}{2}a_0, \quad a_2 = \frac{-1}{20}a_1 = \frac{1}{40}a_0, \quad \ldots$$

and

$$y_2(x) = a_0x^{-1/2}\left(1 - \frac{1}{2}x + \frac{1}{40}x^2 + \cdots\right)$$

The general solution is

$$y = c_1y_1(x) + c_2y_2(x)$$
$$= k_1x^{1/4}\left(1 - \frac{1}{14}x + \frac{1}{616}x^2 + \cdots\right) + k_2x^{-1/2}\left(1 - \frac{1}{2}x + \frac{1}{40}x^2 + \cdots\right)$$

where $k_1 = c_1a_0$ and $k_2 = c_2a_0$.

20.2. Find the general solution near $x = 0$ of $2x^2y'' + 7x(x+1)y' - 3y = 0$.

Here $P(x) = 7(x+1)/2x$ and $Q(x) = -3/2x^2$; hence, $x = 0$ is a regular singular point and the method of Frobenius is applicable. Substituting (*1*), (*2*), and (*3*) of Problem 20.1 into the differential equation and combining, we obtain

$$\begin{aligned} &x^{\lambda}[2\lambda(\lambda-1)a_0 + 7\lambda a_0 - 3a_0] \\ &\quad + x^{\lambda+1}[2(\lambda+1)\lambda a_1 + 7\lambda a_0 + 7(\lambda+1)a_1 - 3a_1] + \cdots \\ &\qquad + x^{\lambda+n}[2(\lambda+n)(\lambda+n-1)a_n + 7(\lambda+n-1)a_{n-1} + 7(\lambda+n)a_n - 3a_n] \\ &\qquad\quad + \cdots \\ &= 0 \end{aligned}$$

Dividing by x^{λ} and simplifying, we have

$$\begin{aligned} &(2\lambda^2 + 5\lambda - 3)a_0 + x[(2\lambda^2 + 9\lambda + 4)a_1 + 7\lambda a_0] + \cdots \\ &\quad + x^n\{[2(\lambda+n)^2 + 5(\lambda+n) - 3]a_n + 7(\lambda+n-1)a_{n-1}\} + \cdots \\ &= 0 \end{aligned}$$

Factoring the coefficient of a_n and equating each coefficient to zero, we find

$$(2\lambda^2 + 5\lambda - 3)a_0 = 0 \tag{1}$$

and, for $n \geqq 1$,

$$[2(\lambda+n) - 1][(\lambda+n) + 3]a_n + 7(\lambda+n-1)a_{n-1} = 0$$

or,

$$a_n = \frac{-7(\lambda+n-1)}{[2(\lambda+n) - 1][(\lambda+n) + 3]}a_{n-1} \tag{2}$$

From (*1*), either $a_0 = 0$ or

$$2\lambda^2 + 5\lambda - 3 = 0 \tag{3}$$

It is convenient to keep a_0 arbitrary; therefore, we require λ to satisfy the indicial equation (*3*). The roots of (*3*) are $\lambda_1 = \frac{1}{2}$ and $\lambda_2 = -3$. Since $\lambda_1 - \lambda_2 = \frac{7}{2}$, the solution is given by (*20.3*) and (*20.4*). Substituting $\lambda = \frac{1}{2}$ into (*2*) and simplifying, we obtain

$$a_n = \frac{-7(2n-1)}{2n(2n+7)}a_{n-1} \qquad (n \geqq 1)$$

Thus,

$$a_1 = -\frac{7}{18}a_0, \quad a_2 = -\frac{21}{44}a_1 = \frac{147}{792}a_0, \quad \ldots$$

and

$$y_1(x) = a_0x^{1/2}\left(1 - \frac{7}{18}x + \frac{147}{792}x^2 + \cdots\right)$$

Substituting $\lambda = -3$ into (*2*) and simplifying, we obtain

$$a_n = \frac{-7(n-4)}{n(2n-7)}a_{n-1} \qquad (n \geqq 1)$$

Thus,

$$a_1 = -\frac{21}{5}a_0, \quad a_2 = -\frac{7}{3}a_1 = \frac{49}{5}a_0, \quad a_3 = -\frac{7}{3}a_2 = -\frac{343}{15}a_0, \quad a_4 = 0$$

and, since $a_4 = 0$, $a_n = 0$ for $n \geqq 4$. Thus,

$$y_2(x) = a_0x^{-3}\left(1 - \frac{21}{5}x + \frac{49}{5}x^2 - \frac{343}{15}x^3\right)$$

The general solution is

$$\begin{aligned} y &= c_1y_1(x) + c_2y_2(x) \\ &= k_1x^{1/2}\left(1 - \frac{7}{18}x + \frac{147}{792}x^2 + \cdots\right) + k_2x^{-3}\left(1 - \frac{21}{5}x + \frac{49}{5}x^2 - \frac{343}{15}x^3\right) \end{aligned}$$

where $k_1 = c_1a_0$ and $k_2 = c_2a_0$.

20.3. Find the general solution near $x=0$ of $3x^2y''-xy'+y=0$.

Here $P(x)=-1/3x$ and $Q(x)=1/3x^2$; hence, $x=0$ is a regular singular point and the method of Frobenius is applicable. Substituting *(1)*, *(2)*, and *(3)* of Problem 20.1 into the differential equation and simplifying, we have

$$x^{\lambda}[3\lambda^2-4\lambda+1]a_0 + x^{\lambda+1}[3\lambda^2+2\lambda]a_1 + \cdots$$
$$+ x^{\lambda+n}[3(\lambda+n)^2-4(\lambda+n)+1]a_n + \cdots = 0$$

Dividing by x^{λ} and equating all coefficients to zero, we find

$$(3\lambda^2-4\lambda+1)a_0 = 0 \tag{1}$$

and

$$[3(\lambda+n)^2-4(\lambda+n)+1]a_n = 0 \qquad (n \geqq 1) \tag{2}$$

From *(1)*, we conclude that the indicial equation is $3\lambda^2-4\lambda+1=0$, which has roots $\lambda_1=1$ and $\lambda_2=\frac{1}{3}$. Since $\lambda_1-\lambda_2=\frac{2}{3}$, the solution is given by *(20.3)* and *(20.4)*. Note that for either value of λ, *(2)* is satisfied by simply choosing $a_n=0$, $n \geqq 1$. Thus,

$$y_1(x) = x^1\sum_{n=0}^{\infty} a_n x^n = a_0 x \qquad y_2(x) = x^{1/3}\sum_{n=0}^{\infty} a_n x^n = a_0 x^{1/3}$$

and the general solution is

$$y = c_1y_1(x)+c_2y_2(x) = k_1x + k_2x^{1/3}$$

where $k_1=c_1a_0$ and $k_2=c_2a_0$.

THE INDICIAL ROOTS ARE EQUAL

20.4. Find the general solution near $x=0$ of $x^2y''+xy'+x^2y=0$.

Here $P(x)=1/x$ and $Q(x)=1$, so $x=0$ is a regular singular point. Substituting *(1)*, *(2)*, and *(3)* of Problem 20.1 into the differential equation and combining, we obtain

$$x^{\lambda}[\lambda^2a_0] + x^{\lambda+1}[(\lambda+1)^2a_1] + x^{\lambda+2}[(\lambda+2)^2a_2+a_0] + \cdots$$
$$+ x^{\lambda+n}[(\lambda+n)^2a_n+a_{n-2}] + \cdots = 0$$

Thus,

$$\lambda^2a_0 = 0 \tag{1}$$

$$(\lambda+1)^2a_1 = 0 \tag{2}$$

and, for $n \geqq 2$, $(\lambda+n)^2a_n+a_{n-2}=0$, or,

$$a_n = \frac{-1}{(\lambda+n)^2}a_{n-2} \qquad (n \geqq 2) \tag{3}$$

The stipulation $n \geqq 2$ is required in *(3)* because a_{n-2} is not defined for $n=0$ or $n=1$. From *(1)*, the indicial equation is $\lambda^2=0$, which has roots $\lambda_1=\lambda_2=0$. Thus, we will obtain only *one* solution of the form *(20.3)*; the second solution, $y_2(x)$, will have the form *(20.5)*.

Substituting $\lambda=0$ into *(2)* and *(3)*, we find that $a_1=0$ and $a_n=-(1/n^2)a_{n-2}$. Since $a_1=0$, it follows that $0=a_3=a_5=a_7=\cdots$. Furthermore,

$$a_2 = -\frac{1}{4}a_0 = -\frac{1}{2^2(1!)^2}a_0 \qquad a_6 = -\frac{1}{36}a_4 = -\frac{1}{2^6(3!)^2}a_0$$

$$a_4 = -\frac{1}{16}a_2 = \frac{1}{2^4(2!)^2}a_0 \qquad a_8 = -\frac{1}{64}a_6 = \frac{1}{2^8(4!)^2}a_0$$

and, in general, $a_{2k} = \frac{(-1)^k}{2^{2k}(k!)^2}a_0$ $(k=1,2,3,\ldots)$. Thus,

$$y_1(x) = a_0x^0\left[1-\frac{1}{2^2(1!)^2}x^2+\frac{1}{2^4(2!)^2}x^4+\cdots+\frac{(-1)^k}{2^{2k}(k!)^2}x^{2k}+\cdots\right]$$

$$= a_0\sum_{n=0}^{\infty}\frac{(-1)^n}{2^{2n}(n!)^2}x^{2n} \tag{4}$$

TO FIND $y_2(x)$ WHEN THE INDICIAL ROOTS ARE EQUAL, keep the recurrence formula in terms of λ and use it to find the coefficients a_n $(n \geqq 1)$ in terms of both λ and a_0, where the coefficient a_0 remains arbitrary. Substitute these a_n into *(20.2)* to obtain a function $y(\lambda, x)$ which depends on the variables λ and x. Then

$$y_2(x) = \left.\frac{\partial y(\lambda, x)}{\partial \lambda}\right|_{\lambda=\lambda_1} \tag{5}$$

For the problem at hand, the recurrence formula is *(3)*, augmented by *(2)* for the special case $n = 1$. From *(2)*, $a_1 = 0$, which implies that $0 = a_3 = a_5 = a_7 = \cdots$. Then, from *(3)*,

$$a_2 = \frac{-1}{(\lambda+2)^2} a_0, \qquad a_4 = \frac{-1}{(\lambda+4)^2} a_2 = \frac{1}{(\lambda+4)^2(\lambda+2)^2} a_0, \qquad \cdots$$

Substituting these values into *(20.2)*, we have

$$y(\lambda, x) = a_0\left[x^\lambda - \frac{1}{(\lambda+2)^2} x^{\lambda+2} + \frac{1}{(\lambda+4)^2(\lambda+2)^2} x^{\lambda+4} + \cdots\right]$$

Recall that $\frac{\partial}{\partial\lambda}(x^{\lambda+k}) = x^{\lambda+k}\ln x$. (When differentiating with respect to λ, x can be thought of as a constant.) Thus,

$$\begin{aligned}\frac{\partial y(\lambda, x)}{\partial\lambda} = a_0\Bigg[& x^\lambda \ln x + \frac{2}{(\lambda+2)^3} x^{\lambda+2} - \frac{1}{(\lambda+2)^2} x^{\lambda+2}\ln x \\ & - \frac{2}{(\lambda+4)^3(\lambda+2)^2} x^{\lambda+4} - \frac{2}{(\lambda+4)^2(\lambda+2)^3} x^{\lambda+4} \\ & + \frac{1}{(\lambda+4)^2(\lambda+2)^2} x^{\lambda+4}\ln x + \cdots\Bigg]\end{aligned}$$

and

$$\begin{aligned} y_2(x) = \left.\frac{\partial y(\lambda, x)}{\partial\lambda}\right|_{\lambda=0} &= a_0\Bigg(\ln x + \frac{2}{2^3}x^2 - \frac{1}{2^2}x^2\ln x \\ &\qquad - \frac{2}{4^3 2^2}x^4 - \frac{2}{4^2 2^3}x^4 + \frac{1}{4^2 2^2}x^4 \ln x + \cdots\Bigg) \\ &= (\ln x)a_0\left[1 - \frac{1}{2^2(1!)}x^2 + \frac{1}{2^4(2!)^2}x^4 + \cdots\right] \\ &\qquad + a_0\left[\frac{x^2}{2^2(1!)^2}(1) - \frac{x^4}{2^4(2!)^2}\left(\frac{1}{2}+1\right) + \cdots\right] \\ &= y_1(x)\ln x + a_0\left[\frac{x^2}{2^2(1!)^2}(1) - \frac{x^4}{2^4(2!)^2}\left(1+\frac{1}{2}\right) + \cdots\right] \end{aligned} \tag{6}$$

which is the form claimed in *(20.5)*. The general solution is $y = c_1y_1(x) + c_2y_2(x)$.

20.5. Find the general solution near $x = 0$ of $x^2y'' - xy' + y = 0$.

Here $P(x) = -1/x$ and $Q(x) = 1/x^2$, so $x = 0$ is a regular singular point. Substituting *(1)*, *(2)*, and *(3)* of Problem 20.1 into the differential equation and combining, we obtain

$$\begin{aligned} & x^\lambda(\lambda-1)^2 a_0 + x^{\lambda+1}[\lambda^2 a_1] + \cdots \\ & \qquad + x^{\lambda+n}[(\lambda+n)^2 - 2(\lambda+n) + 1]a_n + \cdots = 0 \end{aligned}$$

Thus,

$$(\lambda-1)^2 a_0 = 0 \tag{1}$$

and, in general,

$$[(\lambda+n)^2 - 2(\lambda+n) + 1]a_n = 0 \tag{2}$$

From *(1)*, the indicial equation is $(\lambda-1)^2 = 0$, which has roots $\lambda_1 = \lambda_2 = 1$. Substituting $\lambda = 1$ into *(2)*, we obtain $n^2a_n = 0$, which implies that $a_n = 0$, $n \geqq 1$. Thus, $y_1(x) = a_0x$.

To find the second solution, we use the method outlined in Problem 20.4. From (2), the recurrence formula for a_n $(n \geqq 1)$ in terms of λ is still $a_n = 0$. Substituting these results into (20.2), we have $y(\lambda, x) = a_0 x^\lambda$. Thus,

$$\frac{\partial y(\lambda, x)}{\partial \lambda} = a_0 x^\lambda \ln x$$

and

$$y_2(x) = \left.\frac{\partial y(\lambda, x)}{\partial \lambda}\right|_{\lambda=1} = a_0 x \ln x = y_1(x) \ln x$$

which is precisely the form of (20.5), where, for this particular differential equation, $b_n(\lambda_1) = 0$ $(n = 0, 1, 2, \ldots)$. The general solution is

$$y = c_1 y_1(x) + c_2 y_2(x) = k_1 x + k_2 x \ln x$$

where $k_1 = c_1 a_0$ and $k_2 = c_2 a_0$.

THE INDICIAL ROOTS DIFFER BY A POSITIVE INTEGER

20.6. Find the general solution near $x = 0$ of $x^2 y'' + (x^2 - 2x)y' + 2y = 0$.

Here $P(x) = 1 - \dfrac{2}{x}$ and $Q(x) = \dfrac{2}{x^2}$; hence, $x = 0$ is a regular singular point. Substituting (1), (2), and (3) of Problem 20.1 into the differential equation and simplifying, we have

$$x^\lambda[(\lambda^2 - 3\lambda + 2)a_0] + x^{\lambda+1}[(\lambda^2 - \lambda)a_1 + \lambda a_0] + \cdots$$
$$+ x^{\lambda+n}\{[(\lambda+n)^2 - 3(\lambda+n) + 2]a_n + (\lambda + n - 1)a_{n-1}\} + \cdots = 0$$

Dividing by x^λ, factoring the coefficient of a_n, and equating the coefficient of each power of x to zero, we obtain

$$(\lambda^2 - 3\lambda + 2)a_0 = 0 \tag{1}$$

and, in general, $[(\lambda+n) - 2][(\lambda+n) - 1]a_n + (\lambda + n - 1)a_{n-1} = 0$, or,

$$a_n = -\frac{1}{\lambda + n - 2} a_{n-1} \qquad (n \geqq 1) \tag{2}$$

From (1), the indicial equation is $\lambda^2 - 3\lambda + 2 = 0$, which has roots $\lambda_1 = 2$ and $\lambda_2 = 1$. Since $\lambda_1 - \lambda_2 = 1$, a positive integer, the solution is given by (20.3) and (20.6). Substituting $\lambda = 2$ into (2), we have $a_n = -(1/n)a_{n-1}$, from which we obtain

$$a_1 = -a_0$$

$$a_2 = -\tfrac{1}{2}a_1 = \frac{1}{2!}a_0$$

$$a_3 = -\tfrac{1}{3}a_2 = -\frac{1}{3}\frac{1}{2!}a_0 = -\frac{1}{3!}a_0$$

and, in general, $a_k = \dfrac{(-1)^k}{k!}a_0$. Thus,

$$y_1(x) = a_0 x^2 \sum_{n=0}^{\infty} \frac{(-1)^n}{n!} x^n = a_0 x^2 e^{-x}$$

Observe that we can *not* find the second solution $y_2(x)$ by repeating the simple method of Frobenius with *the smaller* root λ_2. In fact, if we substitute $\lambda = 1$ into (2), we obtain

$$a_n = -\frac{1}{n-1} a_{n-1}$$

Now, however, a_1 is undefined, since the denominator is zero when $n = 1$.

TO FIND $y_2(x)$ WHEN THE INDICIAL ROOTS DIFFER BY A POSITIVE INTEGER, first try the simple method of Frobenius with $\lambda = \lambda_2$. If this approach is not applicable, use the method given in Problem 20.4 for equal roots to obtain $y(\lambda, x)$. Then

$$y_2(x) = \left.\frac{\partial}{\partial \lambda}[(\lambda - \lambda_2)y(\lambda, x)]\right|_{\lambda=\lambda_2} \tag{3}$$

For the problem at hand, the general recurrence formula is given by (2). Thus,

$$a_1 = -\frac{1}{\lambda-1}a_0, \quad a_2 = -\frac{1}{\lambda}a_1 = \frac{1}{\lambda(\lambda-1)}a_0, \quad a_3 = \frac{-1}{(\lambda+1)\lambda(\lambda-1)}a_0, \quad \ldots$$

Substituting these values into (20.2), we obtain

$$y(\lambda, x) = a_0\left[x^\lambda - \frac{1}{(\lambda-1)}x^{\lambda+1} + \frac{1}{\lambda(\lambda-1)}x^{\lambda+2} - \frac{1}{(\lambda+1)\lambda(\lambda-1)}x^{\lambda+3} + \cdots\right]$$

and, since $\lambda - \lambda_2 = \lambda - 1$,

$$(\lambda-\lambda_2)y(\lambda, x) = a_0\left[(\lambda-1)x^\lambda - x^{\lambda+1} + \frac{1}{\lambda}x^{\lambda+2} - \frac{1}{\lambda(\lambda+1)}x^{\lambda+3} + \cdots\right]$$

Then

$$\begin{aligned}\frac{\partial}{\partial\lambda}[(\lambda-\lambda_2)y(\lambda,x)] = a_0\Big[& x^\lambda + (\lambda-1)x^\lambda \ln x - x^{\lambda+1}\ln x \\ & - \frac{1}{\lambda^2}x^{\lambda+2} + \frac{1}{\lambda}x^{\lambda+2}\ln x \\ & + \frac{1}{\lambda^2(\lambda+1)}x^{\lambda+3} + \frac{1}{\lambda(\lambda+1)^2}x^{\lambda+3} \\ & - \frac{1}{\lambda(\lambda+1)}x^{\lambda+3}\ln x + \cdots\Big]\end{aligned}$$

and

$$\begin{aligned} y_2(x) &= \frac{\partial}{\partial\lambda}[(\lambda-\lambda_2)y(\lambda,x)]\Big|_{\lambda=\lambda_2=1} \\ &= a_0\left(x + 0 - x^2\ln x - x^3 + x^3\ln x + \frac{1}{2}x^4 + \frac{1}{4}x^4 - \frac{1}{2}x^4\ln x + \cdots\right) \\ &= (-\ln x)a_0\left(x^2 - x^3 + \frac{1}{2}x^4 + \cdots\right) + a_0\left(x - x^3 + \frac{3}{4}x^4 + \cdots\right) \\ &= -y_1(x)\ln x + a_0x\left(1 - x^2 + \frac{3}{4}x^3 + \cdots\right)\end{aligned}$$

This is the form claimed in (20.6), with $d_{-1} = -1$, $d_0 = a_0$, $d_1 = 0$, $d_2 = -a_0$, $d_3 = \frac{3}{4}a_0$, ... The general solution is $y = c_1y_1(x) + c_2y_2(x)$.

20.7. Find the general solution near $x = 0$ of $x^2y'' + xy' + (x^2-1)y = 0$.

Here $P(x) = x^{-1}$ and $Q(x) = 1 - x^{-2}$, so $x = 0$ is a regular singular point. Substituting (1), (2), and (3) of Problem 20.1 into the differential equation, we obtain

$$\begin{aligned} & x^\lambda[(\lambda^2-1)a_0] + x^{\lambda+1}[(\lambda+1)^2-1]a_1 \\ & \quad + x^{\lambda+2}\{[(\lambda+2)^2-1]a_2 + a_0\} + \cdots \\ & \quad + x^{\lambda+n}\{[(\lambda+n)^2-1]a_n + a_{n-2}\} + \cdots = 0\end{aligned}$$

Thus,

$$(\lambda^2-1)a_0 = 0 \tag{1}$$

$$[(\lambda+1)^2-1]a_1 = 0 \tag{2}$$

and, for $n \geqq 2$, $[(\lambda+n)^2-1]a_n + a_{n-2} = 0$, or,

$$a_n = \frac{-1}{(\lambda+n)^2-1}a_{n-2} \qquad (n \geqq 2) \tag{3}$$

From (1), the indicial equation is $\lambda^2 - 1 = 0$, which has roots $\lambda_1 = 1$ and $\lambda_2 = -1$. Since $\lambda_1 - \lambda_2 = 2$, a positive integer, the solution is given by (20.3) and (20.6). Substituting $\lambda = 1$ into (2) and (3), we obtain $a_1 = 0$ and

$$a_n = \frac{-1}{n(n+2)} a_{n-2} \qquad (n \geqq 2)$$

Since $a_1 = 0$, it follows that $0 = a_3 = a_5 = a_7 = \cdots$. Furthermore,

$$a_2 = \frac{-1}{2(4)} a_0 = \frac{-1}{2^2 1!\,2!} a_0, \quad a_4 = \frac{-1}{4(6)} a_2 = \frac{1}{2^4 2!\,3!} a_0, \quad a_6 = \frac{-1}{6(8)} a_4 = \frac{-1}{2^6 3!\,4!} a_0$$

and, in general,

$$a_{2k} = \frac{(-1)^k}{2^{2k} k!\,(k+1)!} a_0 \qquad (k = 1, 2, 3, \ldots)$$

Thus,

$$y_1(x) = a_0 x \sum_{n=0}^{\infty} \frac{(-1)^n}{2^{2n} n!\,(n+1)!} x^{2n} \tag{4}$$

If we substitute $\lambda = \lambda_2 = -1$ into (3), we obtain $a_n = \frac{-1}{n(n-2)} a_{n-2}$, which fails to define a_2. Thus, the simple method of Frobenius does not provide the second solution $y_2(x)$, and we must use the modification described in Problem 20.6. From (2) and (3), $0 = a_1 = a_3 = a_5 = \cdots$ and

$$a_2 = \frac{-1}{(\lambda+3)(\lambda+1)} a_0, \quad a_4 = \frac{1}{(\lambda+5)(\lambda+3)^2(\lambda+1)} a_0, \quad \ldots$$

Thus,

$$y(\lambda, x) = a_0\left[x^\lambda - \frac{1}{(\lambda+3)(\lambda+1)} x^{\lambda+2} + \frac{1}{(\lambda+5)(\lambda+3)^2(\lambda+1)} x^{\lambda+4} + \cdots\right]$$

Since $\lambda - \lambda_2 = \lambda + 1$,

$$(\lambda - \lambda_2) y(\lambda, x) = a_0\left[(\lambda+1)x^\lambda - \frac{1}{(\lambda+3)} x^{\lambda+2} + \frac{1}{(\lambda+5)(\lambda+3)^2} x^{\lambda+4} + \cdots\right]$$

and

$$\begin{aligned}\frac{\partial}{\partial\lambda}[(\lambda-\lambda_2)y(\lambda,x)] = a_0\Bigg[& x^\lambda + (\lambda+1)x^\lambda \ln x + \frac{1}{(\lambda+3)^2} x^{\lambda+2} \\ & - \frac{1}{(\lambda+3)} x^{\lambda+2} \ln x - \frac{1}{(\lambda+5)^2(\lambda+3)^2} x^{\lambda+4} \\ & - \frac{2}{(\lambda+5)(\lambda+3)^3} x^{\lambda+4} + \frac{1}{(\lambda+5)(\lambda+3)^2} x^{\lambda+4} \ln x \\ & + \cdots\Bigg]\end{aligned}$$

Then

$$\begin{aligned} y_2(x) &= \frac{\partial}{\partial\lambda}[(\lambda-\lambda_2)y(\lambda,x)]\Big|_{\lambda=\lambda_2=-1} \\ &= a_0\left(x^{-1} + 0 + \frac{1}{4}x - \frac{1}{2}x \ln x - \frac{1}{64}x^3 \right. \\ &\qquad \left. - \frac{2}{32}x^3 + \frac{1}{16}x^3 \ln x + \cdots\right) \\ &= -\frac{1}{2}(\ln x) a_0 x \left(1 - \frac{1}{8}x^2 + \cdots\right) + a_0\left(x^{-1} + \frac{1}{4}x - \frac{5}{64}x^3 + \cdots\right) \\ &= -\frac{1}{2}(\ln x) y_1(x) + a_0 x^{-1}\left(1 + \frac{1}{4}x^2 - \frac{5}{64}x^4 + \cdots\right) \end{aligned} \tag{5}$$

This is in the form (20.6), with $d_{-1} = -\frac{1}{2}$, $d_0 = a_0$, $d_1 = 0$, $d_2 = \frac{1}{4}a_0$, $d_3 = 0$, $d_4 = \frac{-5}{64}a_0$, ... The general solution is $y = c_1 y_1(x) + c_2 y_2(x)$.

20.8. Find the general solution near $x = 0$ of $x^2 y'' + (x^2 + 2x)y' - 2y = 0$.

Here $P(x) = 1 + (2/x)$ and $Q(x) = -2/x^2$; hence, $x = 0$ is a regular singular point. Substituting (1), (2), and (3) of Problem 20.1 into the differential equation and simplifying, we have

$$x^{\lambda}[(\lambda^2+\lambda-2)a_0] + x^{\lambda+1}[(\lambda^2+3\lambda)a_1+\lambda a_0] + \cdots$$
$$+ x^{\lambda+n}\{[(\lambda+n)^2+(\lambda+n)-2]a_n + (\lambda+n-1)a_{n-1}\}$$
$$+ \cdots = 0$$

Dividing by x^{λ}, factoring the coefficient of a_n, and equating to zero the coefficient of each power of x, we obtain

$$(\lambda^2+\lambda-2)a_0 = 0 \tag{1}$$

and, for $n \geqq 1$,

$$[(\lambda+n)+2][(\lambda+n)-1]a_n + (\lambda+n-1)a_{n-1} = 0$$

which is equivalent to

$$a_n = -\frac{1}{\lambda+n+2}a_{n-1} \qquad (n \geqq 1) \tag{2}$$

From (1), the indicial equation is $\lambda^2+\lambda-2=0$, which has roots $\lambda_1=1$ and $\lambda_2=-2$. Since $\lambda_1-\lambda_2=3$, a positive integer, the solution is given by (20.3) and (20.6). Substituting $\lambda=1$ into (2), we obtain $a_n=[-1/(n+3)]a_{n-1}$, which in turn yields

$$a_1 = -\frac{1}{4}a_0 = -\frac{3!}{4!}a_0$$

$$a_2 = -\frac{1}{5}a_1 = \left(-\frac{1}{5}\right)\left(-\frac{3!}{4!}\right)a_0 = \frac{3!}{5!}a_0$$

$$a_3 = -\frac{1}{6}a_2 = -\frac{3!}{6!}a_0$$

and, in general,

$$a_k = \frac{(-1)^k 3!}{(k+3)!}a_0$$

Hence,
$$y_1(x) = a_0x\left[1 + 3!\sum_{n=1}^{\infty}\frac{(-1)^n x^n}{(n+3)!}\right] = a_0x\sum_{n=0}^{\infty}\frac{(-1)^n 3!\,x^n}{(n+3)!}$$

which can be simplified to

$$y_1(x) = \frac{3a_0}{x^2}(2-2x+x^2-2e^{-x})$$

To find $y_2(x)$, let us try repeating the method of Frobenius with $\lambda=\lambda_2$. Substituting $\lambda=-2$ into (2), we obtain $a_n=(-1/n)a_{n-1}$, $n \geqq 1$. Note that, in contrast to Problems 20.6 and 20.7, no a_n $(n \geqq 1)$ is undefined, so the unmodified method of Frobenius can be used to find $y_2(x)$. We obtain

$$a_1 = -a_0 = -\frac{1}{1!}a_0 \qquad a_2 = -\frac{1}{2}a_2 = \frac{1}{2!}a_0$$

and, in general, $a_k=(-1)^k a_0/k!$. Therefore,

$$y_2(x) = a_0x^{-2}\left[1 - \frac{1}{1!}x + \frac{1}{2!}x^2 + \cdots + \frac{(-1)^k}{k!}x^k + \cdots\right]$$
$$= a_0x^{-2}\sum_{n=0}^{\infty}\frac{(-1)^n x^n}{n!} = a_0x^{-2}e^{-x}$$

This is precisely in the form (20.6), with $d_{-1}=0$ and $d_n=(-1)^n a_0/n!$. The general solution is $y=c_1y_1(x)+c_2y_2(x)$.

MISCELLANEOUS PROBLEMS

20.9. Find a general expression for the indicial equation of (20.1).

Since $x=0$ is a regular singular point, $xP(x)$ and $x^2Q(x)$ are analytic near the origin and can be expanded in Taylor series there. Thus,

$$xP(x) = \sum_{n=0}^{\infty} p_n x^n = p_0 + p_1 x + p_2 x^2 + \cdots$$

$$x^2 Q(x) = \sum_{n=0}^{\infty} q_n x^n = q_0 + q_1 x + q_2 x^2 + \cdots$$

Dividing by x and x^2, respectively, we have

$$P(x) = p_0 x^{-1} + p_1 + p_2 x + \cdots \qquad Q(x) = q_0 x^{-2} + q_1 x^{-1} + q_2 + \cdots$$

Substituting these two results along with (*1*), (*2*), and (*3*) of Problem 20.1 into (*20.1*) and combining, we obtain

$$x^{\lambda-2}[\lambda(\lambda-1)a_0 + \lambda a_0 p_0 + a_0 q_0] + \cdots = 0$$

which can hold only if

$$a_0[\lambda^2 + (p_0 - 1)\lambda + q_0] = 0$$

Since $a_0 \neq 0$ (a_0 is an arbitrary constant, hence can be chosen nonzero), the indicial equation is

$$\lambda^2 + (p_0 - 1)\lambda + q_0 = 0 \tag{1}$$

20.10. Find the indicial equation of $x^2 y'' + xe^x y' + (x^3 - 1)y = 0$ if the solution is required near $x = 0$.

Here $P(x) = \dfrac{e^x}{x}$ and $Q(x) = x - \dfrac{1}{x^2}$, and we have

$$xP(x) = e^x = 1 + x + \frac{x^2}{2!} + \cdots$$

$$x^2 Q(x) = x^3 - 1 = -1 + 0x + 0x^2 + 1x^3 + 0x^4 + \cdots$$

from which $p_0 = 1$ and $q_0 = -1$. Using (*1*) of Problem 20.9, we obtain the indicial equation as $\lambda^2 - 1 = 0$.

20.11. Solve Problem 20.3 by an alternate method.

The given differential equation, $3x^2 y'' - xy' + y = 0$, is a special case of *Euler's equation*

$$b_n x^n y^{(n)} + b_{n-1} x^{n-1} y^{(n-1)} + \cdots + b_2 x^2 y'' + b_1 x y' + b_0 y = \phi(x) \tag{1}$$

where b_j $(j = 0, 1, \ldots, n)$ is a constant. Euler's equation can always be transformed into a linear differential equation with *constant coefficients* by the change of variables

$$z = \ln x \qquad \text{or} \qquad x = e^z \tag{2}$$

It follows from (*2*) and from the chain rule and the product rule of differentiation that

$$\frac{dy}{dx} = \frac{dy}{dz}\frac{dz}{dx} = \frac{1}{x}\frac{dy}{dz} = e^{-z}\frac{dy}{dz} \tag{3}$$

$$\frac{d^2y}{dx^2} = \frac{d}{dx}\left(\frac{dy}{dx}\right) = \frac{d}{dx}\left(e^{-z}\frac{dy}{dz}\right) = \left[\frac{d}{dz}\left(e^{-z}\frac{dy}{dz}\right)\right]\frac{dz}{dx}$$

$$= \left[-e^{-z}\left(\frac{dy}{dz}\right) + e^{-z}\left(\frac{d^2y}{dz^2}\right)\right]e^{-z} = e^{-2z}\left(\frac{d^2y}{dz^2}\right) - e^{-2z}\left(\frac{dy}{dz}\right) \tag{4}$$

Substituting (*2*), (*3*), and (*4*) in the given differential equation and simplifying, we obtain

$$\frac{d^2y}{dz^2} - \frac{4}{3}\frac{dy}{dz} + \frac{1}{3}y = 0$$

Using the method of Chapter 12, we find that the solution of this last equation is $y = c_1 e^z + c_2 e^{(1/3)z}$. Then using (*2*) and noting that $e^{(1/3)z} = (e^z)^{1/3}$, we have, as before,

$$y = c_1 x + c_2 x^{1/3}$$

20.12. Solve Problem 20.5 by an alternate method.

The given differential equation, $x^2y'' - xy' + y = 0$, is a special case of Euler's equation, (*1*) of Problem 20.11. Using the transformations (*2*), (*3*), and (*4*) of Problem 20.11, we reduce the given equation to

$$\frac{d^2y}{dz^2} - 2\frac{dy}{dz} + y = 0$$

The solution to this equation is (see Chapter 12) $y = c_1e^z + c_2ze^z$. Then, using (*2*) of Problem 20.11, we have for the solution of the original differential equation

$$y = c_1x + c_2x \ln x$$

as before.

20.13. Find the general solution near $x = 0$ of the *hypergeometric equation*

$$x(1-x)y'' + [C-(A+B+1)x]y' - ABy = 0$$

where A and B are any real numbers, and C is any real nonintegral number.

Since $x = 0$ is a regular singular point, the method of Frobenius is applicable. Substituting (*1*), (*2*), and (*3*) of Problem 20.1 into the differential equation, simplifying and equating the coefficient of each power of x to zero, we obtain

$$\lambda^2 + (C-1)\lambda = 0 \tag{1}$$

as the indicial equation and

$$a_{n+1} = \frac{(\lambda+n)(\lambda+n+A+B)+AB}{(\lambda+n+1)(\lambda+n+C)}a_n \tag{2}$$

as the recurrence formula. The roots of (*1*) are $\lambda_1 = 0$ and $\lambda_2 = 1-C$; hence, $\lambda_1 - \lambda_2 = C-1$. Since C is not an integer, the solution of the hypergeometric equation is given by (*20.3*) and (*20.4*).

Substituting $\lambda = 0$ into (*2*), we have

$$a_{n+1} = \frac{n(n+A+B)+AB}{(n+1)(n+C)}a_n$$

which is equivalent to

$$a_{n+1} = \frac{(A+n)(B+n)}{(n+1)(n+C)}a_n$$

Thus

$$a_1 = \frac{AB}{C}a_0 = \frac{AB}{1!\,C}a_0$$

$$a_2 = \frac{(A+1)(B+1)}{2(C+1)}a_1 = \frac{A(A+1)B(B+1)}{2!\,C(C+1)}a_0$$

$$a_3 = \frac{(A+2)(B+2)}{3(C+2)}a_2 = \frac{A(A+1)(A+2)B(B+1)(B+2)}{3!\,C(C+1)(C+2)}a_0$$

. . . .

and $y_1(x) = a_0F(A,B;C;x)$, where

$$F(A,B;C;x) = 1 + \frac{AB}{1!\,C}x + \frac{A(A+1)B(B+1)}{2!\,C(C+1)}x^2 + \frac{A(A+1)(A+2)B(B+1)(B+2)}{3!\,C(C+1)(C+2)}x^3 + \cdots$$

The series $F(A,B;C;x)$ is known as the *hypergeometric series*; it can be shown that this series converges for $-1 < x < 1$. It is customary to assign the arbitrary constant a_0 the value 1. Then $y_1(x) = F(A,B;C;x)$ and the hypergeometric series is a solution of the hypergeometric equation.

To find $y_2(x)$, we substitute $\lambda = 1-C$ into (*2*) and obtain

$$a_{n+1} = \frac{(n+1-C)(n+1+A+B-C)+AB}{(n+2-C)(n+1)}a_n$$

or
$$a_{n+1} = \frac{(A-C+n+1)(B-C+n+1)}{(n+2-C)(n+1)} a_n$$

Solving for a_n in terms of a_0, and again setting $a_0 = 1$, it follows that

$$y_2(x) = x^{1-C}F(A-C+1, B-C+1; 2-C; x)$$

The general solution is $y = c_1y_1(x) + c_2y_2(x)$.

Supplementary Problems

In Problems 20.14–20.22, find two linearly independent solutions near $x = 0$ by the method of Frobenius, using where necessary the modifications given in Problems 20.4 and 20.6.

20.14. $2x^2y'' - xy' + (1-x)y = 0.$

20.15. $2x^2y'' + (x^2-x)y' + y = 0.$

20.16. $3x^2y'' - 2xy' - (2+x^2)y = 0.$

20.17. $xy'' + y' - y = 0.$

20.18. $x^2y'' + xy' + x^3y = 0.$

20.19. $x^2y'' + (x-x^2)y' - y = 0.$

20.20. $xy'' - (x+1)y' - y = 0.$

20.21. $4x^2y'' + (4x+2x^2)y' + (3x-1)y = 0.$

20.22. $x^2y'' + (x^2-3x)y' - (x-4)y = 0.$

20.23. Use the method given in Problem 20.11 to find the general solution of

(a) $4x^2y'' + 4xy' - y = 0$ (b) $x^2y'' - 3xy' + 4y = 0$

Answers to Supplementary Problems

RF $\equiv$ recurrence formula

20.14. RF: $a_n = \dfrac{1}{[2(\lambda+n)-1][(\lambda+n)-1]} a_{n-1}$

$$y_1(x) = a_0x\left(1 + \frac{1}{3}x + \frac{1}{30}x^2 + \frac{1}{630}x^3 + \cdots\right)$$

$$y_2(x) = a_0\sqrt{x}\left(1 + x + \frac{1}{6}x^2 + \frac{1}{90}x^3 + \cdots\right)$$

20.15. RF: $a_n = \dfrac{-1}{2(\lambda+n)-1}a_{n-1}$

$$y_1(x) = a_0x\left(1 - \frac{1}{3}x + \frac{1}{15}x^2 - \frac{1}{105}x^3 + \cdots\right)$$

$$y_2(x) = a_0\sqrt{x}\left(1 - \frac{1}{2}x + \frac{1}{8}x^2 - \frac{1}{48}x^3 + \cdots\right)$$

20.16. RF: $a_n = \dfrac{1}{[3(\lambda+n)+1][(\lambda+n)-2]}a_{n-2}$

$$y_1(x) = a_0x^2\left(1 + \frac{1}{26}x^2 + \frac{1}{1976}x^4 + \cdots\right)$$

$$y_2(x) = a_0x^{-1/3}\left(1 - \frac{1}{2}x^2 - \frac{1}{40}x^4 - \frac{1}{2640}x^6 - \cdots\right)$$

20.17. For convenience, first multiply the differential equation by x. Then

RF: $a_n = \dfrac{1}{(\lambda+n)^2}a_{n-1}$

$$y_1(x) = a_0\left(1 + x + \frac{1}{4}x^2 + \frac{1}{36}x^3 + \cdots\right)$$

$$y_2(x) = y_1(x)\ln x + a_0\left(-2x - \frac{3}{4}x^2 + \cdots\right)$$

20.18. RF: $a_n = \dfrac{-1}{(\lambda+n)^2}a_{n-3}$

$$y_1(x) = a_0\left(1 - \frac{1}{9}x^3 + \frac{1}{324}x^6 + \cdots\right)$$

$$y_2(x) = y_1(x)\ln x + a_0\left(\frac{2}{27}x^3 - \frac{1}{324}x^6 + \cdots\right)$$

20.19. RF: $a_n = \dfrac{1}{(\lambda+n)+1}a_{n-1}$

$$y_1(x) = a_0x\left(1 + \frac{1}{3}x + \frac{1}{12}x^2 + \frac{1}{60}x^3 + \cdots\right)$$

$$y_2(x) = a_0x^{-1}\left(1 + x + \frac{1}{2!}x^2 + \frac{1}{3!}x^3 + \cdots\right) = a_0x^{-1}e^x$$

20.20. For convenience, first multiply the differential equation by x. Then

RF: $a_n = \dfrac{1}{(\lambda+n)-2}a_{n-1}$

$$y_1(x) = a_0x^2\left(1 + x + \frac{1}{2!}x^2 + \frac{1}{3!}x^3 + \cdots\right) = a_0x^2e^x$$

$$y_2(x) = -y_1(x)\ln x + a_0(1 - x + x^2 + 0x^3 + \cdots)$$

20.21. RF: $a_n = \dfrac{-1}{2(\lambda+n)-1}\,a_{n-1}$

$$y_1(x) = a_0\sqrt{x}\left(1 - \frac{1}{2}x + \frac{1}{8}x^2 - \frac{1}{48}x^3 + \cdots\right)$$

$$y_2(x) = -\frac{1}{2}y_1(x)\ln x + x^{-1/2}\left(1 - \frac{1}{8}x^2 + \frac{3}{64}x^3 + \cdots\right)$$

20.22. RF: $a_n = \dfrac{-1}{(\lambda+n)-2}\,a_{n-1}$

$$y_1(x) = a_0x^2\left(1 - x + \frac{1}{2!}x^2 - \frac{1}{3!}x^3 + \cdots\right) = a_0x^2e^{-x}$$

$$y_2(x) = y_1(x)\ln x + a_0x^2\left(x - \frac{3}{4}x^2 + \frac{11}{36}x^3 + \cdots\right)$$

20.23. (*a*) $y = c_1x^{1/2} + c_2x^{-1/2}$ (*b*) $y = c_1x^2 + c_2x^2\ln x$

Chapter 21

Gamma Function. Bessel Functions

21.1 GAMMA FUNCTION

The *gamma function*, $\Gamma(p)$, is defined for any positive real number p by

$$\Gamma(p) = \int_0^\infty x^{p-1}e^{-x}\,dx \qquad (21.1)$$

The basic functional equation satisfied by $\Gamma(p)$ is

$$\Gamma(p+1) = p\,\Gamma(p) \qquad (21.2)$$

(see Problem 21.1). From (*21.2*), it can be shown (see Problem 21.3) that when $p = n$, a positive integer, then

$$\Gamma(n+1) = n! \qquad (21.3)$$

Thus, the gamma function (which is defined on all positive real numbers) is an extension of the factorial function (which is defined only on the nonnegative integers).

We extend the definition of $\Gamma(p)$ to negative nonintegral values of p by means of (*21.2*). Thus,

$$\Gamma(p) = \frac{1}{p}\Gamma(p+1) \qquad (p < 0 \text{ and nonintegral}) \qquad (21.4)$$

Since $\Gamma(1) = 1$ (Problem 21.2), we have

$$\lim_{p\to 0+} \Gamma(p) = \lim_{p\to 0+} \frac{\Gamma(p+1)}{p} = \infty \qquad \text{and} \qquad \lim_{p\to 0-} \Gamma(p) = \lim_{p\to 0-} \frac{\Gamma(p+1)}{p} = -\infty$$

Therefore, $\Gamma(0)$ is undefined, and it follows immediately from (*21.4*) that $\Gamma(p)$ is undefined for all negative integer values of p.

The gamma function is tabulated in Appendix A for $1 < p < 2$. Equations (*21.2*) and (*21.4*) can be used with this table to generate other values of $\Gamma(p)$.

Example 21.1. To four decimal places $\Gamma(1.5) = 0.8862$. Find (*a*) $\Gamma(3.5)$ and (*b*) $\Gamma(-0.5)$.

(*a*) From (*21.2*) with $p = 2.5$, we obtain $\Gamma(3.5) = (2.5)\,\Gamma(2.5)$. But also from (*21.2*), with $p = 1.5$, we have $\Gamma(2.5) = (1.5)\,\Gamma(1.5)$. Thus, $\Gamma(3.5) = (2.5)(1.5)\,\Gamma(1.5) = (3.75)(0.8862) = 3.3233$.

(*b*) From (*21.4*) with $p = 0.5$, we obtain $\Gamma(0.5) = 2\Gamma(1.5)$. But also from (*21.4*), with $p = -0.5$, we have $\Gamma(-0.5) = -2\Gamma(0.5)$. Thus, $\Gamma(-0.5) = (-2)(2)\,\Gamma(1.5) = -4(0.8862) = -3.5448$.

21.2 BESSEL FUNCTIONS

Let p represent any real number. The *Bessel function of the first kind of order* p, $J_p(x)$, is

$$J_p(x) = \sum_{k=0}^{\infty} \frac{(-1)^k x^{2k+p}}{2^{2k+p}k!\,\Gamma(p+k+1)} \qquad (21.5)$$

The function $J_p(x)$ is a solution near the regular singular point $x = 0$ of *Bessel's differential equation of order* p:

$$x^2y'' + xy' + (x^2 - p^2)y = 0 \tag{21.6}$$

In fact, $J_p(x)$ is that solution of (*21.6*) guaranteed by Theorem 20.1 (see Problem 21.6).

If p is not an integer, then a second linearly independent solution of (*21.6*) is $J_{-p}(x)$. When $p = n$, an integer, $J_{-p}(x)$ is no longer linearly independent of $J_p(x)$; in that case, the second linearly independent solution, called the *Bessel function of the second kind of order* p, can be found by the methods outlined in Chapter 20 which apply to Cases 2 and 3 of Theorem 20.2. The special cases $p = 0$ and $p = 1$ are considered in Problems 21.7 and 21.8.

Some of the more important properties of $J_p(x)$ are given in Problems 21.11–21.14. Tabular values of $J_0(x)$ and $J_1(x)$ are given in Appendix B.

21.3 ALGEBRAIC OPERATIONS ON INFINITE SERIES

Changing the dummy index. The dummy index in an infinite series can be changed at will without altering the series. For example,

$$\sum_{k=0}^{\infty} \frac{1}{(k+1)!} = \sum_{n=0}^{\infty} \frac{1}{(n+1)!} = \sum_{p=0}^{\infty} \frac{1}{(p+1)!} = \frac{1}{1!} + \frac{1}{2!} + \frac{1}{3!} + \frac{1}{4!} + \frac{1}{5!} + \cdots$$

Change of variables. Consider the infinite series $\sum_{k=0}^{\infty} \frac{1}{(k+1)!}$. If we make the change of variables $j = k+1$, or $k = j-1$, then

$$\sum_{k=0}^{\infty} \frac{1}{(k+1)!} = \sum_{j=1}^{\infty} \frac{1}{j!}$$

Note that a change of variables generally changes the limits on the summation. For instance, if $j = k+1$, it follows that $j = 1$ when $k = 0$, $j = \infty$ when $k = \infty$, and, as k runs from 0 to ∞, j runs from 1 to ∞.

The two operations given above are often used in concert (see Problems 21.9 and 21.10). For example,

$$\sum_{k=0}^{\infty} \frac{1}{(k+1)!} = \sum_{j=2}^{\infty} \frac{1}{(j-1)!} = \sum_{k=2}^{\infty} \frac{1}{(k-1)!}$$

Here, the second series results from the change of variables $j = k+2$ in the first series, while the third series is the result of simply changing the dummy index in the second series from j to k. Note that all three series equal $\frac{1}{1!} + \frac{1}{2!} + \frac{1}{3!} + \frac{1}{4!} + \cdots = e - 1$.

Solved Problems

21.1. Prove that $\Gamma(p+1) = p\,\Gamma(p)$, $p > 0$.

Using (*21.1*) and integration by parts, we have

$$\Gamma(p+1) = \int_0^{\infty} x^{(p+1)-1}e^{-x}\,dx = \lim_{r\to\infty} \int_0^r x^p e^{-x}\,dx$$

$$= \lim_{r\to\infty} \left[-x^p e^{-x} \Big|_0^r + \int_0^r p x^{p-1} e^{-x}\,dx \right]$$

$$= \lim_{r \to \infty} (-r^p e^{-r} + 0) + p \int_0^\infty x^{p-1} e^{-x}\, dx = p\,\Gamma(p)$$

The result $\lim_{r \to \infty} r^p e^{-r} = 0$ is easily obtained by first writing $r^p e^{-r}$ as r^p/e^r and then using L'Hôpital's rule.

21.2. Prove that $\Gamma(1) = 1$.

Using (*21.1*), we find that

$$\Gamma(1) = \int_0^\infty x^{1-1} e^{-x}\, dx = \lim_{r \to \infty} \int_0^r e^{-x}\, dx$$

$$= \lim_{r \to \infty} -e^{-x}\Big|_0^r = \lim_{r \to \infty} (-e^{-r} + 1) = 1$$

21.3. Prove that if $p = n$, a positive integer, then $\Gamma(n+1) = n!$.

The proof is by induction. First we consider $n = 1$. Using Problem 21.1 with $p = 1$ and then Problem 21.2, we have

$$\Gamma(1+1) = 1\,\Gamma(1) = 1(1) = 1 = 1!$$

Next we assume that $\Gamma(n+1) = n!$ holds for $n = k$ and then try to prove its validity for $n = k+1$:

$$\Gamma[(k+1)+1] = (k+1)\,\Gamma(k+1) \qquad \text{(Problem 21.1 with } p = k+1\text{)}$$

$$= (k+1)(k!) \qquad \text{(from the induction hypothesis)}$$

$$= (k+1)!$$

Thus, $\Gamma(n+1) = n!$ is true by induction.

Note that we can now use this equality to define $0!$; that is,

$$0! = \Gamma(0+1) = \Gamma(1) = 1$$

21.4. Prove that $\Gamma(p+k+1) = (p+k)(p+k-1) \cdots (p+2)(p+1)\,\Gamma(p+1)$.

Using Problem 21.1 repeatedly, where first p is replaced by $p+k$, then by $p+k-1$, etc., we obtain

$$\Gamma(p+k+1) = \Gamma[(p+k)+1] = (p+k)\,\Gamma(p+k)$$

$$= (p+k)\,\Gamma[(p+k-1)+1] = (p+k)(p+k-1)\,\Gamma(p+k-1)$$

$$= \cdots = (p+k)(p+k-1)\cdots(p+2)(p+1)\,\Gamma(p+1)$$

21.5. Express $\int_0^\infty e^{-x^2}\, dx$ as a gamma function.

Let $z = x^2$; hence $x = z^{1/2}$ and $dx = \frac{1}{2}z^{-1/2}\, dz$. Substituting these values into the integral and noting that as x goes from 0 to ∞ so does z, we have

$$\int_0^\infty e^{-x^2}\, dx = \int_0^\infty e^{-z}\left(\frac{1}{2}z^{-1/2}\right) dz = \frac{1}{2}\int_0^\infty z^{(1/2)-1} e^{-z}\, dz = \frac{1}{2}\Gamma\left(\frac{1}{2}\right)$$

The last equality follows from (*21.1*), with the dummy variable x replaced by z and with $p = \frac{1}{2}$.

21.6. Use the method of Frobenius to find one solution of Bessel's equation of order p:

$$x^2 y'' + xy' + (x^2 - p^2)y = 0$$

Substituting (1), (2), and (3) of Problem 20.1 into Bessel's equation and simplifying, we find that

$$x^{\lambda}(\lambda^2 - p^2)a_0 + x^{\lambda+1}[(\lambda+1)^2 - p^2]a_1 + x^{\lambda+2}\{[(\lambda+2)^2 - p^2]a_2 + a_0\} + \cdots + x^{\lambda+n}\{[(\lambda+n)^2 - p^2]a_n + a_{n-2}\} + \cdots = 0$$

Thus,
$$(\lambda^2 - p^2)a_0 = 0 \qquad [(\lambda+1)^2 - p^2]a_1 = 0 \tag{1}$$

and, in general, $[(\lambda+n)^2 - p^2]a_n + a_{n-2} = 0$, or,

$$a_n = -\frac{1}{(\lambda+n)^2 - p^2}a_{n-2} \tag{2}$$

The indicial equation is $\lambda^2 - p^2 = 0$, which has the roots $\lambda_1 = p$ and $\lambda_2 = -p$ (p nonnegative). Substituting $\lambda = p$ into (1) and (2) and simplifying, we find that $a_1 = 0$ and

$$a_n = -\frac{1}{n(2p+n)}a_{n-2} \qquad (n \geqq 2)$$

Hence, $0 = a_1 = a_3 = a_5 = a_7 = \cdots$ and

$$a_2 = \frac{-1}{2^2 1!\,(p+1)}a_0$$

$$a_4 = -\frac{1}{2^2 2(p+2)}a_2 = \frac{1}{2^4 2!\,(p+2)(p+1)}a_0$$

$$a_6 = -\frac{1}{2^2 3(p+3)}a_4 = \frac{-1}{2^6 3!\,(p+3)(p+2)(p+1)}a_0$$

and, in general,

$$a_{2k} = \frac{(-1)^k}{2^{2k}k!\,(p+k)(p+k-1)\cdots(p+2)(p+1)}a_0 \qquad (k \geqq 1)$$

Thus,
$$y_1(x) = x^{\lambda}\sum_{n=0}^{\infty} a_n x^n = x^p\left[a_0 + \sum_{k=1}^{\infty} a_{2k}x^{2k}\right]$$

$$= a_0 x^p\left[1 + \sum_{k=1}^{\infty}\frac{(-1)^k x^{2k}}{2^{2k}k!\,(p+k)(p+k-1)\cdots(p+2)(p+1)}\right] \tag{3}$$

It is customary to choose the arbitrary constant a_0 as $a_0 = \dfrac{1}{2^p\,\Gamma(p+1)}$. Then bringing a_0x^p inside the brackets and summation in (3), combining, and finally using Problem 21.4, we obtain

$$y_1(x) = \frac{1}{2^p\,\Gamma(p+1)}x^p + \sum_{k=1}^{\infty}\frac{(-1)^k x^{2k+p}}{2^{2k+p}k!\,\Gamma(p+k+1)}$$

$$= \sum_{k=0}^{\infty}\frac{(-1)^k x^{2k+p}}{2^{2k+p}k!\,\Gamma(p+k+1)} \equiv J_p(x)$$

21.7. Find the general solution to Bessel's equation of order zero.

For $p = 0$, the equation is $x^2y'' + xy' + x^2y = 0$, which was solved in Problem 20.4. By (4) of Problem 20.4, one solution is

$$y_1(x) = a_0\sum_{n=0}^{\infty}\frac{(-1)^n x^{2n}}{2^{2n}(n!)^2}$$

Changing n to k, using Problem 21.3, and letting $a_0 = \dfrac{1}{2^0\,\Gamma(0+1)} = 1$ as indicated in Problem 21.6, it follows that $y_1(x) = J_0(x)$. A second solution is (see (6) of Problem 20.4, with a_0 again chosen to be 1)

$$y_2(x) = J_0(x)\ln x + \left[\frac{x^2}{2^2(1!)^2}(1) - \frac{x^4}{2^4(2!)^2}\left(1+\frac{1}{2}\right) + \frac{x^6}{2^6(3!)^2}\left(1+\frac{1}{2}+\frac{1}{3}\right) - \cdots\right]$$

which is usually designated by $N_0(x)$. Thus, the general solution to Bessel's equation of order zero is $y = c_1 J_0(x) + c_2 N_0(x)$.

Another common form of the general solution is obtained when the second linearly independent solution is not taken to be $N_0(x)$, but a combination of $N_0(x)$ and $J_0(x)$. In particular, if we define

$$Y_0(x) = \frac{2}{\pi}[N_0(x) + (\gamma - \ln 2)J_0(x)] \tag{1}$$

where γ is the *Euler constant* defined by

$$\gamma = \lim_{k\to\infty}\left(1 + \frac{1}{2} + \frac{1}{3} + \cdots + \frac{1}{k} - \ln k\right) \simeq 0.57721566$$

then the general solution to Bessel's equation of order zero can be given as $y = c_1 J_0(x) + c_2 Y_0(x)$. The function $Y_0(x)$ is partially tabulated in Appendix B.

21.8. Find the general solution to Bessel's equation of order one.

For $p = 1$, the equation is $x^2y'' + xy' + (x^2 - 1)y = 0$, which was solved in Problem 20.7. One solution is given by (4) of Problem 20.7; this solution can easily be transformed into $J_1(x)$ if a_0 is chosen to be $1/2\Gamma(2)$ as indicated in Problem 21.6. A second solution is given by (5) of Problem 20.7 as $y_2(x)$. Again, it is customary to take as the second solution not $y_2(x)$, but a special linear combination of $y_2(x)$ and $J_1(x)$ which is denoted by $Y_1(x)$ and is partially tabulated in Appendix B. The general solution of Bessel's equation of order one is $y = c_1 J_1(x) + c_2 Y_1(x)$.

21.9. Prove that

$$\sum_{k=0}^{\infty} \frac{(-1)^k(2k)x^{2k-1}}{2^{2k+p}k!\,\Gamma(p+k+1)} = -\sum_{k=0}^{\infty} \frac{(-1)^k x^{2k+1}}{2^{2k+p+1}k!\,\Gamma(p+k+2)}$$

Writing the $k = 0$ term separately, we have

$$\sum_{k=0}^{\infty} \frac{(-1)^k(2k)x^{2k-1}}{2^{2k+p}k!\,\Gamma(p+k+1)} = 0 + \sum_{k=1}^{\infty} \frac{(-1)^k(2k)x^{2k-1}}{2^{2k+p}k!\,\Gamma(p+k+1)}$$

which, under the change of variables $j = k - 1$, becomes

$$\begin{aligned}\sum_{j=0}^{\infty} \frac{(-1)^{j+1}2(j+1)x^{2(j+1)-1}}{2^{2(j+1)+p}(j+1)!\,\Gamma(p+j+1+1)} &= \sum_{j=0}^{\infty} \frac{(-1)(-1)^j 2(j+1)x^{2j+1}}{2^{2j+p+2}(j+1)!\,\Gamma(p+j+2)} \\ &= -\sum_{j=0}^{\infty} \frac{(-1)^j 2(j+1)x^{2j+1}}{2^{2j+p+1}(2)(j+1)(j!)\,\Gamma(p+j+2)} \\ &= -\sum_{j=0}^{\infty} \frac{(-1)^j x^{2j+1}}{2^{2j+p+1}j!\,\Gamma(p+j+2)}\end{aligned}$$

The desired result follows by changing the dummy variable in the last summation from j to k.

21.10. Prove that

$$-\sum_{k=0}^{\infty} \frac{(-1)^k x^{2k+p+2}}{2^{2k+p+1}k!\,\Gamma(p+k+2)} = \sum_{k=0}^{\infty} \frac{(-1)^k(2k)x^{2k+p}}{2^{2k+p}k!\,\Gamma(p+k+1)}$$

Make the change of variables $j = k + 1$:

$$\begin{aligned}-\sum_{k=0}^{\infty} \frac{(-1)^k x^{2k+p+2}}{2^{2k+p+1}k!\,\Gamma(p+k+2)} &= -\sum_{j=1}^{\infty} \frac{(-1)^{j-1}x^{2(j-1)+p+2}}{2^{2(j-1)+p+1}(j-1)!\,\Gamma(p+j-1+2)} \\ &= \sum_{j=1}^{\infty} \frac{(-1)^j x^{2j+p}}{2^{2j+p-1}(j-1)!\,\Gamma(p+j+1)}\end{aligned}$$

Now, multiply numerator and denominator in the last summation by $2j$, noting that $j(j-1)! = j!$ and $2^{2j+p-1}(2) = 2^{2j+p}$. The result is

$$\sum_{j=1}^{\infty} \frac{(-1)^j(2j)x^{2j+p}}{2^{2j+p}j!\,\Gamma(p+j+1)}$$

Owing to the factor j in the numerator, the last infinite series is not altered if the lower limit in the sum is changed from $j=1$ to $j=0$. Once this is done, the desired result is achieved by simply changing the dummy index from j to k.

21.11. Prove that $\dfrac{d}{dx}[x^{p+1}J_{p+1}(x)] = x^{p+1}J_p(x)$.

We may differentiate the series for the Bessel function term by term. Thus,

$$\begin{aligned}\frac{d}{dx}[x^{p+1}J_{p+1}(x)] &= \frac{d}{dx}\left[x^{p+1}\sum_{k=0}^{\infty}\frac{(-1)^k x^{2k+p+1}}{2^{2k+p+1}k!\,\Gamma(k+p+1+1)}\right] \\ &= \frac{d}{dx}\left[\sum_{k=0}^{\infty}\frac{(-1)^k x^{2k+2p+2}}{2^{2k+p}(2)k!\,\Gamma(k+p+2)}\right] \\ &= \sum_{k=0}^{\infty}\frac{(-1)^k(2k+2p+2)x^{2k+2p+1}}{2^{2k+p}k!\,2\Gamma(k+p+2)}\end{aligned}$$

Noting that $2\Gamma(k+p+2) = 2(k+p+1)\,\Gamma(k+p+1)$ and that the factor $2(k+p+1)$ cancels, we have

$$\frac{d}{dx}[x^{p+1}J_{p+1}(x)] = \sum_{k=0}^{\infty}\frac{(-1)^k x^{2k+2p+1}}{2^{2k+p}k!\,\Gamma(k+p+1)} = x^{p+1}J_p(x)$$

For the particular case $p=0$, it follows that

$$\frac{d}{dx}[xJ_1(x)] = xJ_0(x) \qquad (1)$$

21.12. Prove that $xJ_p'(x) = pJ_p(x) - xJ_{p+1}(x)$.

We have

$$\begin{aligned}pJ_p(x) - xJ_{p+1}(x) &= p\sum_{k=0}^{\infty}\frac{(-1)^k x^{2k+p}}{2^{2k+p}k!\,\Gamma(p+k+1)} - x\sum_{k=0}^{\infty}\frac{(-1)^k x^{2k+p+1}}{2^{2k+p+1}k!\,\Gamma(p+k+2)} \\ &= \sum_{k=0}^{\infty}\frac{(-1)^k p x^{2k+p}}{2^{2k+p}k!\,\Gamma(p+k+1)} - \sum_{k=0}^{\infty}\frac{(-1)^k x^{2k+p+2}}{2^{2k+p+1}k!\,\Gamma(p+k+2)}\end{aligned}$$

Using Problem 21.10 on the last summation, we find

$$\begin{aligned}pJ_p(x) - xJ_{p+1}(x) &= \sum_{k=0}^{\infty}\frac{(-1)^k p x^{2k+p}}{2^{2k+p}k!\,\Gamma(p+k+1)} + \sum_{k=0}^{\infty}\frac{(-1)^k(2k)x^{2k+p}}{2^{2k+p}k!\,\Gamma(p+k+1)} \\ &= \sum_{k=0}^{\infty}\frac{(-1)^k(p+2k)x^{2k+p}}{2^{2k+p}k!\,\Gamma(p+k+1)} = xJ_p'(x)\end{aligned}$$

For the particular case $p=0$, it follows that $xJ_0'(x) = -xJ_1(x)$, or

$$J_0'(x) = -J_1(x) \qquad (1)$$

21.13. Prove that $xJ_p'(x) = -pJ_p(x) + xJ_{p-1}(x)$.

$$-pJ_p(x) + xJ_{p-1}(x) = -p\sum_{k=0}^{\infty}\frac{(-1)^k x^{2k+p}}{2^{2k+p}k!\,\Gamma(p+k+1)} + x\sum_{k=0}^{\infty}\frac{(-1)^k x^{2k+p-1}}{2^{2k+p-1}k!\,\Gamma(p+k)}$$

Multiplying the numerator and denominator in the second summation by $2(p+k)$ and noting that $(p+k)\,\Gamma(p+k) = \Gamma(p+k+1)$, we find

$$-pJ_p(x) + xJ_{p-1}(x) = \sum_{k=0}^{\infty} \frac{(-1)^k(-p)x^{2k+p}}{2^{2k+p}k!\,\Gamma(p+k+1)} + \sum_{k=0}^{\infty} \frac{(-1)^k 2(p+k)x^{2k+p}}{2^{2k+p}k!\,\Gamma(p+k+1)}$$

$$= \sum_{k=0}^{\infty} \frac{(-1)^k[-p+2(p+k)]x^{2k+p}}{2^{2k+p}k!\,\Gamma(p+k+1)}$$

$$= \sum_{k=0}^{\infty} \frac{(-1)^k(2k+p)x^{2k+p}}{2^{2k+p}k!\,\Gamma(p+k+1)} = xJ_p'(x)$$

21.14. Use Problems 21.12 and 21.13 to derive the recurrence formula

$$J_{p+1}(x) = \frac{2p}{x}J_p(x) - J_{p-1}(x)$$

Subtracting the results of Problem 21.13 from the results of Problem 21.12, we find that

$$0 = 2pJ_p(x) - xJ_{p-1}(x) - xJ_{p+1}(x)$$

Upon solving for $J_{p+1}(x)$, we obtain the desired result.

Values of $J_0(x)$ and $J_1(x)$ have been extensively tabulated. (See, for example, Appendix B.) With the aid of the above recurrence formula, values of $J_p(x)$ for any other positive integer p can be obtained easily. For example, setting $p = 1$, we have $J_2(x) = (2/x)J_1(x) - J_0(x)$.

21.15. Show that $y = xJ_1(x)$ is a solution of $xy'' - y' - x^2J_0'(x) = 0$.

First note that $J_1(x)$ is a solution of Bessel's equation of order one:

$$x^2J_1''(x) + xJ_1'(x) + (x^2-1)J_1(x) = 0 \tag{1}$$

Now substitute $y = xJ_1(x)$ into the left side of the given differential equation:

$$x[xJ_1(x)]'' - [xJ_1(x)]' - x^2J_0'(x)$$
$$= x[2J_1'(x) + xJ_1''(x)] - [J_1(x) + xJ_1'(x)] - x^2J_0'(x)$$

But $J_0'(x) = -J_1(x)$ (by (1) of Problem 21.12), so that the right-hand side becomes

$$x^2J_1''(x) + 2xJ_1'(x) - J_1(x) - xJ_1'(x) + x^2J_1(x)$$
$$= x^2J_1''(x) + xJ_1'(x) + (x^2-1)J_1(x) = 0$$

the last equality following from (1).

21.16. Show that $y = \sqrt{x}\,J_{3/2}(x)$ is a solution of $x^2y'' + (x^2-2)y = 0$.

Observe that $J_{3/2}(x)$ is a solution of Bessel's equation of order $\frac{3}{2}$:

$$x^2J_{3/2}''(x) + xJ_{3/2}'(x) + \left(x^2 - \frac{9}{4}\right)J_{3/2}(x) = 0 \tag{1}$$

Now substitute $y = \sqrt{x}\,J_{3/2}(x)$ into the left side of the given differential equation, obtaining

$$x^2[\sqrt{x}\,J_{3/2}(x)]'' + (x^2-2)\sqrt{x}\,J_{3/2}(x)$$

$$= x^2\left[-\frac{1}{4}x^{-3/2}J_{3/2}(x) + x^{-1/2}J_{3/2}'(x) + x^{1/2}J_{3/2}''(x)\right]$$
$$+ (x^2-2)x^{1/2}J_{3/2}(x)$$

$$= \sqrt{x}\left[x^2J_{3/2}''(x) + xJ_{3/2}'(x) + \left(x^2 - \frac{9}{4}\right)J_{3/2}(x)\right]$$

$$= 0$$

the last equality following from (1). Thus $\sqrt{x}\,J_{3/2}(x)$ satisfies the given differential equation.

Supplementary Problems

21.17. $\Gamma(1.6)$ to four decimal places is 0.8935. Find (a) $\Gamma(2.6)$ and (b) $\Gamma(-1.4)$.

21.18. Express $\int_0^\infty e^{-x^3}\,dx$ as a gamma function.

21.19. Evaluate $\int_0^\infty x^3 e^{-x^2}\,dx$.

21.20. Prove that $\sum_{k=0}^{\infty} \frac{(-1)^k(2k)x^{2k-1}}{2^{2k-1}k!\,\Gamma(p+k)} = -\sum_{k=0}^{\infty} \frac{(-1)^k x^{2k+1}}{2^{2k}k!\,\Gamma(p+k+1)}$.

21.21. Prove that $\frac{d}{dx}[x^{-p}J_p(x)] = -x^{-p}J_{p+1}(x)$. (*Hint.* Use Problem 21.9.)

21.22. Prove that $J_{p-1}(x) - J_{p+1}(x) = 2J_p'(x)$.

21.23. (a) Prove that the derivative of $(\frac{1}{2}x^2)[J_0^2(x) + J_1^2(x)]$ is $xJ_0^2(x)$. (*Hint.* Use (*1*) of Problem 21.11 and (*1*) of Problem 21.12.)

(b) Evaluate $\int_0^1 xJ_0^2(x)\,dx$ in terms of Bessel functions.

21.24. Show that $y = xJ_n(x)$ is a solution of $x^2y'' - xy' + (1 + x^2 - n^2)y = 0$.

21.25. Show that $y = x^2J_2(x)$ is a solution of $xy'' - 3y' + xy = 0$.

Answers to Supplementary Problems

21.17. (a) 1.4296 (b) 2.6592 **21.18.** $\frac{1}{3}\Gamma(\frac{1}{3})$ **21.19.** $\frac{1}{2}\Gamma(2) = \frac{1}{2}$

21.20. First separate the $k = 0$ term from the series, then make the change of variables $j = k - 1$, and finally change the dummy index from j to k.

21.23. (b) $\frac{1}{2}[J_0^2(1) + J_1^2(1)]$

Chapter 22

The Laplace Transform

22.1 IMPROPER INTEGRALS

If $g(x)$ is defined for $a \leq x < \infty$, where a is a constant, then the *improper integral* $\int_a^\infty g(x)\,dx$ is defined by

$$\int_a^\infty g(x)\,dx = \lim_{R\to\infty} \int_a^R g(x)\,dx \tag{22.1}$$

if the limit exists. When the limit does exist, the improper integral is said to *converge*; otherwise, the improper integral is said to *diverge*. (See Problems 22.1 through 22.3.)

22.2 DEFINITION OF THE LAPLACE TRANSFORM

Let $f(x)$ be defined for $0 \leq x < \infty$ and let s denote an arbitrary real variable. The *Laplace transform of* $f(x)$, designated by either $\mathcal{L}\{f(x)\}$ or $F(s)$, is

$$\mathcal{L}\{f(x)\} = F(s) = \int_0^\infty e^{-sx} f(x)\,dx \tag{22.2}$$

for all values of s for which the improper integral converges.

When evaluating the integral in (*22.2*), the variable s is treated as a constant, since the integration is with respect to x. (See Problems 22.3 through 22.8.) The Laplace transforms for a number of elementary functions are calculated in Problems 22.4 through 22.8. Additional transforms are given in Table 22-1 and Appendix C.

22.3 CONVERGENCE OF THE LAPLACE TRANSFORM

Not all functions have a Laplace transform. Conditions on $f(x)$ which will guarantee the convergence of the improper integral (*22.2*) are given below.

Definition: A function $f(x)$ is of *exponential order* α if there exist constants α, M, and x_0 such that $e^{-\alpha x}|f(x)| \leq M$ for all $x \geq x_0$.

(See Problems 22.9–22.11.)

Definition: A function $f(x)$ is *piecewise continuous on the open interval* $a < x < b$ if (1) $f(x)$ is continuous everywhere in $a < x < b$ with the possible exception of at most a *finite* number of points $x_1, x_2, \ldots, x_n$, and (2) at these points of discontinuity, the right- and left-hand limits of $f(x)$, respectively $\lim_{\substack{x\to x_j\\ x>x_j}} f(x)$ and $\lim_{\substack{x\to x_j\\ x<x_j}} f(x)$, exist ($j = 1, 2, \ldots, n$).

(Note that a continuous function is piecewise continuous.)

Table 22-1

	$f(x)$	$F(s) = \mathcal{L}\{f(x)\}$
1.	1	$\dfrac{1}{s}$ $(s>0)$
2.	x	$\dfrac{1}{s^2}$ $(s>0)$
3.	x^{n-1} $(n=1,2,\ldots)$	$\dfrac{(n-1)!}{s^n}$ $(s>0)$
4.	$\sqrt{x}$	$\dfrac{1}{2}\sqrt{\pi}\,s^{-3/2}$ $(s>0)$
5.	$1/\sqrt{x}$	$\sqrt{\pi}\,s^{-1/2}$ $(s>0)$
6.	$x^{n-1/2}$ $(n=1,2,\ldots)$	$\dfrac{(1)(3)(5)\cdots(2n-1)\sqrt{\pi}}{2^n}\,s^{-n-1/2}$ $(s>0)$
7.	e^{ax}	$\dfrac{1}{s-a}$ $(s>a)$
8.	$\sin ax$	$\dfrac{a}{s^2+a^2}$ $(s>0)$
9.	$\cos ax$	$\dfrac{s}{s^2+a^2}$ $(s>0)$
10.	$\sinh ax$	$\dfrac{a}{s^2-a^2}$ $(s>\lvert a\rvert)$
11.	$\cosh ax$	$\dfrac{s}{s^2-a^2}$ $(s>\lvert a\rvert)$
12.	$x\sin ax$	$\dfrac{2as}{(s^2+a^2)^2}$ $(s>0)$
13.	$x\cos ax$	$\dfrac{s^2-a^2}{(s^2+a^2)^2}$ $(s>0)$
14.	$x^{n-1}e^{ax}$ $(n=1,2,\ldots)$	$\dfrac{(n-1)!}{(s-a)^n}$ $(s>a)$
15.	$e^{bx}\sin ax$	$\dfrac{a}{(s-b)^2+a^2}$ $(s>b)$
16.	$e^{bx}\cos ax$	$\dfrac{s-b}{(s-b)^2+a^2}$ $(s>b)$
17.	$\sin ax - ax\cos ax$	$\dfrac{2a^3}{(s^2+a^2)^2}$ $(s>0)$

Definition: A function $f(x)$ is *piecewise continuous on the closed interval* $a \leqq x \leqq b$ if (1) it is piecewise continuous on the open interval $a < x < b$, (2) the right-hand limit of $f(x)$ exists at $x = a$, and (3) the left-hand limit of $f(x)$ exists at $x = b$.

(See Problems 22.12 and 22.13.)

Theorem 22.1. If $f(x)$ is piecewise continuous on every finite closed interval $0 \leqq x \leqq b$, $b > 0$, and if $f(x)$ is of exponential order α, then the Laplace transform of $f(x)$ exists for $s > \alpha$.

Solved Problems

22.1. Determine whether the improper integral $\int_2^\infty \frac{1}{x^2}dx$ converges.

Since $\lim_{R\to\infty} \int_2^R \frac{1}{x^2}dx = \lim_{R\to\infty}\left(-\frac{1}{x}\right)\Big|_2^R = \lim_{R\to\infty}\left(-\frac{1}{R}+\frac{1}{2}\right) = \frac{1}{2}$, the improper integral converges to the value $\frac{1}{2}$.

22.2. Determine whether the improper integral $\int_9^\infty \frac{1}{x}dx$ converges.

Since $\lim_{R\to\infty} \int_9^R \frac{1}{x}dx = \lim_{R\to\infty} \ln|x|\Big|_9^R = \lim_{R\to\infty}(\ln R - \ln 9) = \infty$, the improper integral diverges.

22.3. Determine those values of s for which the improper integral $\int_0^\infty e^{-sx}\,dx$ converges.

For $s = 0$,

$$\int_0^\infty e^{-sx}\,dx = \int_0^\infty e^{-(0)(x)}\,dx = \lim_{R\to\infty}\int_0^R (1)\,dx = \lim_{R\to\infty} x\Big|_0^R = \lim_{R\to\infty} R = \infty$$

hence the integral diverges. For $s \neq 0$,

$$\int_0^\infty e^{-sx}\,dx = \lim_{R\to\infty}\int_0^R e^{-sx}\,dx = \lim_{R\to\infty}\left[-\frac{1}{s}e^{-sx}\right]_{x=0}^{x=R}$$

$$= \lim_{R\to\infty}\left(\frac{-1}{s}e^{-sR}+\frac{1}{s}\right)$$

When $s < 0$, $-sR > 0$; hence the limit is ∞ and the integral diverges. When $s > 0$, $-sR < 0$; hence, the limit is $1/s$ and the integral converges.

22.4. Find the Laplace transform of $f(x) \equiv 1$.

Using (*22.2*) and the results of Problem 22.3, we have

$$F(s) = \mathcal{L}\{1\} = \int_0^\infty e^{-sx}(1)\,dx = \frac{1}{s}$$

for $s > 0$.

22.5. Find the Laplace transform of $f(x) = x^2$.

Using (22.2) and integration by parts twice, we find that

$$F(s) = \mathcal{L}\{x^2\} = \int_0^\infty e^{-sx}x^2\,dx = \lim_{R\to\infty}\int_0^R x^2e^{-sx}\,dx$$

$$= \lim_{R\to\infty}\left[-\frac{x^2}{s}e^{-sx} - \frac{2x}{s^2}e^{-sx} - \frac{2}{s^3}e^{-sx}\right]_{x=0}^{x=R}$$

$$= \lim_{R\to\infty}\left(-\frac{R^2}{s}e^{-sR} - \frac{2R}{s^2}e^{-sR} - \frac{2}{s^3}e^{-sR} + \frac{2}{s^3}\right)$$

For $s < 0$, $\lim_{R\to\infty}\left(-\frac{R^2}{s}e^{-sR}\right) = \infty$, and the improper integral diverges. For $s > 0$, it follows from repeated use of L'Hôpital's rule that

$$\lim_{R\to\infty}\left(-\frac{R^2}{s}e^{-sR}\right) = \lim_{R\to\infty}\left(\frac{-R^2}{se^{sR}}\right) = \lim_{R\to\infty}\left(\frac{-2R}{s^2e^{sR}}\right)$$

$$= \lim_{R\to\infty}\left(\frac{-2}{s^3e^{sR}}\right) = 0$$

$$\lim_{R\to\infty}\left(-\frac{2R}{s}e^{-sR}\right) = \lim_{R\to\infty}\left(\frac{-2R}{se^{sR}}\right) = \lim_{R\to\infty}\left(\frac{-2}{s^2e^{sR}}\right) = 0$$

Also, $\lim_{R\to\infty}\left(-\frac{2}{s^3}e^{-sR}\right) = 0$ directly; hence the integral converges, and $F(s) = 2/s^3$. For the special case $s = 0$, we have

$$\int_0^\infty e^{-sx}x^2\,dx = \int_0^\infty e^{-s(0)}x^2\,dx = \lim_{R\to\infty}\int_0^R x^2\,dx = \lim_{R\to\infty}\frac{R^3}{3} = \infty$$

Finally, combining all cases, we obtain $\mathcal{L}\{x^2\} = 2/s^3$, $s > 0$.

22.6. Find $\mathcal{L}\{e^{ax}\}$.

Using (22.2), we obtain

$$F(s) = \mathcal{L}\{e^{ax}\} = \int_0^\infty e^{-sx}e^{ax}\,dx = \lim_{R\to\infty}\int_0^R e^{(a-s)x}\,dx$$

$$= \lim_{R\to\infty}\left[\frac{e^{(a-s)x}}{a-s}\right]_{x=0}^{x=R} = \lim_{R\to\infty}\left[\frac{e^{(a-s)R}-1}{a-s}\right]$$

$$= \frac{1}{s-a}, \quad \text{for } s > a$$

Note that when $s \leq a$, the improper integral diverges.

22.7. Find $\mathcal{L}\{\sin ax\}$.

Using (22.2) and integration by parts twice, we obtain

$$\mathcal{L}\{\sin ax\} = \int_0^\infty e^{-sx}\sin ax\,dx = \lim_{R\to\infty}\int_0^R e^{-sx}\sin ax\,dx$$

$$= \lim_{R\to\infty}\left[\frac{-se^{-sx}\sin ax}{s^2+a^2} - \frac{ae^{-sx}\cos ax}{s^2+a^2}\right]_{x=0}^{x=R}$$

$$= \lim_{R\to\infty}\left[\frac{-se^{-sR}\sin aR}{s^2+a^2} - \frac{ae^{-sR}\cos aR}{s^2+a^2} + \frac{a}{s^2+a^2}\right]$$

$$= \frac{a}{s^2+a^2}, \quad \text{for } s > 0$$

22.8. Find $\mathcal{L}\{f(x)\}$ if $f(x) = \begin{cases} -1 & x \leqq 4 \\ 1 & x > 4 \end{cases}$.

$$\begin{aligned} \mathcal{L}\{f(x)\} &= \int_0^\infty e^{-sx} f(x)\, dx = \int_0^4 e^{-sx}(-1)\, dx + \int_4^\infty e^{-sx}(1)\, dx \\ &= \left.\frac{e^{-sx}}{s}\right|_{x=0}^{x=4} + \lim_{R\to\infty} \int_4^R e^{-sx}\, dx \\ &= \frac{e^{-4s}}{s} - \frac{1}{s} + \lim_{R\to\infty}\left(\frac{-1}{s}e^{-Rs} + \frac{1}{s}e^{-4s}\right) \\ &= \frac{2e^{-4s}}{s} - \frac{1}{s}, \quad \text{for } s > 0 \end{aligned}$$

22.9. Prove that $f(x) = x^2$ is of exponential order α for every $\alpha > 0$.

Using L'Hôpital's rule, we obtain

$$\lim_{x\to\infty} e^{-\alpha x}|x^2| = \lim_{x\to\infty} \frac{x^2}{e^{\alpha x}} = \lim_{x\to\infty} \frac{2x}{\alpha e^{\alpha x}} = \lim_{x\to\infty} \frac{2}{\alpha^2 e^{\alpha x}} = 0$$

Pick $M = 1$ (any other positive number will also do). Then since $\lim_{x\to\infty} e^{-\alpha x}|x^2| = 0$, it follows that there exists an x_0 such that $e^{-\alpha x}|x^2| \leqq 1 = M$ for $x \geqq x_0$.

22.10. Prove that $f(x) = \sin ax$ is of exponential order α for every $\alpha \geqq 0$.

Note that $|\sin ax| \leqq 1$ and that $\lim_{x\to\infty} e^{-\alpha x} = 0$ for $\alpha > 0$. Then, choosing $M = 1$, it follows that there exists an x_0 such that $e^{-\alpha x} \leqq M$ if $x \geqq x_0$. For $\alpha = 0$, $e^{-\alpha x} = M$. Thus,

$$e^{-\alpha x}|\sin ax| \leqq e^{-\alpha x} \leqq M, \quad \text{if } x \geqq x_0$$

22.11. Prove that $f(x) = J_p(\sqrt{x})$ is of exponential order α for every $\alpha > 1$.

Note that for $p \geqq 0$ and for any nonnegative integer k, we have $2^{2k+p} \geqq 1$ and $\Gamma(k+p+1) \geqq \frac{1}{2}$. Then, for $x \geqq 0$,

$$\begin{aligned} |J_p(\sqrt{x})| &= \left|\sum_{k=0}^\infty \frac{(-1)^k(\sqrt{x})^{2k+p}}{2^{2k+p}k!\,\Gamma(k+p+1)}\right| \\ &\leqq \sum_{k=0}^\infty \frac{|(-1)^k|\,|x^{k+p/2}|}{2^{2k+p}k!\,\Gamma(k+p+1)} \\ &= x^{p/2}\sum_{k=0}^\infty \frac{x^k}{2^{2k+p}k!\,\Gamma(k+p+1)} \\ &\leqq x^{p/2}\sum_{k=0}^\infty \frac{x^k}{(1)k!\,(\frac{1}{2})} = 2x^{p/2}e^x \end{aligned}$$

Therefore, $\lim_{x\to\infty} e^{-\alpha x}|J_p(\sqrt{x})| \leqq \lim_{x\to\infty} e^{-\alpha x}2x^{p/2}e^x = \lim_{x\to\infty} \frac{2x^{p/2}}{e^{(\alpha-1)x}} = 0$ if $\alpha > 1$

Finally, choosing $M = 1$, it follows that if $\alpha > 1$ there exists an x_0 such that $e^{-\alpha x}|J_p(\sqrt{x})| \leqq 1 = M$ for $x \geqq x_0$.

22.12. Determine whether $f(x) = \begin{cases} x^2 + 1 & x \geqq 0 \\ 1/x & x < 0 \end{cases}$ is piecewise continuous on $[-1, 1]$.

The given function is continuous everywhere on $[-1,1]$ except at $x=0$. Therefore, if the right- and left-hand limits exist at $x=0$, $f(x)$ will be piecewise continuous on $[-1,1]$. We have

$$\lim_{\substack{x\to 0\\ x>0}} f(x) = \lim_{\substack{x\to 0\\ x>0}} (x^2+1) = 1 \qquad \lim_{\substack{x\to 0\\ x<0}} f(x) = \lim_{\substack{x\to 0\\ x<0}} \frac{1}{x} = -\infty$$

Since the left-hand limit does not exist, $f(x)$ is not piecewise continuous on $[-1,1]$.

22.13. Is
$$f(x) = \begin{cases} \sin \pi x & x>1 \\ 0 & 0 \leqq x \leqq 1 \\ e^x & -1<x<0 \\ x^3 & x \leqq -1 \end{cases}$$
piecewise continuous on $[-2,5]$?

The given function is continuous on $[-2,5]$ except at the two points $x_1=0$ and $x_2=-1$. (Note that $f(x)$ is continuous at $x=1$.) At the two points of discontinuity, we find that

$$\lim_{\substack{x\to 0\\ x>0}} f(x) = \lim_{x\to 0} 0 = 0 \qquad \lim_{\substack{x\to 0\\ x<0}} f(x) = \lim_{x\to 0} e^x = e^0 = 1$$

and

$$\lim_{\substack{x\to -1\\ x>-1}} f(x) = \lim_{x\to -1} e^x = e^{-1} \qquad \lim_{\substack{x\to -1\\ x<-1}} f(x) = \lim_{x\to -1} x^3 = -1$$

Since all required limits exist, $f(x)$ is piecewise continuous on $[-2,5]$.

Supplementary Problems

22.14. Determine whether the following improper integrals converge: (a) $\int_1^\infty x^{-1/2}\,dx$, (b) $\int_5^\infty x^{-3/2}\,dx$.

22.15. Determine values of s, if any, for which the following improper integrals converge:

(a) $\int_0^\infty e^{sx}\,dx$ (b) $\int_0^\infty xe^{-sx}\,dx$

22.16. Find the Laplace transform of (a) x, (b) $\cos bx$, (c) x^3, (d) xe^{ax}.

22.17. Find $\mathcal{L}\{f(x)\}$, if

(a) $f(x) = \begin{cases} x & 0 \leqq x \leqq 2 \\ 2 & x>2 \end{cases}$ (b) $f(x) = \begin{cases} 1 & 0 \leqq x \leqq 1 \\ e^x & 1<x \leqq 4 \\ 0 & x>4 \end{cases}$

22.18. Prove that $f(x) = e^{4x}$ is of exponential order α for every $\alpha \geqq 4$.

22.19. Prove that $f(x) = \cos 7x$ is of exponential order α for every $\alpha \geqq 0$.

22.20. Determine whether the following functions are piecewise continuous on $[-1,5]$:

(a) $f(x) = \begin{cases} x^2 & x \geqq 2 \\ 4 & 0<x<2 \\ x & x \leqq 0 \end{cases}$ (b) $f(x) = \begin{cases} 1/(x-2)^2 & x>2 \\ 5x^2-1 & x \leqq 2 \end{cases}$

(c) $f(x) = \dfrac{1}{(x-2)^2}$ (d) $f(x) = \dfrac{1}{(x+2)^2}$

Answers to Supplementary Problems

22.14. (*a*) diverges (*b*) converges

22.15. (*a*) $s < 0$ (*b*) $s > 0$

22.16. See Table 22-1.

22.17. (*a*) $F(s) = \dfrac{1 - e^{-2s}}{s^2}$

(*b*) $F(s) = \dfrac{1 - e^{-s}}{s} + \dfrac{e^{-(s-1)} - e^{-4(s-1)}}{s - 1}$

22.20. (*a*) yes; (*b*) no, $\lim\limits_{\substack{x \to 2 \\ x > 2}} f(x) = \infty$; (*c*) no, $\lim\limits_{\substack{x \to 2 \\ x > 2}} f(x) = \infty$; (*d*) yes, $f(x)$ is continuous on $[-1, 5]$

Chapter 23

Properties of the Laplace Transform

The six theorems given below are, among other things, very useful in the calculation of Laplace transforms. To simplify the statements of the theorems, we make the

Definition: $f(x) \in E_\alpha$ if (1) $f(x)$ is defined for all $0 \leq x < \infty$; (2) $f(x)$ is piecewise continuous on every closed interval $0 \leq x \leq b$, $b > 0$; and (3) $f(x)$ is of exponential order α.

In other words, we are considering functions that obey the hypotheses of Theorem 22.1 and that, in addition, are well defined at their points of discontinuity. It follows from Theorem 22.1 that if $f(x) \in E_\alpha$ then $F(s) = \mathcal{L}\{f(x)\}$ exists for $s > \alpha$.

Theorem 23.1 (Linearity). If $f_1(x) \in E_\alpha$ and $f_2(x) \in E_\alpha$, then for any two constants c_1 and c_2, $c_1 f_1(x) + c_2 f_2(x) \in E_\alpha$ and

$$\mathcal{L}\{c_1 f_1(x) + c_2 f_2(x)\} = c_1 \mathcal{L}\{f_1(x)\} + c_2 \mathcal{L}\{f_2(x)\} \tag{23.1}$$

(See Problems 23.1 through 23.3.)

Theorem 23.2. If $f(x) \in E_\alpha$, then for any constant a,

$$\mathcal{L}\{e^{ax} f(x)\} = F(s-a) \quad (s > \alpha + a) \tag{23.2}$$

(See Problems 23.4 through 23.6.)

Theorem 23.3. If $f(x) \in E_\alpha$, then for any positive integer n,

$$\mathcal{L}\{x^n f(x)\} = (-1)^n \frac{d^n}{ds^n}[F(s)] \tag{23.3}$$

(See Problems 23.4, 23.7, and 23.8.)

Example 23.1. For the particular cases $n = 1$ and $n = 2$, *(23.3)* reduces to

$$\mathcal{L}\{x\, f(x)\} = -F'(s) \tag{23.4}$$

and

$$\mathcal{L}\{x^2 f(x)\} = F''(s) \tag{23.5}$$

Theorem 23.4. If $f(x) \in E_\alpha$ and if $\lim_{\substack{x \to 0 \\ x > 0}} \frac{f(x)}{x}$ exists, then

$$\mathcal{L}\left\{\frac{1}{x} f(x)\right\} = \int_s^\infty F(t)\, dt \tag{23.6}$$

(See Problem 23.9.)

Theorem 23.5. If $f(x) \in E_\alpha$, then

$$\mathcal{L}\left\{\int_0^x f(t)\,dt\right\} = \frac{1}{s}F(s) \qquad (23.7)$$

(See Problem 23.10.)

Theorem 23.6. If $f(x)$ is periodic with period ω, that is, $f(x+\omega) = f(x)$, then

$$\mathcal{L}\{f(x)\} = \frac{\int_0^\omega e^{-sx} f(x)\,dx}{1 - e^{-\omega s}} \qquad (23.8)$$

(See Problems 23.12 and 23.13.)

Solved Problems

23.1. Find $F(s)$ if $f(x) = 3 + 2x^2$.

From (*23.1*) and Table 22-1,

$$F(s) = \mathcal{L}\{3+2x^2\} = 3\,\mathcal{L}\{1\} + 2\,\mathcal{L}\{x^2\} = 3\left(\frac{1}{s}\right) + 2\left(\frac{2}{s^3}\right)$$

$$= \frac{3}{s} + \frac{4}{s^3}$$

23.2. Find $F(s)$ if $f(x) = 2\sin x + 3\cos 2x$.

From (*23.1*) and Table 22-1,

$$F(s) = \mathcal{L}\{2\sin x + 3\cos 2x\} = 2\,\mathcal{L}\{\sin x\} + 3\,\mathcal{L}\{\cos 2x\}$$

$$= 2\frac{1}{s^2+1} + 3\frac{s}{s^2+4}$$

$$= \frac{2}{s^2+1} + \frac{3s}{s^2+4}$$

Note that $f(x)$ is periodic, so that Theorem 23.6 is also applicable with $\omega = 2\pi$.

23.3. Find $F(s)$ if $f(x) = 2x^2 - 3x + 4$.

Using (*23.1*) repeatedly, we obtain

$$F(s) = \mathcal{L}\{2x^2 - 3x + 4\} = 2\,\mathcal{L}\{x^2\} - 3\,\mathcal{L}\{x\} + 4\,\mathcal{L}\{1\}$$

$$= 2\left(\frac{2}{s^3}\right) - 3\left(\frac{1}{s^2}\right) + 4\left(\frac{1}{s}\right)$$

$$= \frac{4}{s^3} - \frac{3}{s^2} + \frac{4}{s}$$

23.4. Find $\mathcal{L}\{xe^{4x}\}$.

This problem can be done three different ways.

(*a*) Using the defining equation (*22.2*) directly, we obtain $\mathcal{L}\{xe^{4x}\} = 1/(s-4)^2$. See Table 22-1, entry 14.

(b) Using Theorem 23.2 with $a = 4$ and $f(x) = x$, we have

$$F(s) = \mathcal{L}\{f(x)\} = \mathcal{L}\{x\} = \frac{1}{s^2}$$

$$\mathcal{L}\{e^{4x}x\} = F(s-4) = \frac{1}{(s-4)^2}$$

(c) Using Theorem 23.3, where now $f(x) = e^{4x}$ and $n = 1$, we find that

$$F(s) = \mathcal{L}\{f(x)\} = \mathcal{L}\{e^{4x}\} = \frac{1}{s-4}$$

$$\mathcal{L}\{xe^{4x}\} = -F'(s) = -\frac{d}{ds}\left(\frac{1}{s-4}\right) = \frac{1}{(s-4)^2}$$

23.5. Find $\mathcal{L}\{e^{-2x}\sin 5x\}$.

This problem can be done two different ways.

(a) Using the defining equation (*22.2*) directly, we obtain

$$\mathcal{L}\{e^{-2x}\sin 5x\} = \frac{5}{(s+2)^2+25}$$

See Table 22-1, entry 15.

(b) Using Theorem 23.2 with $a = -2$ and $f(x) = \sin 5x$, we have

$$F(s) = \mathcal{L}\{f(x)\} = \mathcal{L}\{\sin 5x\} = \frac{5}{s^2+25}$$

$$\mathcal{L}\{e^{-2x}\sin 5x\} = F(s+2) = \frac{5}{(s+2)^2+25}$$

23.6. Find $\mathcal{L}\{x\cos ax\}$, where a is a constant.

Taking $f(x) = \cos ax$, we have from Table 22-1

$$F(s) = \mathcal{L}\{f(x)\} = \frac{s}{s^2+a^2}$$

Then, using (*23.4*), we obtain

$$\mathcal{L}\{x\cos ax\} = -\frac{d}{ds}\left(\frac{s}{s^2+a^2}\right) = \frac{s^2-a^2}{(s^2+a^2)^2}$$

which agrees with Table 22-1, entry 13.

23.7. Find $\mathcal{L}\{e^{-x}x\cos 2x\}$.

Let $f(x) = x\cos 2x$. From Problem 23.6 with $a = 2$, or Table 22-1, entry 13, we obtain

$$F(s) = \frac{s^2-4}{(s^2+4)^2}$$

Then, from (*23.2*) with $a = -1$,

$$\mathcal{L}\{e^{-x}x\cos 2x\} = F(s+1) = \frac{(s+1)^2-4}{[(s+1)^2+4]^2}$$

23.8. Find $\mathcal{L}\{x^{7/2}\}$.

Define $f(x) \equiv \sqrt{x}$. Then $x^{7/2} = x^3\sqrt{x} = x^3 f(x)$ and, from Table 22-1, entry 4,

$$F(s) = \mathcal{L}\{f(x)\} = \mathcal{L}\{\sqrt{x}\} = \frac{1}{2}\sqrt{\pi}\, s^{-3/2}$$

It now follows from (23.3) that

$$\mathcal{L}\{x^3\sqrt{x}\} = (-1)^3\frac{d^3}{ds^3}\left(\frac{1}{2}\sqrt{\pi}\,s^{-3/2}\right) = \frac{105}{16}\sqrt{\pi}\,s^{-9/2}$$

which agrees with Table 22-1, entry 6, for $n = 4$.

23.9. Find $\mathcal{L}\left\{\dfrac{\sin 3x}{x}\right\}$.

Taking $f(x) = \sin 3x$, we find from Table 22-1

$$F(s) = \frac{3}{s^2+9} \quad \text{or} \quad F(t) = \frac{3}{t^2+9}$$

Then, using (23.6), we obtain

$$\begin{aligned}\mathcal{L}\left\{\frac{\sin 3x}{x}\right\} &= \int_s^\infty \frac{3}{t^2+9}\,dt = \lim_{R\to\infty}\int_s^R \frac{3}{t^2+9}\,dt \\ &= \lim_{R\to\infty} \arctan\frac{t}{3}\bigg|_s^R \\ &= \lim_{R\to\infty}\left(\arctan\frac{R}{3} - \arctan\frac{s}{3}\right) \\ &= \frac{\pi}{2} - \arctan\frac{s}{3}\end{aligned}$$

23.10. Find $\mathcal{L}\left\{\int_0^x \sinh 2t\,dt\right\}$.

Taking $f(t) = \sinh 2t$, we have $f(x) = \sinh 2x$. It now follows from Table 22-1 that $F(s) = 2/(s^2-4)$, and then, from (23.7), that

$$\mathcal{L}\left\{\int_0^x \sinh 2t\,dt\right\} = \frac{1}{s}\left(\frac{2}{s^2-4}\right) = \frac{2}{s(s^2-4)}$$

23.11. Prove that if $f(x+\omega) = -f(x)$, then

$$\mathcal{L}\{f(x)\} = \frac{\int_0^\omega e^{-sx}f(x)\,dx}{1+e^{-\omega s}} \tag{1}$$

Since

$$f(x+2\omega) = f[(x+\omega)+\omega] = -f(x+\omega) = -[-f(x)] = f(x)$$

$f(x)$ is periodic with period 2ω. Then, using Theorem 23.6 with ω replaced by 2ω, we have

$$\mathcal{L}\{f(x)\} = \frac{\int_0^{2\omega} e^{-sx}f(x)\,dx}{1-e^{-2\omega s}} = \frac{\int_0^{\omega} e^{-sx}f(x)\,dx + \int_\omega^{2\omega} e^{-sx}f(x)\,dx}{1-e^{-2\omega s}}$$

Substituting $y = x - \omega$ into the second integral, we find that

$$\begin{aligned}\int_\omega^{2\omega} e^{-sx}f(x)\,dx &= \int_0^\omega e^{-s(y+\omega)}f(y+\omega)\,dy = e^{-\omega s}\int_0^\omega e^{-sy}[-f(y)]\,dy \\ &= -e^{-\omega s}\int_0^\omega e^{-sy}f(y)\,dy\end{aligned}$$

The last integral, upon changing the dummy variable of integration back to x, equals

$$-e^{-\omega s}\int_0^\omega e^{-sx}f(x)\,dx$$

Thus,

$$\mathcal{L}\{f(x)\} = \frac{(1-e^{-\omega s})\int_0^\omega e^{-sx} f(x)\,dx}{1-e^{-2\omega s}}$$

$$= \frac{(1-e^{-\omega s})\int_0^\omega e^{-sx} f(x)\,dx}{(1-e^{-\omega s})(1+e^{-\omega s})} = \frac{\int_0^\omega e^{-sx} f(x)\,dx}{1+e^{-\omega s}}$$

23.12. Find $\mathcal{L}\{f(x)\}$ for the square wave shown in Fig. 23-1.

This problem can be done two different ways.

(a) Note that $f(x)$ is periodic with period $\omega = 2$, and in the interval $0 < x \leqq 2$ it can be defined analytically by

$$f(x) = \begin{cases} 1 & 0 < x \leqq 1 \\ -1 & 1 < x \leqq 2 \end{cases}$$

From (*23.8*), we have

$$\mathcal{L}\{f(x)\} = \frac{\int_0^2 e^{-sx} f(x)\,dx}{1-e^{-2s}}$$

Since

$$\int_0^2 e^{-sx} f(x)\,dx = \int_0^1 e^{-sx}(1)\,dx + \int_1^2 e^{-sx}(-1)\,dx$$

$$= \frac{1}{s}(e^{-2s} - 2e^{-s} + 1) = \frac{1}{s}(e^{-s}-1)^2$$

it follows that

$$F(s) = \frac{(e^{-s}-1)^2}{s(1-e^{-2s})} = \frac{(1-e^{-s})^2}{s(1-e^{-s})(1+e^{-s})} = \frac{1-e^{-s}}{s(1+e^{-s})}$$

$$= \left[\frac{e^{s/2}}{e^{s/2}}\right]\left[\frac{1-e^{-s}}{s(1+e^{-s})}\right] = \frac{e^{s/2}-e^{-s/2}}{s(e^{s/2}+e^{-s/2})} = \frac{1}{s}\tanh\frac{s}{2}$$

(b) The square wave $f(x)$ also satisfies the equation $f(x+1) = -f(x)$. Thus, using (*1*) of Problem 23.11 with $\omega = 1$, we obtain

$$\mathcal{L}\{f(x)\} = \frac{\int_0^1 e^{-sx} f(x)\,dx}{1+e^{-s}} = \frac{\int_0^1 e^{-sx}(1)\,dx}{1+e^{-s}}$$

$$= \frac{(1/s)(1-e^{-s})}{1+e^{-s}} = \frac{1}{s}\tanh\frac{s}{2}$$

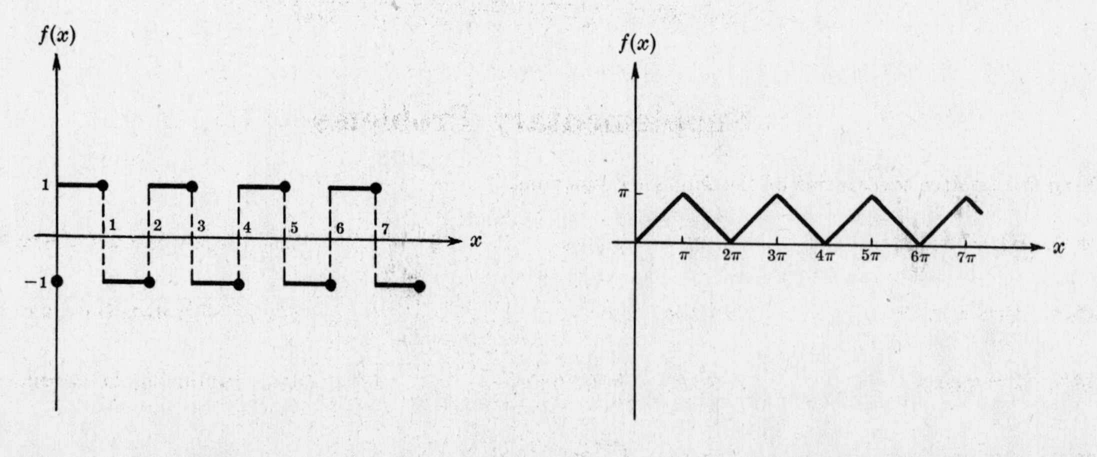

Fig. 23-1

Fig. 23-2

23.13. Find the Laplace transform of the function graphed in Fig. 23-2.

Note that $f(x)$ is periodic with period $\omega = 2\pi$, and in the interval $0 \leq x < 2\pi$ it can be defined analytically by

$$f(x) = \begin{cases} x & 0 \leq x \leq \pi \\ 2\pi - x & \pi \leq x < 2\pi \end{cases}$$

From (*23.8*), we have

$$\mathcal{L}\{f(x)\} = \frac{\int_0^{2\pi} e^{-sx} f(x)\, dx}{1 - e^{-2\pi s}}$$

Since

$$\int_0^{2\pi} e^{-sx} f(x)\, dx = \int_0^{\pi} e^{-sx} x\, dx + \int_{\pi}^{2\pi} e^{-sx}(2\pi - x)\, dx$$

$$= \frac{1}{s^2}(e^{-2\pi s} - 2e^{-\pi s} + 1) = \frac{1}{s^2}(e^{-\pi s} - 1)^2$$

it follows that

$$\mathcal{L}\{f(x)\} = \frac{(1/s^2)(e^{-\pi s} - 1)^2}{1 - e^{-2\pi s}} = \frac{(1/s^2)(e^{-\pi s} - 1)^2}{(1 - e^{-\pi s})(1 + e^{-\pi s})}$$

$$= \frac{1}{s^2}\left(\frac{1 - e^{-\pi s}}{1 + e^{-\pi s}}\right) = \frac{1}{s^2} \tanh \frac{\pi s}{2}$$

23.14. Find $\mathcal{L}\left\{e^{4x} x \int_0^x \frac{1}{t} e^{-4t} \sin 3t\, dt\right\}$.

Using (*23.2*) with $a = -4$ on the results of Problem 23.9, we obtain

$$\mathcal{L}\left\{\frac{1}{x} e^{-4x} \sin 3x\right\} = \frac{\pi}{2} - \arctan \frac{s+4}{3}$$

It now follows from (*23.7*) that

$$\mathcal{L}\left\{\int_0^x \frac{1}{t} e^{-4t} \sin 3t\, dt\right\} = \frac{\pi}{2s} - \frac{1}{s} \arctan \frac{s+4}{3}$$

and then from (*23.4*) that

$$\mathcal{L}\left\{x \int_0^x \frac{1}{t} e^{-4t} \sin 3t\, dt\right\} = \frac{\pi}{2s^2} - \frac{1}{s^2} \arctan \frac{s+4}{3} + \frac{3}{s[9 + (s+4)^2]}$$

Finally, using (*23.2*) with $a = 4$, we conclude that the required transform is

$$\frac{\pi}{2(s-4)^2} - \frac{1}{(s-4)^2} \arctan \frac{s}{3} + \frac{3}{(s-4)(s^2+9)}$$

Supplementary Problems

Find the Laplace transforms of the following functions.

23.15. $x^3 + 3 \cos 2x$.

23.16. $5e^{2x} + 7e^{-x}$.

23.17. $2x^2 \cosh x$.

23.18. $2x^2 e^{-x} \cosh x$.

23.19. $x^2 \sin 4x$.

23.20. $\sqrt{x}\, e^{2x}$.

23.21. $\int_0^x t \sinh t\, dt$.

23.22. $\int_0^x e^{3t} \cos t\, dt$.

23.23. $f(x)$ in Fig. 23-3, page 144.

23.24. $f(x)$ in Fig. 23-4, page 144.

23.25. $f(x)$ in Fig. 23-5, page 144.

Answers to Supplementary Problems

23.15. $\dfrac{6}{s^4} + \dfrac{3s}{s^2+4}$

23.16. $\dfrac{5}{s-2} + \dfrac{7}{s+1}$

23.17. $\dfrac{4s(s^2+3)}{(s^2-1)^3}$

23.18. $\dfrac{4(s+1)[(s+1)^2+3]}{[(s+1)^2-1]^3}$

23.19. $\dfrac{8(3s^2-16)}{(s^2+16)^3}$

23.20. $\dfrac{1}{2}\sqrt{\pi}\,(s-2)^{-3/2}$

23.21. $\dfrac{2}{(s^2-1)^2}$

23.22. $\dfrac{1}{s}\left[\dfrac{s-3}{(s-3)^2+1}\right]$

23.23. $\dfrac{1}{s(1+e^{-s})}$

23.24. $\dfrac{1-e^{-s}-se^{-2s}}{s^2(1-e^{-2s})}$

23.25. $\dfrac{(s+1)e^{-2s}+s-1}{s^2(1-e^{-2s})}$

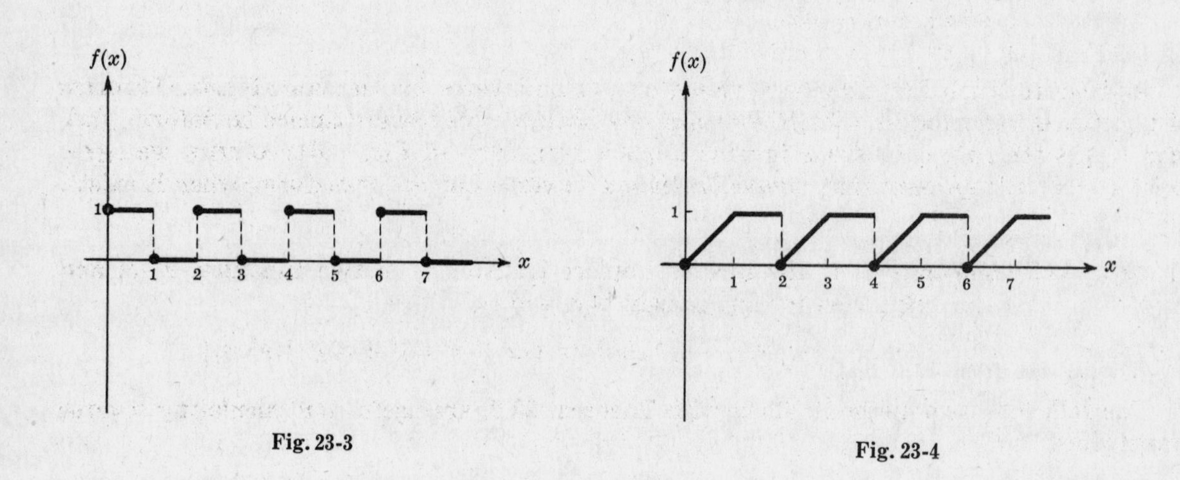

Fig. 23-3

Fig. 23-4

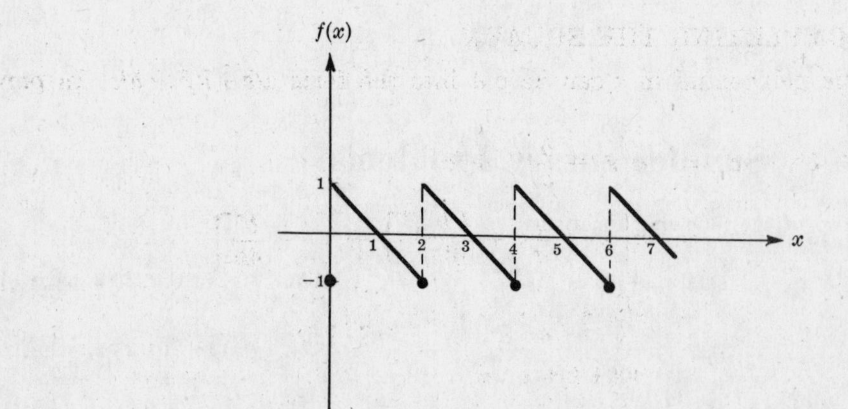

Fig. 23-5

Chapter 24

Inverse Laplace Transforms

24.1 DEFINITION. UNIQUENESS THEOREM

An *inverse Laplace transform* of a given function $F(s)$, designated by $\mathcal{L}^{-1}\{F(s)\}$, is another function $f(x)$ having the property that $\mathcal{L}\{f(x)\} = F(s)$.

Example 24.1. If $F(s) = 1/s$, then $\mathcal{L}^{-1}\{F(s)\} = 1$, since $\mathcal{L}\{1\} = 1/s$. If $F(s) = 1/(s^2+1)$, then $\mathcal{L}^{-1}\{F(s)\} = \sin x$, since $\mathcal{L}\{\sin x\} = 1/(s^2+1)$.

Theorem 24.1. If $\mathcal{L}\{f(x)\} \equiv \mathcal{L}\{g(x)\}$ and if $f(x)$ and $g(x)$ are both continuous in $0 \leqq x < \infty$, then $f(x) \equiv g(x)$.

A given function $F(s)$ may have many, one, or no inverse Laplace transforms. Theorem 24.1, however, guarantees that if $F(s)$ has one *continuous* inverse Laplace transform, $f(x)$, then $f(x)$ is the *only continuous* inverse Laplace transform of $F(s)$. Henceforth, we agree to let $\mathcal{L}^{-1}\{F(s)\}$ represent that *unique continuous* inverse Laplace transform, when it exists.

Theorem 24.2 (Linearity). If the inverse Laplace transforms of two functions $F_1(s)$ and $F_2(s)$ exist, then for any constants c_1 and c_2,

$$\mathcal{L}^{-1}\{c_1F_1(s) + c_2F_2(s)\} = c_1\mathcal{L}^{-1}\{F_1(s)\} + c_2\mathcal{L}^{-1}\{F_2(s)\}$$

The following two methods, along with Theorem 23.2, are useful for calculating inverse transforms.

24.2 METHOD OF COMPLETING THE SQUARE

Every real quadratic polynomial in s can be put into the form $a(s+k)^2 + h^2$. In particular,

$$\begin{aligned} as^2 + bs + c &= a\left(s^2 + \frac{b}{a}s\right) + c \\ &= a\left[s^2 + \frac{b}{a}s + \left(\frac{b}{2a}\right)^2\right] + \left[c - \frac{b^2}{4a}\right] \\ &= a\left(s + \frac{b}{2a}\right)^2 + \left(c - \frac{b^2}{4a}\right) \\ &\equiv a(s+k)^2 + h^2 \end{aligned}$$

where $k = \dfrac{b}{2a}$ and $h = \sqrt{c - \dfrac{b^2}{4a}}$. (See Problems 24.3 through 24.5.)

24.3 METHOD OF PARTIAL FRACTIONS

Every function of the form $a(s)/b(s)$, where both $a(s)$ and $b(s)$ are polynomials in s, can be reduced to the sum of other fractions such that the denominator of each new fraction is either a first-degree or a quadratic polynomial raised to some power. The method requires only that (1) the degree of $a(s)$ be less than the degree of $b(s)$ (if this is not the case, first perform long division, and consider the remainder term) and (2) $b(s)$ be factored into the product of distinct linear and quadratic polynomials raised to various powers.

The method is carried out as follows. To each factor of $b(s)$ of the form $(s-a)^m$, assign a sum of m fractions, of the form

$$\frac{A_1}{s-a} + \frac{A_2}{(s-a)^2} + \cdots + \frac{A_m}{(s-a)^m}$$

To each factor of $b(s)$ of the form $(s^2+bs+c)^p$, assign a sum of p fractions, of the form

$$\frac{B_1s+C_1}{s^2+bs+c} + \frac{B_2s+C_2}{(s^2+bs+c)^2} + \cdots + \frac{B_ps+C_p}{(s^2+bs+c)^p}$$

Here A_i, B_j, and C_k $(i=1,2,\ldots,m;\ j,k=1,2,\ldots,p)$ are constants which still must be determined.

Set the original fraction $a(s)/b(s)$ equal to the sum of the new fractions just constructed. Clear the resulting equation of fractions and then equate coefficients of like powers of s, thereby obtaining a set of simultaneous linear equations in the unknown constants A_i, B_j, and C_k. Finally, solve these equations for A_i, B_j and C_k. (See Problems 24.6 through 24.9.)

Solved Problems

24.1. Find $\mathcal{L}^{-1}\left\{\frac{2s}{(s^2+1)^2}\right\}$.

From Table 22-1, entry 12, with $a=1$,

$$\mathcal{L}\{x \sin x\} = \frac{2s}{(s^2+1)^2}$$

Therefore,

$$\mathcal{L}^{-1}\left\{\frac{2s}{(s^2+1)^2}\right\} = x \sin x$$

24.2. Find $\mathcal{L}^{-1}\{1/\sqrt{s}\}$.

From Table 22-1, entry 5, $\mathcal{L}\{1/\sqrt{x}\} = \sqrt{\pi}/\sqrt{s}$; hence $\mathcal{L}^{-1}\{\sqrt{\pi}/\sqrt{s}\} = 1/\sqrt{x}$. Then, using Theorem 24.2, we obtain

$$\mathcal{L}^{-1}\left\{\frac{1}{\sqrt{s}}\right\} = \frac{\sqrt{\pi}}{\sqrt{\pi}}\mathcal{L}^{-1}\left\{\frac{1}{\sqrt{s}}\right\} = \frac{1}{\sqrt{\pi}}\mathcal{L}^{-1}\left\{\frac{\sqrt{\pi}}{\sqrt{s}}\right\} = \frac{1}{\sqrt{\pi}}\frac{1}{\sqrt{x}}$$

24.3. Find $\mathcal{L}^{-1}\left\{\frac{1}{s^2-2s+9}\right\}$.

No function of this form appears in Table 22-1. But, by completing the square, we obtain

$$s^2 - 2s + 9 = (s^2-2s+1) + (9-1) = (s-1)^2 + (\sqrt{8})^2$$

Hence,

$$\frac{1}{s^2-2s+9} = \frac{1}{(s-1)^2+(\sqrt{8})^2} = \left(\frac{1}{\sqrt{8}}\right)\frac{\sqrt{8}}{(s-1)^2+(\sqrt{8})^2}$$

Then, using Theorem 24.2 and Table 22-1, entry 15, with $a = \sqrt{8}$ and $b = 1$, we find that

$$\mathcal{L}^{-1}\left\{\frac{1}{s^2 - 2s + 9}\right\} = \frac{1}{\sqrt{8}}\mathcal{L}^{-1}\left\{\frac{\sqrt{8}}{(s-1)^2 + (\sqrt{8})^2}\right\} = \frac{1}{\sqrt{8}}e^x \sin\sqrt{8}\,x$$

24.4. Find $\mathcal{L}^{-1}\left\{\dfrac{s+4}{s^2 + 4s + 8}\right\}$.

No function of this form appears in Table 22-1. Completing the square in the denominator, we have

$$s^2 + 4s + 8 = (s^2 + 4s + 4) + (8 - 4) = (s+2)^2 + (2)^2$$

Hence,

$$\frac{s+4}{s^2 + 4s + 8} = \frac{s+4}{(s+2)^2 + (2)^2}$$

This expression also is not found in Table 22-1. However, if we rewrite the numerator as $s + 4 = (s+2) + 2$ and then decompose the fraction, we have

$$\frac{s+4}{s^2 + 4s + 8} = \frac{s+2}{(s+2)^2 + (2)^2} + \frac{2}{(s+2)^2 + (2)^2}$$

Then, from entries 15 and 16 of Table 22-1,

$$\mathcal{L}^{-1}\left\{\frac{s+4}{s^2 + 4s + 8}\right\} = \mathcal{L}^{-1}\left\{\frac{s+2}{(s+2)^2 + (2)^2}\right\} + \mathcal{L}^{-1}\left\{\frac{2}{(s+2)^2 + (2)^2}\right\}$$

$$= e^{-2x}\cos 2x + e^{-2x}\sin 2x$$

24.5. Find $\mathcal{L}^{-1}\left\{\dfrac{s+2}{s^2 - 3s + 4}\right\}$.

No function of this form appears in Table 22-1. Completing the square in the denominator, we obtain

$$s^2 - 3s + 4 = \left(s^2 - 3s + \frac{9}{4}\right) + \left(4 - \frac{9}{4}\right) = \left(s - \frac{3}{2}\right)^2 + \left(\frac{\sqrt{7}}{2}\right)^2$$

so that

$$\frac{s+2}{s^2 - 3s + 4} = \frac{s+2}{\left(s - \frac{3}{2}\right)^2 + \left(\frac{\sqrt{7}}{2}\right)^2}$$

We now rewrite the numerator as

$$s + 2 = \left(s - \frac{3}{2}\right) + \frac{7}{2} = \left(s - \frac{3}{2}\right) + \sqrt{7}\left(\frac{\sqrt{7}}{2}\right)$$

so that

$$\frac{s+2}{s^2 - 3s + 4} = \frac{s - \frac{3}{2}}{\left(s - \frac{3}{2}\right)^2 + \left(\frac{\sqrt{7}}{2}\right)^2} + \sqrt{7}\,\frac{\frac{\sqrt{7}}{2}}{\left(s - \frac{3}{2}\right)^2 + \left(\frac{\sqrt{7}}{2}\right)^2}$$

Then,

$$\mathcal{L}^{-1}\left\{\frac{s+2}{s^2 - 3s + 4}\right\} = \mathcal{L}^{-1}\left\{\frac{s - \frac{3}{2}}{\left(s - \frac{3}{2}\right)^2 + \left(\frac{\sqrt{7}}{2}\right)^2}\right\} + \sqrt{7}\,\mathcal{L}^{-1}\left\{\frac{\frac{\sqrt{7}}{2}}{\left(s - \frac{3}{2}\right)^2 + \left(\frac{\sqrt{7}}{2}\right)^2}\right\}$$

$$= e^{(3/2)x}\cos\frac{\sqrt{7}}{2}x + \sqrt{7}\,e^{(3/2)x}\sin\frac{\sqrt{7}}{2}x$$

24.6. Use partial fractions to decompose $\dfrac{1}{(s+1)(s^2+1)}$.

To the linear factor $s+1$, we associate the fraction $A/(s+1)$; whereas to the quadratic factor s^2+1, we associate the fraction $(Bs+C)/(s^2+1)$. We then set

$$\frac{1}{(s+1)(s^2+1)} \equiv \frac{A}{s+1} + \frac{Bs+C}{s^2+1} \tag{1}$$

Clearing fractions, we obtain

$$1 \equiv A(s^2+1) + (Bs+C)(s+1) \tag{2}$$

or

$$s^2(0) + s(0) + 1 \equiv s^2(A+B) + s(B+C) + (A+C)$$

Equating coefficients of like powers of s, we conclude that $A+B=0$, $B+C=0$, and $A+C=1$. The solution of this set of equations is $A=\frac{1}{2}$, $B=-\frac{1}{2}$ and $C=\frac{1}{2}$. Substituting these values into (1), we obtain the partial-fractions decomposition

$$\frac{1}{(s+1)(s^2+1)} \equiv \frac{\frac{1}{2}}{s+1} + \frac{-\frac{1}{2}s+\frac{1}{2}}{s^2+1}$$

The following is an alternate procedure for finding the constants A, B, and C in (1). Since (2) must hold for all s, it must in particular hold for $s=-1$. Substituting this value into (2), we immediately find $A=\frac{1}{2}$. Equation (2) must also hold for $s=0$. Substituting this value along with $A=\frac{1}{2}$ into (2), we obtain $C=\frac{1}{2}$. Finally, substituting any other value of s into (2), we find that $B=-\frac{1}{2}$.

24.7. Use partial fractions to decompose $\dfrac{1}{(s^2+1)(s^2+4s+8)}$.

To the quadratic factors s^2+1 and s^2+4s+8, we associate the fractions $(As+B)/(s^2+1)$ and $(Cs+D)/(s^2+4s+8)$. We set

$$\frac{1}{(s^2+1)(s^2+4s+8)} \equiv \frac{As+B}{s^2+1} + \frac{Cs+D}{s^2+4s+8} \tag{1}$$

and clear fractions to obtain

$$1 \equiv (As+B)(s^2+4s+8) + (Cs+D)(s^2+1)$$

or

$$s^3(0) + s^2(0) + s(0) + 1 \equiv s^3(A+C) + s^2(4A+B+D) + s(8A+4B+C) + (8B+D)$$

Equating coefficients of like powers of s, we obtain

$$A+C=0,\quad 4A+B+D=0,\quad 8A+4B+C=0,\quad \text{and}\quad 8B+D=1$$

The solution of this set of equations is

$$A=-\frac{4}{65}\qquad B=\frac{7}{65}\qquad C=\frac{4}{65}\qquad D=\frac{9}{65}$$

Therefore,

$$\frac{1}{(s^2+1)(s^2+4s+8)} \equiv \frac{-\frac{4}{65}s+\frac{7}{65}}{s^2+1} + \frac{\frac{4}{65}s+\frac{9}{65}}{s^2+4s+8}$$

24.8. Use partial fractions to decompose $\dfrac{s+3}{(s-2)(s+1)}$.

To the linear factors $s-2$ and $s+1$, we associate respectively the fractions $A/(s-2)$ and $B/(s+1)$. We set

$$\frac{s+3}{(s-2)(s+1)} \equiv \frac{A}{s-2} + \frac{B}{s+1}$$

and, upon clearing fractions, obtain

$$s+3 \equiv A(s+1) + B(s-2) \tag{1}$$

To find A and B, we use the alternate procedure suggested in Problem 24.6. Substituting $s=-1$ and then $s=2$ into (1), we immediately obtain $A=5/3$ and $B=-2/3$. Thus,

$$\frac{s+3}{(s-2)(s+1)} \equiv \frac{5/3}{s-2} - \frac{2/3}{s+1}$$

24.9. Use partial fractions to decompose $\dfrac{8}{s^3(s^2-s-2)}$.

Note that s^2-s-2 factors into $(s-2)(s+1)$. To the factor $s^3=(s-0)^3$, which is a linear polynomial raised to the third power, we associate the sum $A_1/s+A_2/s^2+A_3/s^3$. To the linear factors $(s-2)$ and $(s+1)$, we associate the fractions $B/(s-2)$ and $C/(s+1)$. Then

$$\frac{8}{s^3(s^2-s-2)} \equiv \frac{A_1}{s} + \frac{A_2}{s^2} + \frac{A_3}{s^3} + \frac{B}{s-2} + \frac{C}{s+1}$$

or, clearing fractions,

$$8 \equiv A_1s^2(s-2)(s+1) + A_2s(s-2)(s+1)$$
$$+ A_3(s-2)(s+1) + Bs^3(s+1) + Cs^3(s-2)$$

Letting $s=-1$, 2, and 0, consecutively, we obtain, respectively, $C=\frac{8}{3}$, $B=\frac{1}{3}$, and $A_3=-4$. Then choosing $s=1$ and $s=-2$, and simplifying, we obtain the equations $A_1+A_2=-1$ and $2A_1-A_2=-8$, which have the solution $A_1=-3$ and $A_2=2$. Note that any other two values for s (not -1, 2, or 0) will also do; the resulting equations may be different, but the solution will be identical. Finally,

$$\frac{2}{s^3(s^2-s-2)} \equiv -\frac{3}{s} + \frac{2}{s^2} - \frac{4}{s^3} + \frac{1/3}{s-2} + \frac{8/3}{s+1}$$

24.10. Find $\mathcal{L}^{-1}\left\{\dfrac{s+3}{(s-2)(s+1)}\right\}$.

No function of this form appears in Table 22-1. Using the results of Problem 24.8 and Theorem 24.2, we obtain

$$\mathcal{L}^{-1}\left\{\frac{s+3}{(s-2)(s+1)}\right\} = \frac{5}{3}\mathcal{L}^{-1}\left\{\frac{1}{s-2}\right\} - \frac{2}{3}\mathcal{L}^{-1}\left\{\frac{1}{s+1}\right\}$$
$$= \frac{5}{3}e^{2x} - \frac{2}{3}e^{-x}$$

24.11. Find $\mathcal{L}^{-1}\left\{\dfrac{8}{s^3(s^2-s-2)}\right\}$.

No function of this form appears in Table 22-1. Using the results of Problem 24.9 and Theorem 24.2, we obtain

$$\mathcal{L}^{-1}\left\{\frac{8}{s^3(s^2-s-2)}\right\} = -3\mathcal{L}^{-1}\left\{\frac{1}{s}\right\} + 2\mathcal{L}^{-1}\left\{\frac{1}{s^2}\right\}$$
$$-2\mathcal{L}^{-1}\left\{\frac{2}{s^3}\right\} + \frac{1}{3}\mathcal{L}^{-1}\left\{\frac{1}{s-2}\right\} + \frac{8}{3}\mathcal{L}^{-1}\left\{\frac{1}{s+1}\right\}$$
$$= -3 + 2x - 2x^2 + \frac{1}{3}e^{2x} + \frac{8}{3}e^{-x}$$

24.12. Find $\mathcal{L}^{-1}\left\{\dfrac{1}{(s+1)(s^2+1)}\right\}$.

Using the results of Problem 24.6, and noting that

$$\frac{-\frac{1}{2}s+\frac{1}{2}}{s^2+1} = -\frac{1}{2}\left(\frac{s}{s^2+1}\right) + \frac{1}{2}\left(\frac{1}{s^2+1}\right)$$

we find that

$$\mathcal{L}^{-1}\left\{\frac{1}{(s+1)(s^2+1)}\right\} = \frac{1}{2}\mathcal{L}^{-1}\left\{\frac{1}{s+1}\right\} - \frac{1}{2}\mathcal{L}^{-1}\left\{\frac{s}{s^2+1}\right\} + \frac{1}{2}\mathcal{L}^{-1}\left\{\frac{1}{s^2+1}\right\}$$

$$= \frac{1}{2}e^{-x} - \frac{1}{2}\cos x + \frac{1}{2}\sin x$$

24.13. Find $\mathcal{L}^{-1}\left\{\dfrac{1}{(s^2+1)(s^2+4s+8)}\right\}$.

From Problem 24.7, we have

$$\mathcal{L}^{-1}\left\{\frac{1}{(s^2+1)(s^2+4s+8)}\right\} = \mathcal{L}^{-1}\left\{\frac{-\frac{4}{65}s+\frac{7}{65}}{s^2+1}\right\} + \mathcal{L}^{-1}\left\{\frac{\frac{4}{65}s+\frac{9}{65}}{s^2+4s+8}\right\}$$

The first term can be evaluated easily if we note that

$$\frac{-\frac{4}{65}s+\frac{9}{65}}{s^2+1} = \left(-\frac{4}{65}\right)\frac{s}{s^2+1} + \left(\frac{7}{65}\right)\frac{1}{s^2+1}$$

To evaluate the second term, we must first complete the square in the denominator, $s^2+4s+8 = (s+2)^2+(2)^2$, and then note that

$$\frac{\frac{4}{65}s+\frac{9}{65}}{s^2+4s+8} = \frac{4}{65}\left[\frac{s+2}{(s+2)^2+(2)^2}\right] + \frac{1}{130}\left[\frac{2}{(s+2)^2+(2)^2}\right]$$

Therefore,

$$\mathcal{L}^{-1}\left\{\frac{1}{(s^2+1)(s^2+4s+8)}\right\}$$

$$= -\frac{4}{65}\mathcal{L}^{-1}\left\{\frac{s}{s^2+1}\right\} + \frac{7}{65}\mathcal{L}^{-1}\left\{\frac{1}{s^2+1}\right\}$$

$$+ \frac{4}{65}\mathcal{L}^{-1}\left\{\frac{s+2}{(s+2)^2+(2)^2}\right\} + \frac{1}{130}\mathcal{L}^{-1}\left\{\frac{2}{(s+2)^2+(2)^2}\right\}$$

$$= -\frac{4}{65}\cos x + \frac{7}{65}\sin x + \frac{4}{65}e^{-2x}\cos 2x + \frac{1}{130}e^{-2x}\sin 2x$$

24.14. Find $\mathcal{L}^{-1}\left\{\dfrac{1}{s(s^2+4)}\right\}$.

By the method of partial fractions, we obtain

$$\frac{1}{s(s^2+4)} \equiv \frac{1/4}{s} + \frac{(-1/4)s}{s^2+4}$$

Thus, $$\mathcal{L}^{-1}\left\{\frac{1}{s(s^2+4)}\right\} = \frac{1}{4}\mathcal{L}^{-1}\left\{\frac{1}{s}\right\} - \frac{1}{4}\mathcal{L}^{-1}\left\{\frac{s}{s^2+4}\right\} = \frac{1}{4} - \frac{1}{4}\cos 2x$$

Supplementary Problems

24.15. Using Table 22-1, find the inverse Laplace transform of

(a) $\dfrac{1}{s+2}$ (c) $\dfrac{2}{(s-2)^2+9}$ (e) $\dfrac{2s+1}{(s-1)^2+7}$

(b) $\dfrac{1}{s^2+4}$ (d) $\dfrac{s}{(s+1)^2+5}$ (f) $\dfrac{1}{2s^2+1}$

24.16. By first completing the square, find the inverse Laplace transform of

(a) $\dfrac{1}{s^2-2s+2}$ (b) $\dfrac{s+3}{s^2+2s+5}$ (c) $\dfrac{s}{s^2-s+17/4}$ (d) $\dfrac{s+1}{s^2+3s+5}$

24.17. Use the method of partial fractions to decompose

(a) $\dfrac{2s^2}{(s-1)(s^2+1)}$ (b) $\dfrac{1}{s^2-1}$ (c) $\dfrac{2}{(s^2+1)(s-1)^2}$

24.18. Find the inverse Laplace transforms of the functions given in Problem 24.17.

24.19. Find the inverse Laplace transforms of

(a) $\dfrac{2s-13}{s(s^2-4s+13)}$

(b) $\dfrac{2(s-1)}{s^2-s+1}$

(c) $\dfrac{s}{(s^2+9)^2}$

(d) $\dfrac{1}{2(s-1)(s^2-s-1)} = \dfrac{1/2}{(s-1)(s^2-s-1)}$

(e) $\dfrac{s}{2s^2+4s+5/2} = \dfrac{\frac{1}{2}s}{s^2+2s+5/4}$

Answers to Supplementary Problems

24.15. (a) e^{-2x}

(b) $\dfrac{1}{2}\sin 2x$

(c) $\dfrac{2}{3}e^{2x}\sin 3x$

(d) $e^{-x}\cos\sqrt{5}\,x - \dfrac{1}{\sqrt{5}}e^{-x}\sin\sqrt{5}\,x$

(e) $2e^{x}\cos\sqrt{7}\,x + \dfrac{3}{\sqrt{7}}e^{x}\sin\sqrt{7}\,x$

(f) $\dfrac{1}{\sqrt{2}}\sin\dfrac{1}{\sqrt{2}}x$

24.16. (a) $e^{x}\sin x$

(b) $e^{-x}\cos 2x + e^{-x}\sin 2x$

(c) $e^{(1/2)x}\cos 2x + \dfrac{1}{4}e^{(1/2)x}\sin 2x$

(d) $e^{-(3/2)x}\cos\dfrac{\sqrt{11}}{2}x - \dfrac{1}{\sqrt{11}}e^{-(3/2)x}\sin\dfrac{\sqrt{11}}{2}x$

24.17. (a) $\dfrac{1}{s-1} + \dfrac{s+1}{s^2+1}$ (b) $\dfrac{\frac{1}{2}}{s-1} + \dfrac{-\frac{1}{2}}{s+1}$ (c) $\dfrac{s}{s^2+1} + \dfrac{-1}{s-1} + \dfrac{1}{(s-1)^2}$

24.18. (a) $e^{x} + \cos x + \sin x$ (b) $\dfrac{1}{2}e^{x} - \dfrac{1}{2}e^{-x}$ (c) $\cos x - e^{x} + xe^{x}$

24.19. (a) $-1 + e^{2x}\cos 3x$

(b) $2e^{(1/2)x}\cos\dfrac{\sqrt{3}}{2}x - \dfrac{2}{\sqrt{3}}e^{(1/2)x}\sin\dfrac{\sqrt{3}}{2}x$

(c) $\dfrac{1}{6}x\sin 3x$ (Table 22-1, entry 12)

(d) $-\dfrac{1}{2}e^{x} + \dfrac{1}{2}e^{(1/2)x}\cosh\dfrac{\sqrt{5}}{2}x + \dfrac{1}{2\sqrt{5}}e^{(1/2)x}\sinh\dfrac{\sqrt{5}}{2}x$

(e) $\dfrac{1}{2}e^{-x}\cos\dfrac{1}{2}x - e^{-x}\sin\dfrac{1}{2}x$

Chapter 25

Convolutions and the Unit Step Function

25.1 CONVOLUTIONS

Let $f(x) \in E_\alpha$ and $g(x) \in E_\alpha$ (see Definition in Chapter 23). The *convolution* (or *faltung*) of $f(x)$ and $g(x)$ is

$$f(x) * g(x) = \int_0^x f(t)\, g(x-t)\, dt \tag{25.1}$$

Example 25.1. If $f(x) = e^{3x}$ and $g(x) = e^{2x}$, then $f(t) = e^{3t}$, $g(x-t) = e^{2(x-t)}$, and

$$f(x) * g(x) = \int_0^x e^{3t}e^{2(x-t)}\, dt = \int_0^x e^{3t}e^{2x}e^{-2t}\, dt$$

$$= e^{2x}\int_0^x e^t\, dt = e^{2x}\Big[e^t\Big]_{t=0}^{t=x} = e^{2x}(e^x - 1) = e^{3x} - e^{2x}$$

Theorem 25.1. $f(x) * g(x) = g(x) * f(x)$.
(See Problem 25.3.)

Theorem 25.2 (Convolution theorem). If $\mathcal{L}\{f(x)\} = F(s)$ and $\mathcal{L}\{g(x)\} = G(s)$, then

$$\mathcal{L}\{f(x) * g(x)\} = \mathcal{L}\{f(x)\}\,\mathcal{L}\{g(x)\} = F(s)\, G(s)$$

For applications, it is useful to express Theorem 25.2 as

$$\mathcal{L}^{-1}\{F(s)\, G(s)\} = f(x) * g(x) \tag{25.2}$$

It is sometimes easier to evaluate $g(x) * f(x)$ than $f(x) * g(x)$. One then rewrites (*25.2*), using Theorem 25.1, as $\mathcal{L}^{-1}\{F(s)\, G(s)\} = g(x) * f(x)$. (See Problem 25.4.)

25.2 UNIT STEP FUNCTION

The *unit step function* $u(x)$ is defined as

$$u(x) = \begin{cases} 0 & x < 0 \\ 1 & x \geq 0 \end{cases}$$

As an immediate consequence of the definition, we have for any number c,

$$u(x-c) = \begin{cases} 0 & x < c \\ 1 & x \geq c \end{cases}$$

The graph of $u(x-c)$ is given in Fig. 25-1.

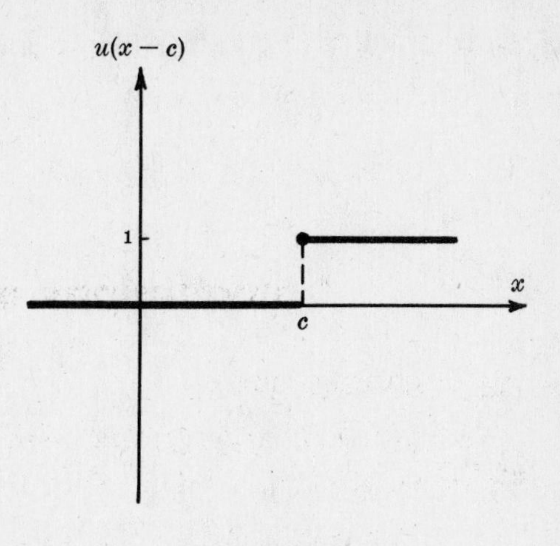

Fig. 25-1

Theorem 25.3. $\mathcal{L}\{u(x-c)\} = \frac{1}{s}e^{-cs}$.
(See Problem 25.7.)

Given a function $f(x)$ defined for $x \geqq 0$, the function

$$u(x-c)\,f(x-c) = \begin{cases} 0 & x < c \\ f(x-c) & x \geqq c \end{cases}$$

represents a shift, or translation, of the function $f(x)$ by c units in the positive x-direction. For example, if $f(x)$ is given graphically by Fig. 25-2, then $u(x-c)\,f(x-c)$ is given graphically by Fig. 25-3.

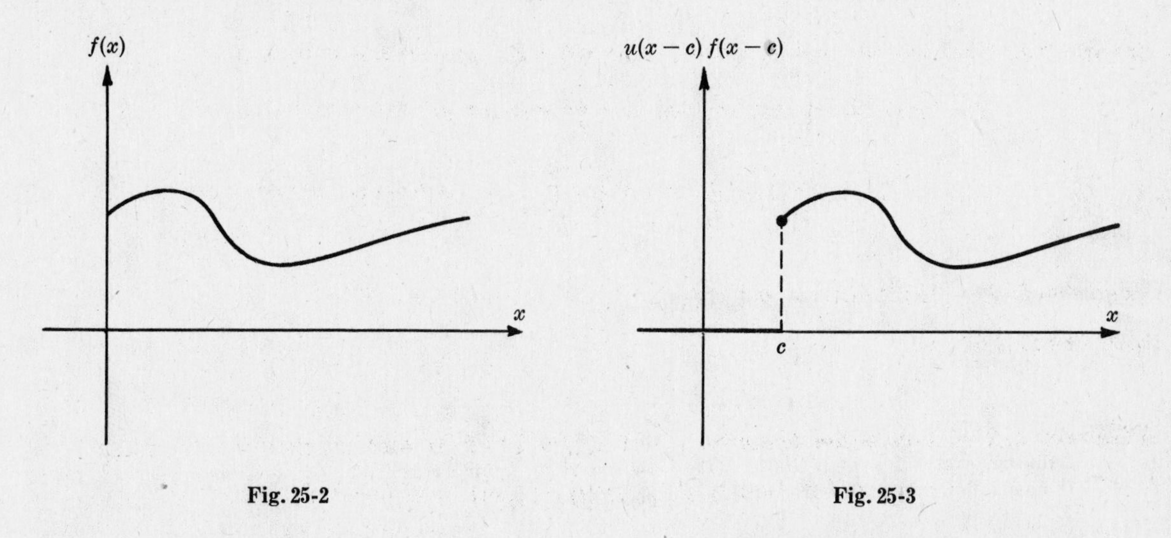

Fig. 25-2 Fig. 25-3

Theorem 25.4. If $f(x) \in E_\alpha$, and $F(s) = \mathcal{L}\{f(x)\}$, then $\mathcal{L}\{u(x-c)\,f(x-c)\} = e^{-cs}F(s)$. Conversely,

$$\mathcal{L}^{-1}\{e^{-cs}F(s)\} = u(x-c)\,f(x-c) = \begin{cases} 0 & x < c \\ f(x-c) & x \geqq c \end{cases}$$

Solved Problems

25.1. For $g(x)$ and $f(x)$ as in Example 25.1, find $g(x) * f(x)$ and thus verify Theorem 25.1.

With $f(x-t) = e^{3(x-t)}$ and $g(t) = e^{2t}$,

$$\begin{aligned} g(x) * f(x) &= \int_0^x g(t)\,f(x-t)\,dt = \int_0^x e^{2t}e^{3(x-t)}\,dt \\ &= e^{3x}\int_0^x e^{-t}\,dt = e^{3x}\Big[-e^{-t}\Big]_{t=0}^{t=x} \\ &= e^{3x}(-e^{-x}+1) = e^{3x} - e^{2x} \end{aligned}$$

which, from Example 25.1, equals $f(x) * g(x)$.

25.2. Find $f(x) * g(x)$ if $f(x) = x$ and $g(x) = x^2$.

Here $f(t) = t$ and $g(x-t) = (x-t)^2 = x^2 - 2xt + t^2$. Thus,

$$f(x) * g(x) = \int_0^x t(x^2 - 2xt + t^2)\,dt$$

$$= x^2 \int_0^x t\,dt - 2x \int_0^x t^2\,dt + \int_0^x t^3\,dt$$

$$= x^2 \frac{x^2}{2} - 2x\frac{x^3}{3} + \frac{x^4}{4} = \frac{1}{12}x^4$$

25.3. Prove Theorem 25.1.

Making the substitution $\tau = x - t$ in the right-hand side of (*25.1*), we have

$$f(x) * g(x) = \int_0^x f(t)\,g(x-t)\,dt = \int_x^0 f(x-\tau)\,g(\tau)\,(-d\tau)$$

$$= -\int_x^0 g(\tau)\,f(x-\tau)\,d\tau = \int_0^x g(\tau)\,f(x-\tau)\,d\tau$$

$$= g(x) * f(x)$$

25.4. Find $\mathcal{L}^{-1}\left\{\frac{1}{s(s^2+4)}\right\}$ by convolutions.

Note that

$$\frac{1}{s(s^2+4)} = \frac{1}{s}\,\frac{1}{s^2+4}$$

Defining $F(s) = 1/s$ and $G(s) = 1/(s^2+4)$, we have, from Table 22-1, $f(x) = 1$ and $g(x) = \frac{1}{2}\sin 2x$. It now follows from Theorems 25.2 and 25.1 that

$$\mathcal{L}^{-1}\left\{\frac{1}{s(s^2+4)}\right\} = \mathcal{L}^{-1}\{F(s)\,G(s)\} = g(x) * f(x)$$

$$= \int_0^x g(t)\,f(x-t)\,dt = \int_0^x \left(\frac{1}{2}\sin 2t\right)(1)\,dt$$

$$= \frac{1}{4}(1 - \cos 2x)$$

See also Problem 24.14.

25.5. Find $\mathcal{L}^{-1}\left\{\frac{1}{(s-1)^2}\right\}$ by convolutions.

If we define $F(s) = G(s) = 1/(s-1)$, then $f(x) = g(x) = e^x$ and

$$\mathcal{L}^{-1}\left\{\frac{1}{(s-1)^2}\right\} = \mathcal{L}^{-1}\{F(s)\,G(s)\} = f(x) * g(x)$$

$$= \int_0^x f(t)\,g(x-t)\,dt = \int_0^x e^t e^{x-t}\,dt$$

$$= e^x \int_0^x (1)\,dt = xe^x$$

25.6. Prove that $f(x) * [g(x) + h(x)] = f(x) * g(x) + f(x) * h(x)$.

$$\begin{aligned} f(x) * [g(x) + h(x)] &= \int_0^x f(t)\,[g(x-t) + h(x-t)]\,dt \\ &= \int_0^x [f(t)\,g(x-t) + f(t)\,h(x-t)]\,dt \\ &= \int_0^x f(t)\,g(x-t)\,dt + \int_0^x f(t)\,h(x-t)\,dt \\ &= f(x) * g(x) + f(x) * h(x) \end{aligned}$$

25.7. Prove Theorem 25.3.

$$\begin{aligned} \mathcal{L}\{u(x-c)\} &= \int_0^\infty e^{-sx} u(x-c)\,dx = \int_0^c e^{-sx}(0)\,dx + \int_c^\infty e^{-sx}(1)\,dx \\ &= \int_c^\infty e^{-sx}\,dx = \lim_{R\to\infty} \int_c^R e^{-sx}\,dx = \lim_{R\to\infty} \frac{e^{-sR} - e^{-sc}}{-s} \\ &= \frac{1}{s} e^{-sc} \qquad (\text{if } s > 0) \end{aligned}$$

25.8. Graph the function $f(x) = u(x-2) - u(x-3)$.

Note that

$$u(x-2) = \begin{cases} 0 & x < 2 \\ 1 & x \geqq 2 \end{cases} \qquad \text{and} \qquad u(x-3) = \begin{cases} 0 & x < 3 \\ 1 & x \geqq 3 \end{cases}$$

Thus,
$$f(x) = u(x-2) - u(x-3) = \begin{cases} 0 - 0 = 0 & x < 2 \\ 1 - 0 = 1 & 2 \leqq x < 3 \\ 1 - 1 = 0 & x \geqq 3 \end{cases}$$

the graph of which is given in Fig. 25-4.

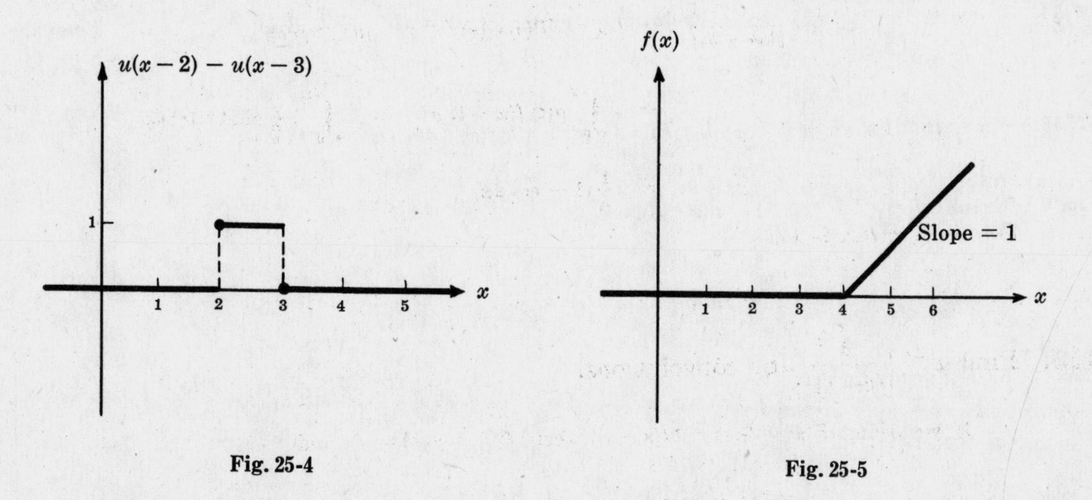

Fig. 25-4

Fig. 25-5

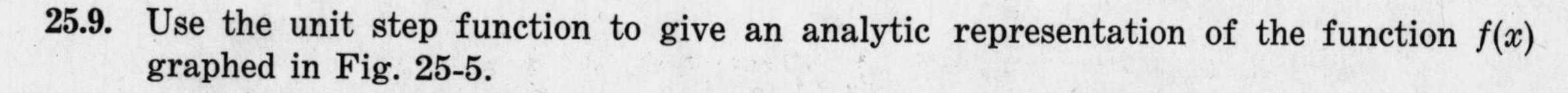

25.9. Use the unit step function to give an analytic representation of the function $f(x)$ graphed in Fig. 25-5.

Note that $f(x)$ is the function $g(x) = x$, $x \geqq 0$, translated four units in the positive x-direction. Thus, $f(x) = u(x-4)g(x-4) = u(x-4)(x-4)$.

25.10. Find $\mathcal{L}\{g(x)\}$ if $g(x) = \begin{cases} 0 & x < 4 \\ (x-4)^2 & x \geqq 4 \end{cases}$.

If we define $f(x) = x^2$, then $g(x)$ can be given compactly as $g(x) = u(x-4)\,f(x-4) = u(x-4)(x-4)^2$. Then, noting that $\mathcal{L}\{f(x)\} = F(s) = 2/s^3$ and using Theorem 25.4, we conclude that

$$\mathcal{L}\{g(x)\} = \mathcal{L}\{u(x-4)(x-4)^2\} = e^{-4s}\frac{2}{s^3}$$

25.11. Find $\mathcal{L}\{g(x)\}$ if $g(x) = \begin{cases} 0 & x < 4 \\ x^2 & x \geqq 4 \end{cases}$.

We first determine a function $f(x)$ such that $f(x-4) = x^2$. Once this has been done, $g(x)$ can be written as $g(x) = u(x-4)f(x-4)$ and Theorem 25.4 can be applied. Now, $f(x-4) = x^2$ only if

$$f(x) = f(x+4-4) = (x+4)^2 = x^2 + 8x + 16$$

Since

$$\mathcal{L}\{f(x)\} = \mathcal{L}\{x^2\} + 8\,\mathcal{L}\{x\} + 16\,\mathcal{L}\{1\} = \frac{2}{s^3} + \frac{8}{s^2} + \frac{16}{s}$$

it follows that

$$\mathcal{L}\{g(x)\} = \mathcal{L}\{u(x-4)\,f(x-4)\} = e^{-4s}\left(\frac{2}{s^3} + \frac{8}{s^2} + \frac{16}{s}\right)$$

Supplementary Problems

25.12. Determine $f(x) * g(x)$ and $g(x) * f(x)$ if $f(x) = 4x$ and $g(x) = e^{2x}$.

25.13. Use convolutions to find the inverse Laplace transform of

(a) $\dfrac{1}{(s-1)(s-2)}$ (b) $\dfrac{1}{(s)(s)}$ (c) $\dfrac{2}{s(s+1)}$

25.14. Find $\mathcal{L}^{-1}\left\{\dfrac{1}{s(s^2+4)}\right\}$ by convolutions, taking $F(s) = 1/s^2$ and $G(s) = s/(s^2+4)$. Compare your result with Problem 25.4.

25.15. Prove that for any constant k, $[kf(x)] * g(x) = k[f(x) * g(x)]$.

25.16. Graph $f(x) = 2u(x-2) - u(x-4)$.

25.17. Graph $f(x) = u(x-2) - 2u(x-3) + u(x-4)$.

25.18. Find $\mathcal{L}\{g(x)\}$ for

(a) $g(x) = \begin{cases} 0 & x < 1 \\ \sin(x-1) & x \geqq 1 \end{cases}$ (b) $g(x) = \begin{cases} 0 & x < 2 \\ x^3 + 1 & x \geqq 2 \end{cases}$

25.19. Determine $\mathcal{L}^{-1}\{F(s)\} = f(x)$ for

(a) $F(s) = \dfrac{s}{s^2+4}e^{-\pi s}$ (b) $F(s) = \dfrac{1}{s^3}e^{-s}$

25.20. Graph the functions $f(x)$ found in Problem 25.19.

Answers to Supplementary Problems

25.12. $e^{2x} - (2x+1)$

25.13. (a) $e^{2x} - e^{x}$ (b) x (c) $2(1-e^{-x})$

25.16. See Fig. 25-6.

25.17. See Fig. 25-7.

25.18. (a) $e^{-s}\dfrac{1}{s^2+1}$

(b) $g(x) = u(x-2)\,f(x-2)$ if

$f(x) = x^3 + 6x^2 + 12x + 9;$

$$G(s) = e^{-2s}\left(\frac{6}{s^4} + \frac{12}{s^3} + \frac{12}{s^2} + \frac{9}{s}\right)$$

25.19. (a) $u(x-\pi)\cos 2(x-\pi)$ (b) $\dfrac{1}{2}(x-1)^2 u(x-1)$

25.20. (a) See Fig. 25-8. (b) See Fig. 25-9.

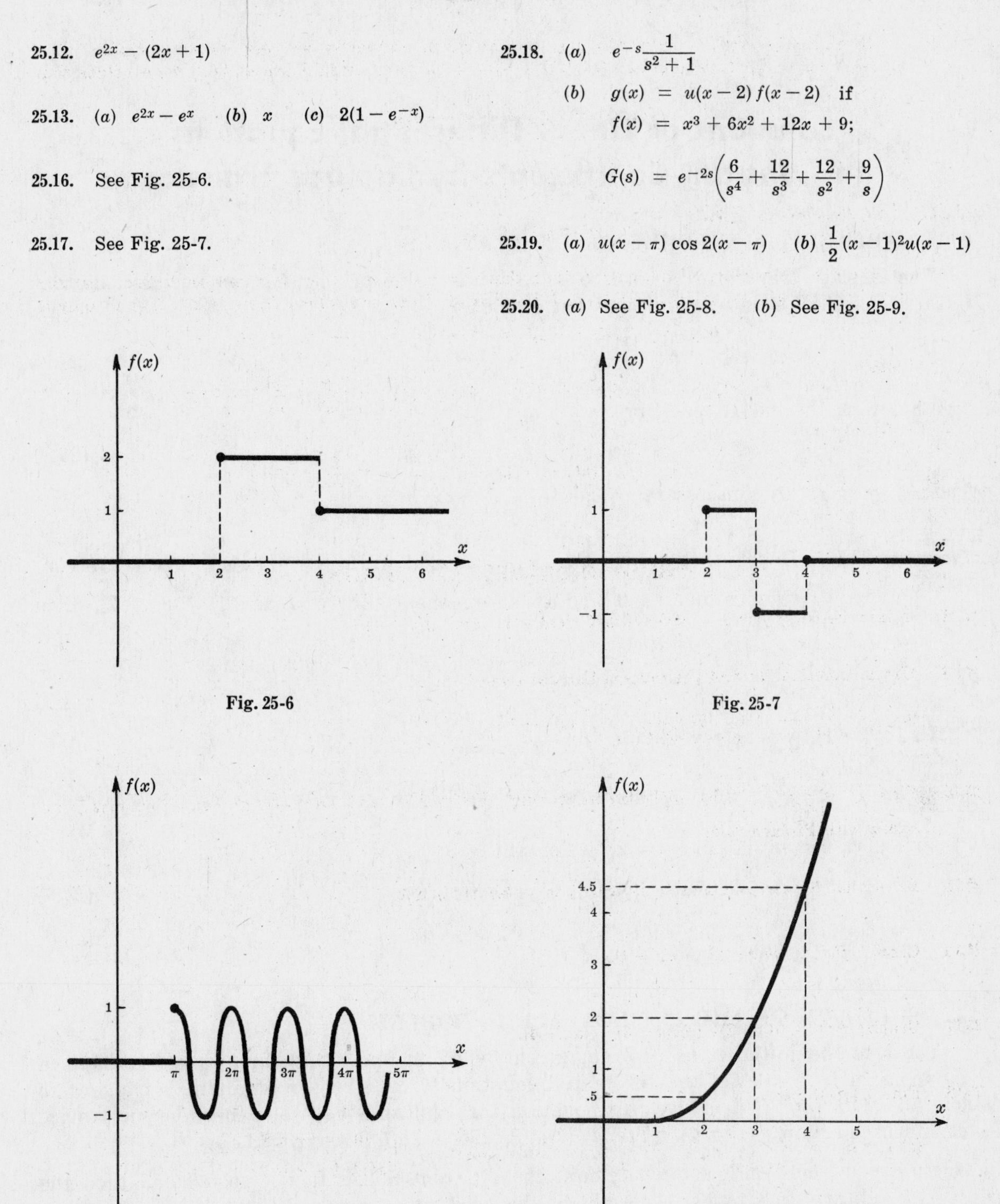

Fig. 25-6

Fig. 25-7

Fig. 25-8

Fig. 25-9

Chapter 26

Solutions of Linear Differential Equations with Constant Coefficients by Laplace Transforms

26.1 LAPLACE TRANSFORMS OF DERIVATIVES

The Laplace transform is used to solve initial-value problems given by the nth-order linear differential equation with constant coefficients

$$b_n \frac{d^n y}{dx^n} + b_{n-1}\frac{d^{n-1}y}{dx^{n-1}} + \cdots + b_1 \frac{dy}{dx} + b_0 y = g(x) \tag{26.1}$$

together with the initial conditions

$$y(0) = c_0, \quad y'(0) = c_1, \quad \ldots, \quad y^{(n-1)}(0) = c_{n-1} \tag{26.2}$$

The following result is necessary.

Theorem 26.1. Denote $\mathcal{L}\{y(x)\}$ by $Y(s)$. If $y(x)$ and its first $n-1$ derivatives are continuous for $x \geqq 0$ and are of exponential order α, and if $\dfrac{d^n y}{dx^n} \in E_\alpha$, then

$$\mathcal{L}\left\{\frac{d^n y}{dx^n}\right\} = s^n Y(s) - s^{n-1}y(0) - s^{n-2}y'(0) - \cdots - sy^{(n-2)}(0) - y^{(n-1)}(0) \tag{26.3}$$

Using (*26.2*), we can rewrite (*26.3*) as

$$\mathcal{L}\left\{\frac{d^n y}{dx^n}\right\} = s^n Y(s) - c_0 s^{n-1} - c_1 s^{n-2} - \cdots - c_{n-2}s - c_{n-1} \tag{26.4}$$

In particular, for $n = 1$ and $n = 2$, we obtain

$$\mathcal{L}\{y'(x)\} = sY(s) - c_0 \tag{26.5}$$

$$\mathcal{L}\{y''(x)\} = s^2 Y(s) - c_0 s - c_1 \tag{26.6}$$

26.2 SOLUTION OF THE INITIAL-VALUE PROBLEM

To solve the initial-value problem given by (*26.1*) and (*26.2*), first take the Laplace transform of both sides of the differential equation (*26.1*), thereby obtaining an algebraic equation involving $Y(s)$. Solve this equation for $Y(s)$ and then take the inverse Laplace transform to obtain $y(x) = \mathcal{L}^{-1}\{Y(s)\}$. (See Problems 26.1 through 26.12.)

Unlike previous methods, where first the differential equation is solved and then the initial conditions are applied to evaluate the arbitrary constants, the Laplace transform method usually solves the entire initial-value problem in one step. One exception is when the initial conditions are not given at $x = 0$. A procedure for handling this case will be found in Problems 26.3 and 26.9.

Solved Problems

26.1. Solve $y' - 5y = 0;\ y(0) = 2.$

Taking the Laplace transform of both sides of this differential equation and using Theorem 23.1, we obtain $\mathcal{L}\{y'\} - 5\,\mathcal{L}\{y\} = \mathcal{L}\{0\}$. Then, using (26.5) with $c_0 = 2$, we find

$$[sY(s) - 2] - 5Y(s) = 0 \qquad \text{from which} \qquad Y(s) = \frac{2}{s-5}$$

Finally, taking the inverse Laplace transform of $Y(s)$, we obtain

$$y(x) = \mathcal{L}^{-1}\{Y(s)\} = \mathcal{L}^{-1}\left\{\frac{2}{s-5}\right\} = 2\,\mathcal{L}^{-1}\left\{\frac{1}{s-5}\right\} = 2e^{5x}$$

26.2. Solve $y' - 5y = e^{5x};\ y(0) = 0.$

Taking the Laplace transform of both sides of this differential equation and using Theorem 23.1, we find that $\mathcal{L}\{y'\} - 5\,\mathcal{L}\{y\} = \mathcal{L}\{e^{5x}\}$. Then, using Table 22-1 and (26.5) with $c_0 = 0$, we obtain

$$[sY(s) - 0] - 5Y(s) = \frac{1}{s-5} \qquad \text{from which} \qquad Y(s) = \frac{1}{(s-5)^2}$$

Finally, taking the inverse Laplace transform of $Y(s)$, we obtain

$$y(x) = \mathcal{L}^{-1}\{Y(s)\} = \mathcal{L}^{-1}\left\{\frac{1}{(s-5)^2}\right\} = xe^{5x}$$

(see Table 22-1, entry 14).

26.3. Solve $y' - 5y = 0;\ y(\pi) = 2.$

Taking the Laplace transform of both sides of the differential equation, we obtain

$$\mathcal{L}\{y'\} - 5\,\mathcal{L}\{y\} = \mathcal{L}\{0\}$$

Then, using (26.5) with $c_0 = y(0)$ kept arbitrary, we have

$$[sY(s) - c_0] - 5Y(s) = 0 \qquad \text{or} \qquad Y(s) = \frac{c_0}{s-5}$$

Taking the inverse Laplace transform, we find that

$$y(x) = \mathcal{L}^{-1}\{Y(s)\} = c_0\,\mathcal{L}^{-1}\left\{\frac{1}{s-5}\right\} = c_0 e^{5x}$$

Now we use the initial condition to solve for c_0 (see Chapter 16). The result is $c_0 = 2e^{-5\pi}$, and $y(x) = 2e^{5(x-\pi)}$.

26.4. Solve $y' + y = \sin x;\ y(0) = 1.$

Taking the Laplace transform of both sides of the differential equation, we obtain

$$\mathcal{L}\{y'\} + \mathcal{L}\{y\} = \mathcal{L}\{\sin x\} \qquad \text{or} \qquad [sY(s) - 1] + Y(s) = \frac{1}{s^2+1}$$

Solving for $Y(s)$, we find

$$Y(s) = \frac{1}{(s+1)(s^2+1)} + \frac{1}{s+1}$$

Taking the inverse Laplace transform, and using the result of Problem 24.12, we obtain

$$y(x) = \mathcal{L}^{-1}\{Y(s)\} = \mathcal{L}^{-1}\left\{\frac{1}{(s+1)(s^2+1)}\right\} + \mathcal{L}^{-1}\left\{\frac{1}{s+1}\right\}$$

$$= \left(\frac{1}{2}e^{-x} - \frac{1}{2}\cos x + \frac{1}{2}\sin x\right) + e^{-x} = \frac{3}{2}e^{-x} - \frac{1}{2}\cos x + \frac{1}{2}\sin x$$

26.5. Solve $y'' + 4y = 0;\ y(0) = 2,\ y'(0) = 2.$

Taking Laplace transforms, we have $\mathcal{L}\{y''\} + 4\,\mathcal{L}\{y\} = \mathcal{L}\{0\}$. Then, using *(26.6)* with $c_0 = 2$ and $c_1 = 2$, we obtain

$$[s^2Y(s) - 2s - 2] + 4Y(s) = 0$$

or

$$Y(s) = \frac{2s+2}{s^2+4} = \frac{2s}{s^2+4} + \frac{2}{s^2+4}$$

Finally, taking the inverse Laplace transform, we obtain

$$y(x) = \mathcal{L}^{-1}\{Y(s)\} = 2\,\mathcal{L}^{-1}\left\{\frac{s}{s^2+4}\right\} + \mathcal{L}^{-1}\left\{\frac{2}{s^2+4}\right\} = 2\cos 2x + \sin 2x$$

(See Problem 12.5.)

26.6. Solve $y'' - 3y' + 4y = 0;\ y(0) = 1,\ y'(0) = 5.$

Taking Laplace transforms, we obtain $\mathcal{L}\{y''\} - 3\,\mathcal{L}\{y'\} + 4\,\mathcal{L}\{y\} = \mathcal{L}\{0\}$. Then, using *both (26.5) and (26.6)* with $c_0 = 1$ and $c_1 = 5$, we have

$$[s^2Y(s) - s - 5] - 3[sY(s) - 1] + 4Y(s) = 0$$

or

$$Y(s) = \frac{s+2}{s^2 - 3s + 4}$$

Finally, taking the inverse Laplace transform and using the result of Problem 24.5, we obtain

$$y(x) = e^{(3/2)x}\cos\frac{\sqrt{7}}{2}x + \sqrt{7}\,e^{(3/2)x}\sin\frac{\sqrt{7}}{2}x$$

26.7. Solve $y'' - y' - 2y = 4x^2;\ y(0) = 1,\ y'(0) = 4.$

Taking Laplace transforms, we have $\mathcal{L}\{y''\} - \mathcal{L}\{y'\} - 2\,\mathcal{L}\{y\} = 4\,\mathcal{L}\{x^2\}$. Then, using *both (26.5) and (26.6)* with $c_0 = 1$ and $c_1 = 4$, we obtain

$$[s^2Y(s) - s - 4] - [sY(s) - 1] - 2Y(s) = \frac{8}{s^3}$$

or, upon solving for $Y(s)$,

$$Y(s) = \frac{s+3}{s^2 - s - 2} + \frac{8}{s^3(s^2 - s - 2)}$$

Finally, taking the inverse Laplace transform and using the results of Problems 24.10 and 24.11, we obtain

$$\begin{aligned} y(x) &= \left(\frac{5}{3}e^{2x} - \frac{2}{3}e^{-x}\right) + \left(-3 + 2x - 2x^2 + \frac{1}{3}e^{2x} + \frac{8}{3}e^{-x}\right) \\ &= 2e^{2x} + 2e^{-x} - 2x^2 + 2x - 3 \end{aligned}$$

(See Problem 16.1.)

26.8. Solve $y'' + 4y' + 8y = \sin x;\ y(0) = 1,\ y'(0) = 0.$

Taking Laplace transforms, we obtain $\mathcal{L}\{y''\} + 4\,\mathcal{L}\{y'\} + 8\,\mathcal{L}\{y\} = \mathcal{L}\{\sin x\}$. Since $c_0 = 1$ and $c_1 = 0$, this becomes

$$[s^2Y(s) - s - 0] + 4[sY(s) - 1] + 8Y(s) = \frac{1}{s^2+1}$$

Thus,

$$Y(s) = \frac{s+4}{s^2+4s+8} + \frac{1}{(s^2+1)(s^2+4s+8)}$$

Finally, taking the inverse Laplace transform and using the results of Problems 24.4 and 24.13, we obtain

$$\begin{aligned} y(x) &= (e^{-2x}\cos 2x + e^{-2x}\sin 2x) \\ &\quad + \left(-\frac{4}{65}\cos x + \frac{7}{65}\sin x + \frac{4}{65}e^{-2x}\cos 2x + \frac{1}{130}e^{-2x}\sin 2x\right) \\ &= e^{-2x}\left(\frac{69}{65}\cos 2x + \frac{131}{130}\sin 2x\right) + \frac{7}{65}\sin x - \frac{4}{65}\cos x \end{aligned}$$

(See Problem 16.3.)

26.9. Solve $y'' - 3y' + 2y = e^{-x};\ y(1) = 0,\ y'(1) = 0.$

Taking Laplace transforms, we have $\mathcal{L}\{y''\} - 3\mathcal{L}\{y'\} + 2\mathcal{L}\{y\} = \mathcal{L}\{e^{-x}\}$, or

$$[s^2Y(s) - sc_0 - c_1] - 3[sY(s) - c_0] + 2[Y(s)] = 1/(s+1)$$

Here c_0 and c_1 must remain arbitrary, since they represent $y(0)$ and $y'(0)$, respectively, which are still unknown. Thus,

$$Y(s) = c_0\frac{s-3}{s^2-3s+2} + c_1\frac{1}{s^2-3s+2} + \frac{1}{(s+1)(s^2-3s+2)}$$

Using the method of partial fractions (Section 24.3) and noting that $s^2 - 3s + 2 = (s-1)(s-2)$, we obtain

$$\begin{aligned} y(x) &= c_0\mathcal{L}^{-1}\left\{\frac{2}{s-1} + \frac{-1}{s-2}\right\} + c_1\mathcal{L}^{-1}\left\{\frac{-1}{s-1} + \frac{1}{s-2}\right\} + \mathcal{L}^{-1}\left\{\frac{1/6}{s+1} + \frac{-1/2}{s-1} + \frac{1/3}{s-2}\right\} \\ &= c_0(2e^x - e^{2x}) + c_1(-e^x + e^{2x}) + \left(\frac{1}{6}e^{-x} - \frac{1}{2}e^x + \frac{1}{3}e^{2x}\right) \\ &= \left(2c_0 - c_1 - \frac{1}{2}\right)e^x + \left(-c_0 + c_1 + \frac{1}{3}\right)e^{2x} + \frac{1}{6}e^{-x} \\ &\equiv d_0e^x + d_1e^{2x} + \frac{1}{6}e^{-x} \end{aligned}$$

where $d_0 = 2c_0 - c_1 - \frac{1}{2}$ and $d_1 = -c_0 + c_1 + \frac{1}{3}$.

Applying the initial conditions to this last equation, we find that $d_0 = -\frac{1}{2}e^{-2}$ and $d_1 = \frac{1}{3}e^{-3}$; hence,

$$y(x) = -\frac{1}{2}e^{x-2} + \frac{1}{3}e^{2x-3} + \frac{1}{6}e^{-x}$$

26.10. Solve $y'' - 2y' + y = f(x);\ y(0) = 0,\ y'(0) = 0.$

In this equation $f(x)$ is unspecified. Taking Laplace transforms and designating $\mathcal{L}\{f(x)\}$ by $F(s)$, we obtain

$$[s^2Y(s) - (0)s - 0] - 2[sY(s) - 0] + Y(s) = F(s) \quad \text{or} \quad Y(s) = \frac{F(s)}{(s-1)^2}$$

From Table 22-1, entry 14, $\mathcal{L}^{-1}\{1/(s-1)^2\} = xe^x$. Thus, taking the inverse transform of $Y(s)$ and using convolutions, we conclude that

$$y(x) = xe^x * f(x) = \int_0^x te^t f(x-t)\,dt$$

26.11. Solve $y'' + y = f(x);\ y(0) = 0,\ y'(0) = 0$ if $f(x) = \begin{cases} 0 & x < 1 \\ 2 & x \geqq 1 \end{cases}$.

Note that $f(x) = 2u(x-1)$. Taking Laplace transforms, we obtain

$$[s^2Y(s) - (0)s - 0] + Y(s) = \mathcal{L}\{f(x)\} = 2\mathcal{L}\{u(x-1)\} = 2e^{-s}/s$$

or
$$Y(s) = e^{-s}\frac{2}{s(s^2+1)}$$

Since
$$\mathcal{L}^{-1}\left\{\frac{2}{s(s^2+1)}\right\} = 2\mathcal{L}^{-1}\left\{\frac{1}{s}\right\} - 2\mathcal{L}^{-1}\left\{\frac{s}{s^2+1}\right\} = 2 - 2\cos x$$

it follows from Theorem 25.4 that

$$y(x) = \mathcal{L}^{-1}\left\{e^{-s}\frac{2}{s(s^2+1)}\right\} = [2 - 2\cos(x-1)]u(x-1)$$

26.12. Solve $y''' + y' = e^x;\ y(0) = y'(0) = y''(0) = 0.$

Taking Laplace transforms, we obtain $\mathcal{L}\{y'''\} + \mathcal{L}\{y'\} = \mathcal{L}\{e^x\}$. Then, using Theorem 26.1 with $n = 3$ and *(26.5)*, we have

$$[s^3Y(s) - (0)s^2 - (0)s - 0] + [sY(s) - 0] = \frac{1}{s-1} \quad \text{or} \quad Y(s) = \frac{1}{(s-1)(s^3+s)}$$

Finally, using the method of partial fractions and taking the inverse transform, we obtain

$$y(x) = \mathcal{L}^{-1}\left\{-\frac{1}{s} + \frac{\frac{1}{2}}{s-1} + \frac{\frac{1}{2}s - \frac{1}{2}}{s^2+1}\right\} = -1 + \frac{1}{2}e^x + \frac{1}{2}\cos x - \frac{1}{2}\sin x$$

Supplementary Problems

Use Laplace transforms to solve the following initial-value problems.

26.13. $y' + 2y = 0;\ y(0) = 1.$

26.14. $y' + 2y = 2;\ y(0) = 1.$

26.15. $y' + 2y = e^x;\ y(0) = 1.$

26.16. $y' + 2y = 0;\ y(1) = 1.$

26.17. $y' + 5y = 0;\ y(1) = 0.$

26.18. $y'' - y = 0;\ y(0) = 1,\ y'(0) = 1.$

26.19. $y'' - y = \sin x;\ y(0) = 0,\ y'(0) = 1.$

26.20. $y'' - y = e^x;\ y(0) = 1,\ y'(0) = 0.$

26.21. $y'' + 2y' - 3y = \sin 2x;\ y(0) = y'(0) = 0.$

26.22. $y'' + y = \sin x;\ y(0) = 0,\ y'(0) = 2.$

26.23. $y'' + y' + y = 0;\ y(0) = 4,\ y'(0) = -3.$

26.24. $y'' + 2y' + 5y = 3e^{-2x};\ y(0) = 1,\ y'(0) = 1.$

26.25. $y'' + 5y' - 3y = u(x-4);\ y(0) = 0,\ y'(0) = 0.$

26.26. $y'' + y = 0;\ y(\pi) = 0,\ y'(\pi) = -1.$

26.27. $y''' - y = 5;\ y(0) = 0,\ y'(0) = 0,\ y''(0) = 0.$

26.28. $y^{(4)} - y = 0;\ y(0) = 1,\ y'(0) = 0,\ y''(0) = 0,\ y'''(0) = 0.$

Answers to Supplementary Problems

26.13. $y = e^{-2x}$

26.14. $y = 1$

26.15. $y = \frac{2}{3}e^{-2x} + \frac{1}{3}e^{x}$

26.16. $y = e^{-2(x-1)}$

26.17. $y = 0$

26.18. $y = e^{x}$

26.19. $y = \frac{3}{4}e^{x} - \frac{3}{4}e^{-x} - \frac{1}{2}\sin x$

26.20. $y = \frac{1}{4}e^{x} + \frac{3}{4}e^{-x} + \frac{1}{2}xe^{x}$

26.21. $y = \frac{1}{10}e^{x} - \frac{1}{26}e^{-3x} - \frac{4}{65}\cos 2x - \frac{7}{65}\sin 2x$

26.22. $y = \frac{5}{2}\sin x - \frac{1}{2}x\cos x$

26.23. $y = 4e^{-(1/2)x}\cos\frac{\sqrt{3}}{2}x - \frac{2}{\sqrt{3}}e^{-(1/2)x}\sin\frac{\sqrt{3}}{2}x$

26.24. $y = \frac{3}{5}e^{-2x} + \frac{2}{5}e^{-x}\cos 2x + \frac{13}{10}e^{-x}\sin 2x$

26.25. $y = \left[-\frac{1}{3} + \frac{1}{3}e^{-(5/2)(x-4)}\cosh\frac{\sqrt{37}}{2}(x-4) + \frac{5}{3\sqrt{37}}e^{-(5/2)(x-4)}\sinh\frac{\sqrt{37}}{2}(x-4)\right]u(x-4)$

26.26. $y = \sin x$

26.27. $y = -5 + \frac{5}{3}e^{x} + \frac{10}{3}e^{-(1/2)x}\cos\frac{\sqrt{3}}{2}x$

26.28. $y = \frac{1}{4}e^{x} + \frac{1}{4}e^{-x} + \frac{1}{2}\cos x$

Chapter 27

Solutions of Systems of Linear Differential Equations with Constant Coefficients by Laplace Transforms

Laplace transforms are also used to solve systems of linear differential equations with constant coefficients. The method is identical to the one given in Chapter 26, except that instead of solving one linear algebraic equation, we must now solve a system of simultaneous linear algebraic equations.

Solved Problems

27.1. Solve the system

$$y' + z = x$$
$$z' + 4y = 0;$$
$$y(0) = 1, \quad z(0) = -1$$

Denote $\mathcal{L}\{y(x)\}$ and $\mathcal{L}\{z(x)\}$ by $Y(s)$ and $Z(s)$, respectively. Then, taking Laplace transforms of both differential equations and using Theorem 26.1, we obtain

$$[sY(s)-1] + Z(s) = \frac{1}{s^2} \qquad\qquad sY(s) + Z(s) = \frac{s^2+1}{s^2}$$

or

$$[sZ(s)+1] + 4Y(s) = 0 \qquad\qquad 4Y(s) + sZ(s) = -1$$

The solution to this last set of simultaneous linear equations is

$$Y(s) = \frac{s^2+s+1}{s(s^2-4)} \qquad\qquad Z(s) = -\frac{s^3+4s^2+4}{s^2(s^2-4)}$$

Finally, using the method of partial fractions and taking inverse transforms, we obtain

$$y(x) = \mathcal{L}^{-1}\{Y(s)\} = \mathcal{L}^{-1}\left\{-\frac{1/4}{s} + \frac{7/8}{s-2} + \frac{3/8}{s+2}\right\}$$

$$= -\frac{1}{4} + \frac{7}{8}e^{2x} + \frac{3}{8}e^{-2x}$$

$$z(x) = \mathcal{L}^{-1}\{Z(s)\} = \mathcal{L}^{-1}\left\{\frac{1}{s^2} - \frac{7/4}{s-2} + \frac{3/4}{s+2}\right\}$$

$$= x - \frac{7}{4}e^{2x} + \frac{3}{4}e^{-2x}$$

27.2. Solve the system

$$w' + y = \sin x$$
$$y' - z = e^x$$
$$z' + w + y = 1;$$
$$w(0) = 0, \quad y(0) = 1, \quad z(0) = 1$$

Denote $\mathcal{L}\{w(x)\}$, $\mathcal{L}\{y(x)\}$, and $\mathcal{L}\{z(x)\}$ by $W(s)$, $Y(s)$, and $Z(s)$, respectively. Then, taking Laplace transforms of all three differential equations, we have

$$[sW(s) - 0] + Y(s) = \frac{1}{s^2+1} \qquad\qquad sW(s) + Y(s) = \frac{1}{s^2+1}$$

$$[sY(s) - 1] - Z(s) = \frac{1}{s-1} \qquad \text{or} \qquad sY(s) - Z(s) = \frac{s}{s-1}$$

$$[sZ(s) - 1] + W(s) + Y(s) = \frac{1}{s} \qquad\qquad W(s) + Y(s) + sZ(s) = \frac{s+1}{s}$$

The solution to this last system of simultaneous linear equations is

$$W(s) = \frac{-1}{s(s-1)} \qquad Y(s) = \frac{s^2+s}{(s-1)(s^2+1)} \qquad Z(s) = \frac{s}{s^2+1}$$

Using the method of partial fractions and then taking inverse transforms, we obtain

$$w(x) = \mathcal{L}^{-1}\{W(s)\} = \mathcal{L}^{-1}\left\{\frac{1}{s} - \frac{1}{s-1}\right\} = 1 - e^x$$

$$y(x) = \mathcal{L}^{-1}\{Y(s)\} = \mathcal{L}^{-1}\left\{\frac{1}{s-1} + \frac{1}{s^2+1}\right\} = e^x + \sin x$$

$$z(x) = \mathcal{L}^{-1}\{Z(s)\} = \mathcal{L}^{-1}\left\{\frac{s}{s^2+1}\right\} = \cos x$$

27.3. Solve the system

$$y'' + z + y = 0$$
$$z' + y' = 0;$$
$$y(0) = 0, \quad y'(0) = 0, \quad z(0) = 1$$

Taking Laplace transforms of both differential equations, we obtain

$$[s^2Y(s) - (0)s - (0)] + Z(s) + Y(s) = 0 \qquad\qquad (s^2+1)Y(s) + Z(s) = 0$$

$$[sZ(s) - 1] + [sY(s) - 0] = 0 \qquad \text{or} \qquad Y(s) + Z(s) = \frac{1}{s}$$

Solving this last system for $Y(s)$ and $Z(s)$, we find that

$$Y(s) = -\frac{1}{s^3} \qquad Z(s) = \frac{1}{s} + \frac{1}{s^3}$$

Thus, taking inverse transforms, we conclude that

$$y(x) = -\frac{1}{2}x^2 \qquad z(x) = 1 + \frac{1}{2}x^2$$

27.4. Solve the system

$$z'' + y' = \cos x$$
$$y'' - z = \sin x;$$
$$z(0) = -1, \quad z'(0) = -1, \quad y(0) = 1, \quad y'(0) = 0$$

Taking Laplace transforms of both differential equations, we obtain

$$[s^2Z(s) + s + 1] + [sY(s) - 1] = \frac{s}{s^2+1}$$
$$[s^2Y(s) - s - 0] - Z(s) = \frac{1}{s^2+1}$$

or

$$s^2Z(s) + sY(s) = -\frac{s^3}{s^2+1}$$
$$-Z(s) + s^2Y(s) = \frac{s^3+s+1}{s^2+1}$$

Solving this last system for $Z(s)$ and $Y(s)$, we find that

$$Z(s) = -\frac{s+1}{s^2+1} \qquad Y(s) = \frac{s}{s^2+1}$$

Finally, taking inverse transforms, we obtain

$$z(x) = -\cos x - \sin x \qquad y(x) = \cos x$$

27.5. Solve the system

$$w'' - y + 2z = 3e^{-x}$$
$$-2w' + 2y' + z = 0$$
$$2w' - 2y + z' + 2z'' = 0;$$
$$w(0) = 1, \quad w'(0) = 1, \quad y(0) = 2, \quad z(0) = 2, \quad z'(0) = -2$$

Taking Laplace transforms of all three differential equations, we find that

$$[s^2W(s) - s - 1] - Y(s) + 2Z(s) = \frac{3}{s+1}$$
$$-2[sW(s) - 1] + 2[sY(s) - 2] + Z(s) = 0$$
$$2[sW(s) - 1] - 2Y(s) + [sZ(s) - 2] + 2[s^2Z(s) - 2s + 2] = 0$$

or

$$s^2W(s) - Y(s) + 2Z(s) = \frac{s^2+2s+4}{s+1}$$
$$-2sW(s) + 2sY(s) + Z(s) = 2$$
$$2sW(s) - 2Y(s) + (2s^2+s)Z(s) = 4s$$

The solution to this system is

$$W(s) = \frac{1}{s-1} \qquad Y(s) = \frac{2s}{(s-1)(s+1)} \qquad Z(s) = \frac{2}{s+1}$$

Hence,

$$w(x) = e^x \qquad y(x) = \mathcal{L}^{-1}\left\{\frac{1}{s-1} + \frac{1}{s+1}\right\} = e^x + e^{-x} \qquad z(x) = 2e^{-x}$$

Supplementary Problems

Use Laplace transforms to solve the following systems.

27.6. $y' + z = x$
$z' - y = 0;$
$y(0) = 1, \quad z(0) = 0$

27.7. $y' - z = 0$
$y - z' = 0;$
$y(0) = 1, \quad z(0) = 1$

27.8. $w' - w - 2y = 1$
$y' - 4w - 3y = -1;$
$w(0) = 1, \quad y(0) = 2$

27.9. $w' - y = 0$
$w + y' + z = 1$
$w - y + z' = 2 \sin x;$
$w(0) = 1, \quad y(0) = 1, \quad z(0) = 1$

27.10. $u'' + v = 0$
$u'' - v' = -2e^x;$
$u(0) = 0, \quad u'(0) = -2, \quad v(0) = 0, \quad v'(0) = 2$

27.11. $u'' - 2v = 2$
$u + v' = 5e^{2x} + 1;$
$u(0) = 2, \quad u'(0) = 2, \quad v(0) = 1$

27.12. $w'' - 2z = 0$
$w' + y' - z = 2x$
$w' - 2y + z'' = 0;$
$w(0) = 0, \quad w'(0) = 0, \quad y(0) = 0,$
$z(0) = 1, \quad z'(0) = 0$

27.13. $w'' + y + z = -1$
$w + y'' - z = 0$
$-w' - y' + z'' = 0;$
$w(0) = 0, \quad w'(0) = 1, \quad y(0) = 0,$
$y'(0) = 0, \quad z(0) = -1, \quad z'(0) = 1$

Answers to Supplementary Problems

27.6. $y(x) = 1 \qquad z(x) = x$

27.7. $y(x) = e^x \qquad z(x) = e^x$

27.8. $w(x) = e^{5x} - e^{-x} + 1$
$y(x) = 2e^{5x} + e^{-x} - 1$

27.9. $w(x) = \cos x + \sin x$
$y(x) = \cos x - \sin x$
$z(x) = 1$

27.10. $u(x) = -e^x + e^{-x} \qquad v(x) = e^x - e^{-x}$

27.11. $u(x) = e^{2x} + 1 \qquad v(x) = 2e^{2x} - 1$

27.12. $w(x) = x^2 \qquad y(x) = x \qquad z(x) = 1$

27.13. $w(x) = \sin x$
$y(x) = -1 + \cos x$
$z(x) = \sin x - \cos x$

Chapter 28

Matrices

28.1 MATRICES AND VECTORS

A *matrix* (designated by an uppercase boldface letter) is a rectangular array of elements arranged in horizontal rows and vertical columns. In this book, the elements of matrices will always be numbers or functions of the variable t. If all the elements are numbers, then the matrix is called a *constant matrix*.

Example 28.1.

$$\begin{bmatrix} 1 & 2 \\ 3 & 4 \end{bmatrix}, \quad \begin{bmatrix} 1 & e^t & 2 \\ t & -1 & 1 \end{bmatrix}, \quad \text{and} \quad [1 \quad t^2 \quad \cos t]$$

are all matrices. In particular, the first matrix is a constant matrix, whereas the last two are not.

A general matrix $\mathbf{A}$ having p rows and n columns is given by

$$\mathbf{A} = [a_{ij}] = \begin{bmatrix} a_{11} & a_{12} & \dots & a_{1n} \\ a_{21} & a_{22} & \dots & a_{2n} \\ \vdots & \vdots & & \vdots \\ a_{p1} & a_{p2} & \dots & a_{pn} \end{bmatrix}$$

where a_{ij} represents that element appearing in the ith row and jth column. A matrix is *square* if it has the same number of rows and columns.

A *vector* (designated by a lowercase boldface letter) is a matrix having only one column or one row. (The third matrix given in Example 28.1 is a vector.)

28.2 MATRIX ADDITION

The *sum* $\mathbf{A}+\mathbf{B}$ of two matrices $A = [a_{ij}]$ and $B = [b_{ij}]$ having the same number of rows and the same number of columns is the matrix obtained by adding the corresponding elements of $\mathbf{A}$ and $\mathbf{B}$. That is,

$$\mathbf{A}+\mathbf{B} = [a_{ij}] + [b_{ij}] = [a_{ij} + b_{ij}]$$

(See Problem 28.1.)

Matrix addition is both associative and commutative. Thus, $\mathbf{A}+(\mathbf{B}+\mathbf{C}) = (\mathbf{A}+\mathbf{B})+\mathbf{C}$ and $\mathbf{A}+\mathbf{B} = \mathbf{B}+\mathbf{A}$.

28.3 SCALAR AND MATRIX MULTIPLICATION

If λ is either a number or a function of t, then $\lambda\mathbf{A}$ (or, equivalently, $\mathbf{A}\lambda$) is defined to be the matrix obtained by multiplying every element of $\mathbf{A}$ by λ. That is,

$$\lambda\mathbf{A} = \lambda[a_{ij}] = [\lambda a_{ij}]$$

(See Problem 28.2.)

Let $\mathbf{A} = [a_{ij}]$ and $\mathbf{B} = [b_{ij}]$ be two matrices such that **A** has r rows and n columns and **B** has n rows and p columns. Then the *product* **AB** is defined to be the matrix $\mathbf{C} = [c_{ij}]$ given by

$$c_{ij} = \sum_{k=1}^{n} a_{ik} b_{kj} \qquad (i = 1, 2, \ldots, r;\ j = 1, 2, \ldots, p)$$

According to this definition, c_{ij} is obtained by multiplying the elements of the ith row of **A** into the corresponding elements of the jth column of **B** and summing the results. (See Problems 28.3 through 28.6.)

Matrix multiplication is associative and distributes over addition; in general, however, it is *not* commutative. Thus,

$$\mathbf{A(BC)} = \mathbf{(AB)C}, \quad \mathbf{A(B+C)} = \mathbf{AB} + \mathbf{AC}, \quad \text{and} \quad \mathbf{(B+C)A} = \mathbf{BA} + \mathbf{CA}$$

but, in general, $\mathbf{AB} \neq \mathbf{BA}$.

28.4 IDENTITY AND ZERO MATRICES

An *identity matrix* **I** is a square matrix of the form

$$\mathbf{I} = \begin{bmatrix} 1 & 0 & 0 & \ldots & 0 & 0 \\ 0 & 1 & 0 & \ldots & 0 & 0 \\ 0 & 0 & 1 & \ldots & 0 & 0 \\ \vdots & & & \ddots & & \vdots \\ 0 & 0 & 0 & \ldots & 1 & 0 \\ 0 & 0 & 0 & \ldots & 0 & 1 \end{bmatrix}$$

By definition of matrix multiplication, $\mathbf{AI} = \mathbf{IA} = \mathbf{A}$ for any square matrix **A**.

A *zero matrix* **0** is a matrix having all its elements equal to zero.

28.5 POWERS OF A SQUARE MATRIX

If n is a positive integer and **A** is a square matrix, then

$$\mathbf{A}^n = \underbrace{\mathbf{AA}\cdots\mathbf{A}}_{n \text{ times}}$$

In particular, $\mathbf{A}^2 = \mathbf{AA}$ and $\mathbf{A}^3 = \mathbf{AAA}$. By definition, $\mathbf{A}^0 = \mathbf{I}$.

28.6 DIFFERENTIATION AND INTEGRATION OF MATRICES

The *derivative* of $\mathbf{A} = [a_{ij}]$ is the matrix obtained by differentiating each element of **A**; that is,

$$\frac{d\mathbf{A}}{dt} = \left[\frac{da_{ij}}{dt}\right]$$

Similarly, the *integral* of **A**, either definite or indefinite, is obtained by integrating each element of **A**. Thus,

$$\int_a^b \mathbf{A}\,dt = \left[\int_a^b a_{ij}\,dt\right] \qquad \int \mathbf{A}\,dt = \left[\int a_{ij}\,dt\right]$$

(See Problems 28.7 through 28.10.)

28.7 THE CHARACTERISTIC EQUATION

The *characteristic equation* of a square matrix $\mathbf{A}$ is the polynomial equation in λ given by

$$\det(\mathbf{A} - \lambda\mathbf{I}) = 0 \tag{28.1}$$

where det( ) stands for "the determinant of ( )." Those values of λ which satisfy (*28.1*), that is, the roots of (*28.1*), are the *eigenvalues* of $\mathbf{A}$, a k-fold root being called an *eigenvalue of multiplicity* k. (See Problems 28.11 through 28.14.)

Theorem 28.1 ***(Cayley-Hamilton theorem).*** Any square matrix satisfies its own characteristic equation. That is, if

$$\det(\mathbf{A} - \lambda\mathbf{I}) = b_n\lambda^n + b_{n-1}\lambda^{n-1} + \cdots + b_2\lambda^2 + b_1\lambda + b_0$$

then

$$b_n\mathbf{A}^n + b_{n-1}\mathbf{A}^{n-1} + \cdots + b_2\mathbf{A}^2 + b_1\mathbf{A} + b_0\mathbf{I} = \mathbf{0}$$

(See Problems 28.15 and 28.16.)

Solved Problems

28.1. Show that $\mathbf{A}+\mathbf{B} = \mathbf{B}+\mathbf{A}$ for

$$\mathbf{A} = \begin{bmatrix} 1 & 2 \\ 3 & 4 \end{bmatrix} \qquad \mathbf{B} = \begin{bmatrix} 5 & 6 \\ 7 & 8 \end{bmatrix}$$

$$\mathbf{A} + \mathbf{B} = \begin{bmatrix} 1 & 2 \\ 3 & 4 \end{bmatrix} + \begin{bmatrix} 5 & 6 \\ 7 & 8 \end{bmatrix} = \begin{bmatrix} 1+5 & 2+6 \\ 3+7 & 4+8 \end{bmatrix} = \begin{bmatrix} 6 & 8 \\ 10 & 12 \end{bmatrix}$$

$$\mathbf{B} + \mathbf{A} = \begin{bmatrix} 5 & 6 \\ 7 & 8 \end{bmatrix} + \begin{bmatrix} 1 & 2 \\ 3 & 4 \end{bmatrix} = \begin{bmatrix} 5+1 & 6+2 \\ 7+3 & 8+4 \end{bmatrix} = \begin{bmatrix} 6 & 8 \\ 10 & 12 \end{bmatrix}$$

Since the corresponding elements of the resulting matrices are equal, the desired equality follows.

28.2. Find $3\mathbf{A} - \frac{1}{2}\mathbf{B}$ for the matrices given in Problem 28.1.

$$3\mathbf{A} - \tfrac{1}{2}\mathbf{B} = 3\begin{bmatrix} 1 & 2 \\ 3 & 4 \end{bmatrix} + \left(-\tfrac{1}{2}\right)\begin{bmatrix} 5 & 6 \\ 7 & 8 \end{bmatrix}$$

$$= \begin{bmatrix} 3 & 6 \\ 9 & 12 \end{bmatrix} + \begin{bmatrix} -\frac{5}{2} & -3 \\ -\frac{7}{2} & -4 \end{bmatrix}$$

$$= \begin{bmatrix} 3 + \left(-\frac{5}{2}\right) & 6 + (-3) \\ 9 + \left(-\frac{7}{2}\right) & 12 + (-4) \end{bmatrix} = \begin{bmatrix} \frac{1}{2} & 3 \\ \frac{11}{2} & 8 \end{bmatrix}$$

28.3. Find **AB** and **BA** for the matrices given in Problem 28.1.

$$\mathbf{AB} = \begin{bmatrix} 1 & 2 \\ 3 & 4 \end{bmatrix}\begin{bmatrix} 5 & 6 \\ 7 & 8 \end{bmatrix} = \begin{bmatrix} 1(5)+2(7) & 1(6)+2(8) \\ 3(5)+4(7) & 3(6)+4(8) \end{bmatrix} = \begin{bmatrix} 19 & 22 \\ 43 & 50 \end{bmatrix}$$

$$\mathbf{BA} = \begin{bmatrix} 5 & 6 \\ 7 & 8 \end{bmatrix}\begin{bmatrix} 1 & 2 \\ 3 & 4 \end{bmatrix} = \begin{bmatrix} 5(1)+6(3) & 5(2)+6(4) \\ 7(1)+8(3) & 7(2)+8(4) \end{bmatrix} = \begin{bmatrix} 23 & 34 \\ 31 & 46 \end{bmatrix}$$

Note that for these matrices, $\mathbf{AB} \neq \mathbf{BA}$.

28.4. Find **AB** and **BA** for $\mathbf{A} = \begin{bmatrix} 1 & 2 & 3 \\ 4 & 5 & 6 \end{bmatrix}$, $\mathbf{B} = \begin{bmatrix} 7 & 0 \\ 8 & -1 \end{bmatrix}$.

Since **A** has three columns and **B** has two rows, the product **AB** is not defined. But

$$\mathbf{BA} = \begin{bmatrix} 7 & 0 \\ 8 & -1 \end{bmatrix}\begin{bmatrix} 1 & 2 & 3 \\ 4 & 5 & 6 \end{bmatrix} = \begin{bmatrix} 7(1)+(0)(4) & 7(2)+(0)(5) & 7(3)+(0)(6) \\ 8(1)+(-1)(4) & 8(2)+(-1)(5) & 8(3)+(-1)(6) \end{bmatrix}$$

$$= \begin{bmatrix} 7 & 14 & 21 \\ 4 & 11 & 18 \end{bmatrix}$$

28.5. Find **AB** and **AC** if

$$\mathbf{A} = \begin{bmatrix} 4 & 2 & 0 \\ 2 & 1 & 0 \\ -2 & -1 & 1 \end{bmatrix}, \quad \mathbf{B} = \begin{bmatrix} 2 & 3 & 1 \\ 2 & -2 & -2 \\ -1 & 2 & 1 \end{bmatrix}, \quad \mathbf{C} = \begin{bmatrix} 3 & 1 & -3 \\ 0 & 2 & 6 \\ -1 & 2 & 1 \end{bmatrix}$$

$$\mathbf{AB} = \begin{bmatrix} 4(2)+2(2)+(0)(-1) & 4(3)+2(-2)+(0)(2) & 4(1)+2(-2)+(0)(1) \\ 2(2)+1(2)+(0)(-1) & 2(3)+1(-2)+(0)(2) & 2(1)+1(-2)+(0)(1) \\ -2(2)+(-1)(2)+1(-1) & -2(3)+(-1)(-2)+1(2) & -2(1)+(-1)(-2)+1(1) \end{bmatrix}$$

$$= \begin{bmatrix} 12 & 8 & 0 \\ 6 & 4 & 0 \\ -7 & -2 & 1 \end{bmatrix}$$

$$\mathbf{AC} = \begin{bmatrix} 4(3)+2(0)+(0)(-1) & 4(1)+2(2)+(0)(2) & 4(-3)+2(6)+(0)(1) \\ 2(3)+1(0)+(0)(-1) & 2(1)+1(2)+(0)(2) & 2(-3)+1(6)+(0)(1) \\ -2(3)+(-1)(0)+1(-1) & -2(1)+(-1)(2)+1(2) & -2(-3)+(-1)(6)+1(1) \end{bmatrix}$$

$$= \begin{bmatrix} 12 & 8 & 0 \\ 6 & 4 & 0 \\ -7 & -2 & 1 \end{bmatrix}$$

Note that for these matrices $\mathbf{AB} = \mathbf{AC}$ and yet $\mathbf{B} \neq \mathbf{C}$. Therefore, the cancellation law is not valid for matrix multiplication.

28.6. Find $\mathbf{Ax}$ if

$$\mathbf{A} = \begin{bmatrix} 1 & 2 & 3 & 4 \\ 5 & 6 & 7 & 8 \end{bmatrix} \qquad \mathbf{x} = \begin{bmatrix} 9 \\ -1 \\ -2 \\ 0 \end{bmatrix}$$

$$\mathbf{Ax} = \begin{bmatrix} 1(9) + 2(-1) + 3(-2) + 4(0) \\ 5(9) + 6(-1) + 7(-2) + 8(0) \end{bmatrix} = \begin{bmatrix} 1 \\ 25 \end{bmatrix}$$

28.7. Find $\dfrac{d\mathbf{A}}{dt}$ if $\mathbf{A} = \begin{bmatrix} t^2+1 & e^{2t} \\ \sin t & 45 \end{bmatrix}$.

$$\frac{d\mathbf{A}}{dt} = \begin{bmatrix} \dfrac{d}{dt}(t^2+1) & \dfrac{d}{dt}(e^{2t}) \\ \dfrac{d}{dt}(\sin t) & \dfrac{d}{dt}(45) \end{bmatrix} = \begin{bmatrix} 2t & 2e^{2t} \\ \cos t & 0 \end{bmatrix}$$

28.8. Find $\dfrac{d\mathbf{x}}{dt}$ if $\mathbf{x} = \begin{bmatrix} x_1(t) \\ x_2(t) \\ x_3(t) \end{bmatrix}$.

$$\frac{d\mathbf{x}}{dt} = \begin{bmatrix} \dfrac{dx_1(t)}{dt} \\ \dfrac{dx_2(t)}{dt} \\ \dfrac{dx_3(t)}{dt} \end{bmatrix} = \begin{bmatrix} \dot{x}_1(t) \\ \dot{x}_2(t) \\ \dot{x}_3(t) \end{bmatrix}$$

28.9. Find $\int \mathbf{A}\,dt$ for $\mathbf{A}$ as given in Problem 28.7.

$$\int \mathbf{A}\,dt = \begin{bmatrix} \int (t^2+1)\,dt & \int e^{2t}\,dt \\ \int \sin t\,dt & \int 45\,dt \end{bmatrix} = \begin{bmatrix} \frac{1}{3}t^3 + t + c_1 & \frac{1}{2}e^{2t} + c_2 \\ -\cos t + c_3 & 45t + c_4 \end{bmatrix}$$

28.10. Find $\int_0^1 \mathbf{x}\,dt$ if $\mathbf{x} = \begin{bmatrix} 1 \\ e^t \\ 0 \end{bmatrix}$.

$$\int_0^1 \mathbf{x}\,dt = \begin{bmatrix} \int_0^1 1\,dt \\ \int_0^1 e^t\,dt \\ \int_0^1 0\,dt \end{bmatrix} = \begin{bmatrix} 1 \\ e-1 \\ 0 \end{bmatrix}$$

28.11. Find the eigenvalues of $\mathbf{A} = \begin{bmatrix} 1 & 3 \\ 4 & 2 \end{bmatrix}$.

We have

$$\mathbf{A} - \lambda\mathbf{I} = \begin{bmatrix} 1 & 3 \\ 4 & 2 \end{bmatrix} + (-\lambda)\begin{bmatrix} 1 & 0 \\ 0 & 1 \end{bmatrix}$$

$$= \begin{bmatrix} 1 & 3 \\ 4 & 2 \end{bmatrix} + \begin{bmatrix} -\lambda & 0 \\ 0 & -\lambda \end{bmatrix} = \begin{bmatrix} 1-\lambda & 3 \\ 4 & 2-\lambda \end{bmatrix}$$

Hence,

$$\det(\mathbf{A} - \lambda\mathbf{I}) = \det\begin{bmatrix} 1-\lambda & 3 \\ 4 & 2-\lambda \end{bmatrix}$$

$$= (1-\lambda)(2-\lambda) - (3)(4) = \lambda^2 - 3\lambda - 10$$

The characteristic equation of **A** is $\lambda^2 - 3\lambda - 10 = 0$, which can be factored into $(\lambda - 5)(\lambda + 2) = 0$. The roots of this equation are $\lambda_1 = 5$ and $\lambda_2 = -2$, which are the eigenvalues of **A**.

28.12. Find the eigenvalues of $\mathbf{A}t$ if $\mathbf{A} = \begin{bmatrix} 2 & 5 \\ -1 & -2 \end{bmatrix}$.

$$\mathbf{A}t - \lambda\mathbf{I} = \begin{bmatrix} 2 & 5 \\ -1 & -2 \end{bmatrix} t + (-\lambda)\begin{bmatrix} 1 & 0 \\ 0 & 1 \end{bmatrix}$$

$$= \begin{bmatrix} 2t & 5t \\ -t & -2t \end{bmatrix} + \begin{bmatrix} -\lambda & 0 \\ 0 & -\lambda \end{bmatrix} = \begin{bmatrix} 2t-\lambda & 5t \\ -t & -2t-\lambda \end{bmatrix}$$

Then,

$$\det(\mathbf{A} - \lambda\mathbf{I}) = \det\begin{bmatrix} 2t-\lambda & 5t \\ -t & -2t-\lambda \end{bmatrix}$$

$$= (2t-\lambda)(-2t-\lambda) - (5t)(-t) = \lambda^2 + t^2$$

and the characteristic equation of $\mathbf{A}t$ is $\lambda^2 + t^2 = 0$. The roots of this equation, which are the eigenvalues of $\mathbf{A}t$, are $\lambda_1 = it$ and $\lambda_2 = -it$, where $i = \sqrt{-1}$.

28.13. Find the eigenvalues of $\mathbf{A} = \begin{bmatrix} 4 & 1 & 0 \\ -1 & 2 & 0 \\ 2 & 1 & -3 \end{bmatrix}$.

$$\mathbf{A} - \lambda\mathbf{I} = \begin{bmatrix} 4 & 1 & 0 \\ -1 & 2 & 0 \\ 2 & 1 & -3 \end{bmatrix} - \lambda\begin{bmatrix} 1 & 0 & 0 \\ 0 & 1 & 0 \\ 0 & 0 & 1 \end{bmatrix}$$

$$= \begin{bmatrix} 4-\lambda & 1 & 0 \\ -1 & 2-\lambda & 0 \\ 2 & 1 & -3-\lambda \end{bmatrix}$$

Thus,

$$\det(\mathbf{A}-\lambda\mathbf{I}) = \det\begin{bmatrix} 4-\lambda & 1 & 0 \\ -1 & 2-\lambda & 0 \\ 2 & 1 & -3-\lambda \end{bmatrix}$$

$$= (-3-\lambda)[(4-\lambda)(2-\lambda)-(1)(-1)]$$

$$= (-3-\lambda)(\lambda-3)(\lambda-3)$$

The characteristic equation of **A** is

$$(-3-\lambda)(\lambda-3)(\lambda-3) = 0$$

hence, the eigenvalues of **A** are $\lambda_1 = -3$, $\lambda_2 = 3$, and $\lambda_3 = 3$. Here, $\lambda = 3$ is an eigenvalue of multiplicity two, while $\lambda = -3$ is an eigenvalue of multiplicity one.

28.14. Find the eigenvalues of

$$\mathbf{A} = \begin{bmatrix} 5 & 7 & 0 & 0 \\ -3 & -5 & 0 & 0 \\ 0 & 0 & -2 & 1 \\ 0 & 0 & 0 & -2 \end{bmatrix}$$

$$\mathbf{A} - \lambda\mathbf{I} = \begin{bmatrix} 5-\lambda & 7 & 0 & 0 \\ -3 & -5-\lambda & 0 & 0 \\ 0 & 0 & -2-\lambda & 1 \\ 0 & 0 & 0 & -2-\lambda \end{bmatrix}$$

and
$$\det(\mathbf{A}-\lambda\mathbf{I}) = [(5-\lambda)(-5-\lambda)-(-3)(7)](-2-\lambda)(-2-\lambda)$$

$$= (\lambda^2-4)(-2-\lambda)(-2-\lambda)$$

The characteristic equation of **A** is

$$(\lambda^2-4)(-2-\lambda)(-2-\lambda) = 0$$

which has roots $\lambda_1 = 2$, $\lambda_2 = -2$, $\lambda_3 = -2$, and $\lambda_4 = -2$. Thus, $\lambda = -2$ is an eigenvalue of multiplicity three, whereas $\lambda = 2$ is an eigenvalue of multiplicity one.

28.15. Verify the Cayley-Hamilton theorem for $\mathbf{A} = \begin{bmatrix} 2 & -7 \\ 3 & 6 \end{bmatrix}$.

For this matrix, we have $\det(\mathbf{A}-\lambda\mathbf{I}) = \lambda^2 - 8\lambda + 33$; hence

$$\mathbf{A}^2 - 8\mathbf{A} + 33\mathbf{I} = \begin{bmatrix} 2 & -7 \\ 3 & 6 \end{bmatrix}\begin{bmatrix} 2 & -7 \\ 3 & 6 \end{bmatrix} - 8\begin{bmatrix} 2 & -7 \\ 3 & 6 \end{bmatrix} + 33\begin{bmatrix} 1 & 0 \\ 0 & 1 \end{bmatrix}$$

$$= \begin{bmatrix} -17 & -56 \\ 24 & 15 \end{bmatrix} - \begin{bmatrix} 16 & -56 \\ 24 & 48 \end{bmatrix} + \begin{bmatrix} 33 & 0 \\ 0 & 33 \end{bmatrix}$$

$$= \begin{bmatrix} 0 & 0 \\ 0 & 0 \end{bmatrix}$$

28.16. Verify the Cayley-Hamilton theorem for the matrix of Problem 28.13.

For that matrix we found $\det(\mathbf{A}-\lambda\mathbf{I}) = -(\lambda+3)(\lambda-3)^2$; hence

$$-(\mathbf{A}+3\mathbf{I})(\mathbf{A}-3\mathbf{I})^2 = -\begin{bmatrix} 7 & 1 & 0 \\ -1 & 5 & 0 \\ 2 & 1 & 0 \end{bmatrix}\begin{bmatrix} 1 & 1 & 0 \\ -1 & -1 & 0 \\ 2 & 1 & -6 \end{bmatrix}^2$$

$$= -\begin{bmatrix} 7 & 1 & 0 \\ -1 & 5 & 0 \\ 2 & 1 & 0 \end{bmatrix}\begin{bmatrix} 0 & 0 & 0 \\ 0 & 0 & 0 \\ -11 & -5 & 36 \end{bmatrix} = \begin{bmatrix} 0 & 0 & 0 \\ 0 & 0 & 0 \\ 0 & 0 & 0 \end{bmatrix}$$

Supplementary Problems

In Problems 28.17 through 28.25 let

$$\mathbf{A} = \begin{bmatrix} 2 & 3 \\ -1 & -2 \end{bmatrix} \qquad \mathbf{B} = \begin{bmatrix} 1 & -4 \\ 3 & 1 \end{bmatrix} \qquad \mathbf{C} = \begin{bmatrix} 3 & 5 & 0 \\ -2 & -3 & 0 \\ 1 & 1 & 1 \end{bmatrix}$$

$$\mathbf{D} = \begin{bmatrix} 1 & 0 & 2 \\ 1 & 0 & 1 \\ 2 & 0 & 4 \end{bmatrix} \qquad \mathbf{x} = \begin{bmatrix} 1 \\ -2 \end{bmatrix} \qquad \mathbf{y} = \begin{bmatrix} 1 \\ 1 \\ 2 \end{bmatrix}$$

28.17. Find (*a*) $\mathbf{A}+\mathbf{B}$, (*b*) $3\mathbf{A}-2\mathbf{B}$, and (*c*) $\mathbf{C}-\mathbf{D}$.

28.18. Find (*a*) $\mathbf{AB}$ and (*b*) $\mathbf{BA}$.

28.19. Find (*a*) $\mathbf{CD}$ and (*b*) $\mathbf{DC}$.

28.20. Find (*a*) $\mathbf{Ax}$ and (*b*) $\mathbf{xA}$.

28.21. Find $(\mathbf{C}+\mathbf{D})\mathbf{y}$.

28.22. Find the characteristic equation and the eigenvalues of **A**.

28.23. Find the characteristic equation and the eigenvalues of **B**.

28.24. Find the characteristic equation and the eigenvalues of **C**. Determine the multiplicity of each eigenvalue.

28.25. Find the characteristic equation and the eigenvalues of **D**. Determine the multiplicity of each eigenvalue.

28.26. Find the characteristic equation and the eigenvalues of $\mathbf{A} = \begin{bmatrix} t & t^2 \\ 1 & 2t \end{bmatrix}$.

28.27. Find the characteristic equation and the eigenvalues of $\mathbf{A} = \begin{bmatrix} t & 6t & 0 \\ 4t & -t & 0 \\ 0 & 1 & 5t \end{bmatrix}$.

28.28. Find $\dfrac{d\mathbf{A}}{dt}$ for $\mathbf{A}$ as given in Problem 28.26.

28.29. Find $\dfrac{d\mathbf{A}}{dt}$ for $\mathbf{A} = \begin{bmatrix} \cos 2t \\ te^{3t^2} \end{bmatrix}$.

28.30. Find $\displaystyle\int_0^1 \mathbf{A}\,dt$ for $\mathbf{A}$ as given in Problem 28.29.

Answers to Supplementary Problems

28.17. (a) $\begin{bmatrix} 3 & -1 \\ 2 & -1 \end{bmatrix}$ (b) $\begin{bmatrix} 4 & 17 \\ -9 & -8 \end{bmatrix}$ (c) $\begin{bmatrix} 2 & 5 & -2 \\ -3 & -3 & -1 \\ -1 & 1 & -3 \end{bmatrix}$

28.18. (a) $\begin{bmatrix} 11 & -5 \\ -7 & 2 \end{bmatrix}$ (b) $\begin{bmatrix} 6 & 11 \\ 5 & 7 \end{bmatrix}$

28.19. (a) $\begin{bmatrix} 8 & 0 & 11 \\ -5 & 0 & -7 \\ 4 & 0 & 7 \end{bmatrix}$ (b) $\begin{bmatrix} 5 & 7 & 2 \\ 4 & 6 & 1 \\ 10 & 14 & 4 \end{bmatrix}$

28.20. (a) $\begin{bmatrix} -4 \\ 3 \end{bmatrix}$ (b) undefined

28.21. $\begin{bmatrix} 13 \\ -2 \\ 14 \end{bmatrix}$

28.22. $\lambda^2 - 1 = 0;\ \lambda_1 = 1,\ \lambda_2 = -1$

28.23. $\lambda^2 - 2\lambda + 13 = 0;\ \lambda_1 = 1 + 2\sqrt{3}\,i,\ \lambda_2 = 1 - 2\sqrt{3}\,i$

28.24. $(1-\lambda)(\lambda^2+1) = 0;\ \lambda_1 = 1,\ \lambda_2 = i,\ \lambda_3 = -i$
Each eigenvalue has multiplicity one.

28.25. $(-\lambda)(\lambda^2 - 5\lambda) = 0;\ \lambda_1 = 0,\ \lambda_2 = 0,\ \lambda_3 = 5$
The eigenvalue $\lambda = 0$ has multiplicity two, while $\lambda = 5$ has multiplicity one.

28.26. $\lambda^2 - 3t\lambda + t^2 = 0;\ \lambda_1 = \left(\frac{3}{2} + \frac{1}{2}\sqrt{5}\right)t,\ \lambda_2 = \left(\frac{3}{2} - \frac{1}{2}\sqrt{5}\right)t$

28.27. $(5t - \lambda)(\lambda^2 - 25t^2) = 0;\ \lambda_1 = 5t,\ \lambda_2 = 5t,\ \lambda_3 = -5t$

28.28. $\begin{bmatrix} 1 & 2t \\ 0 & 2 \end{bmatrix}$ **28.29.** $\begin{bmatrix} -2\sin 2t \\ (1+6t^2)e^{3t^2} \end{bmatrix}$ **28.30.** $\begin{bmatrix} \frac{1}{2}\sin 2 \\ \frac{1}{6}(e^3 - 1) \end{bmatrix}$

Chapter 29

$\mathbf{e}^{\mathbf{A}t}$

29.1 DEFINITION

For a square matrix **A**,

$$e^{\mathbf{A}t} \equiv \mathbf{I} + \frac{1}{1!}\mathbf{A}t + \frac{1}{2!}\mathbf{A}^2t^2 + \cdots = \sum_{n=0}^{\infty} \frac{1}{n!}\mathbf{A}^n t^n \tag{29.1}$$

The infinite series (*29.1*) converges for every **A** and t, so that $e^{\mathbf{A}t}$ is defined for all square matrices.

29.2 COMPUTATION OF $e^{\mathbf{A}t}$

For actually computing the elements of $e^{\mathbf{A}t}$, (*29.1*) is not generally useful. However, it follows (with some effort) from Theorem 28.1, applied to the matrix $\mathbf{A}t$, that the infinite series can be reduced to a polynomial in t. Thus:

Theorem 29.1. If **A** is a matrix having n rows and n columns, then

$$e^{\mathbf{A}t} = \alpha_{n-1}\mathbf{A}^{n-1}t^{n-1} + \alpha_{n-2}\mathbf{A}^{n-2}t^{n-2} + \cdots + \alpha_2\mathbf{A}^2t^2 + \alpha_1\mathbf{A}t + \alpha_0\mathbf{I} \tag{29.2}$$

where $\alpha_0, \alpha_1, \ldots, \alpha_{n-1}$ are functions of t which must be determined for each **A**.

Example 29.1. When **A** has two rows and two columns, then $n = 2$ and

$$e^{\mathbf{A}t} = \alpha_1\mathbf{A}t + \alpha_0\mathbf{I} \tag{29.3}$$

When **A** has three rows and three columns, then $n = 3$ and

$$e^{\mathbf{A}t} = \alpha_2\mathbf{A}^2t^2 + \alpha_1\mathbf{A}t + \alpha_0\mathbf{I} \tag{29.4}$$

Theorem 29.2. Let **A** be as in Theorem 29.1, and define

$$r(\lambda) \equiv \alpha_{n-1}\lambda^{n-1} + \alpha_{n-2}\lambda^{n-2} + \cdots + \alpha_2\lambda^2 + \alpha_1\lambda + \alpha_0 \tag{29.5}$$

Then if λ_i is an eigenvalue of $\mathbf{A}t$,

$$e^{\lambda_i} = r(\lambda_i) \tag{29.6}$$

Furthermore, if λ_i is an eigenvalue of multiplicity k, $k > 1$, then the following equations are also valid:

$$\begin{aligned} e^{\lambda_i} &= \frac{d}{d\lambda}r(\lambda)\bigg|_{\lambda=\lambda_i} \\ e^{\lambda_i} &= \frac{d^2}{d\lambda^2}r(\lambda)\bigg|_{\lambda=\lambda_i} \\ &\cdots\cdots\cdots\cdots \\ e^{\lambda_i} &= \frac{d^{k-1}}{d\lambda^{k-1}}r(\lambda)\bigg|_{\lambda=\lambda_i} \end{aligned} \tag{29.7}$$

Note that Theorem 29.2 involves the eigenvalues of $\mathbf{A}t$; these are t-times the eigenvalues of $\mathbf{A}$. (See Problem 29.11.) When computing the various derivatives in (*29.7*), one first calculates the appropriate derivatives of the expression (*29.5*) with respect to λ, and then substitutes $\lambda = \lambda_i$. The reverse procedure of first substituting $\lambda = \lambda_i$ (a function of t) into (*29.5*), and then calculating the derivatives with respect to t, can give erroneous results.

Example 29.2. Let $\mathbf{A}$ have four rows and four columns and let $\lambda = 5t$ and $\lambda = 2t$ be eigenvalues of $\mathbf{A}t$ of multiplicities three and one, respectively. Then $n = 4$ and

$$r(\lambda) = \alpha_3\lambda^3 + \alpha_2\lambda^2 + \alpha_1\lambda + \alpha_0$$

$$r'(\lambda) = 3\alpha_3\lambda^2 + 2\alpha_2\lambda + \alpha_1$$

$$r''(\lambda) = 6\alpha_3\lambda + 2\alpha_2$$

Since $\lambda = 5t$ is an eigenvalue of multiplicity three, it follows that $e^{5t} = r(5t)$, $e^{5t} = r'(5t)$ and $e^{5t} = r''(5t)$. Thus,

$$e^{5t} = \alpha_3(5t)^3 + \alpha_2(5t)^2 + \alpha_1(5t) + \alpha_0$$

$$e^{5t} = 3\alpha_3(5t)^2 + 2\alpha_2(5t) + \alpha_1$$

$$e^{5t} = 6\alpha_3(5t) + 2\alpha_2$$

Also, since $\lambda = 2t$ is an eigenvalue of multiplicity one, it follows that $e^{2t} = r(2t)$, or

$$e^{2t} = \alpha_3(2t)^3 + \alpha_2(2t)^2 + \alpha_1(2t) + \alpha_0$$

Notice that we now have four equations in the four unknown α's.

***Method of computation*:** For each eigenvalue λ_i of $\mathbf{A}t$, apply Theorem 29.2 to obtain a set of linear equations. When this is done for each eigenvalue, the set of all equations so obtained can be solved for $\alpha_0, \alpha_1, \ldots, \alpha_{n-1}$. These values are then substituted into (*29.2*), which, in turn, is used to compute $e^{\mathbf{A}t}$.

Solved Problems

29.1. Find $e^{\mathbf{A}t}$ for $\mathbf{A} = \begin{bmatrix} 1 & 2 \\ 4 & 3 \end{bmatrix}$.

Here, $n = 2$. From (*29.3*),

$$e^{\mathbf{A}t} = \alpha_1\mathbf{A}t + \alpha_0\mathbf{I} = \begin{bmatrix} \alpha_1 t + \alpha_0 & 2\alpha_1 t \\ 4\alpha_1 t & 3\alpha_1 t + \alpha_0 \end{bmatrix} \qquad (1)$$

and from (*29.5*), $r(\lambda) = \alpha_1\lambda + \alpha_0$. The eigenvalues of $\mathbf{A}t$ are $\lambda_1 = -t$ and $\lambda_2 = 5t$, which are both of multiplicity one. Substituting these values successively into (*29.6*), we obtain the two equations

$$e^{-t} = \alpha_1(-t) + \alpha_0 \qquad e^{5t} = \alpha_1(5t) + \alpha_0$$

Solving these equations for α_1 and α_0, we find that

$$\alpha_1 = \frac{1}{6t}(e^{5t} - e^{-t}) \qquad \alpha_0 = \frac{1}{6}(e^{5t} + 5e^{-t})$$

Finally, substituting these values into (*1*) and simplifying, we have

$$e^{\mathbf{A}t} = \frac{1}{6}\begin{bmatrix} 2e^{5t} + 4e^{-t} & 2e^{5t} - 2e^{-t} \\ 4e^{5t} - 4e^{-t} & 4e^{5t} + 2e^{-t} \end{bmatrix}$$

29.2. Find $e^{\mathbf{A}t}$ for $\mathbf{A} = \begin{bmatrix} 0 & 1 \\ 8 & -2 \end{bmatrix}$.

Since $n = 2$, it follows from (*29.3*) and (*29.5*) that

$$e^{\mathbf{A}t} = \alpha_1 \mathbf{A}t + \alpha_0 \mathbf{I} = \begin{bmatrix} \alpha_0 & \alpha_1 t \\ 8\alpha_1 t & -2\alpha_1 t + \alpha_0 \end{bmatrix} \quad (1)$$

and $r(\lambda) = \alpha_1\lambda + \alpha_0$. The eigenvalues of $\mathbf{A}t$ are $\lambda_1 = 2t$ and $\lambda_2 = -4t$, which are both of multiplicity one. Substituting these values successively into (*29.6*), we obtain

$$e^{2t} = \alpha_1(2t) + \alpha_0 \qquad e^{-4t} = \alpha_1(-4t) + \alpha_0$$

Solving these equations for α_1 and α_0, we find that

$$\alpha_1 = \frac{1}{6t}(e^{2t} - e^{-4t}) \qquad \alpha_0 = \frac{1}{3}(2e^{2t} + e^{-4t})$$

Substituting these values into (*1*) and simplifying, we have

$$e^{\mathbf{A}t} = \frac{1}{6}\begin{bmatrix} 4e^{2t} + 2e^{-4t} & e^{2t} - e^{-4t} \\ 8e^{2t} - 8e^{-4t} & 2e^{2t} + 4e^{-4t} \end{bmatrix}$$

29.3. Find $e^{\mathbf{A}t}$ for $\mathbf{A} = \begin{bmatrix} 0 & 1 \\ -1 & 0 \end{bmatrix}$.

Here $n = 2$; hence,

$$e^{\mathbf{A}t} = \alpha_1 \mathbf{A}t + \alpha_0 \mathbf{I} = \begin{bmatrix} \alpha_0 & \alpha_1 t \\ -\alpha_1 t & \alpha_0 \end{bmatrix} \quad (1)$$

and $r(\lambda) = \alpha_1\lambda + \alpha_0$. The eigenvalues of $\mathbf{A}t$ are $\lambda_1 = it$ and $\lambda^2 = -it$, which are both of multiplicity one. Substituting these values successively into (*29.6*), we obtain

$$e^{it} = \alpha_1(it) + \alpha_0 \qquad e^{-it} = \alpha_1(-it) + \alpha_0$$

Solving these equations for α_1 and α_0 and using Euler's relations (page 64), we find that

$$\alpha_1 = \frac{1}{2it}(e^{it} - e^{-it}) = \frac{\sin t}{t}$$

$$\alpha_0 = \frac{1}{2}(e^{it} + e^{-it}) = \cos t$$

Substituting these values into (*1*), we obtain

$$e^{\mathbf{A}t} = \begin{bmatrix} \cos t & \sin t \\ -\sin t & \cos t \end{bmatrix}$$

29.4. Find $e^{\mathbf{A}t}$ for $\mathbf{A} = \begin{bmatrix} 3 & 1 & 0 \\ 0 & 3 & 1 \\ 0 & 0 & 3 \end{bmatrix}$.

Here $n = 3$. From (*29.4*) and (*29.5*) we have

$$e^{\mathbf{A}t} = \alpha_2 \mathbf{A}^2 t^2 + \alpha_1 \mathbf{A}t + \alpha_0 \mathbf{I}$$

$$= \alpha_2 \begin{bmatrix} 9 & 6 & 1 \\ 0 & 9 & 6 \\ 0 & 0 & 9 \end{bmatrix} t^2 + \alpha_1 \begin{bmatrix} 3 & 1 & 0 \\ 0 & 3 & 1 \\ 0 & 0 & 3 \end{bmatrix} t + \alpha_0 \begin{bmatrix} 1 & 0 & 0 \\ 0 & 1 & 0 \\ 0 & 0 & 1 \end{bmatrix}$$

$$= \begin{bmatrix} 9\alpha_2t^2 + 3\alpha_1t + \alpha_0 & 6\alpha_2t^2 + \alpha_1t & \alpha_2t^2 \\ 0 & 9\alpha_2t^2 + 3\alpha_1t + \alpha_0 & 6\alpha_2t^2 + \alpha_1t \\ 0 & 0 & 9\alpha_2t^2 + 3\alpha_1t + \alpha_0 \end{bmatrix} \qquad (1)$$

and $r(\lambda) = \alpha_2\lambda^2 + \alpha_1\lambda + \alpha_0$. Thus,

$$\frac{dr(\lambda)}{d\lambda} = 2\alpha_2\lambda + \alpha_1 \qquad \frac{d^2r(\lambda)}{d\lambda^2} = 2\alpha_2$$

Since the eigenvalues of $\mathbf{A}t$ are $\lambda_1 = \lambda_2 = \lambda_3 = 3t$, an eigenvalue of multiplicity three, it follows from Theorem 29.2 that

$$e^{3t} = \alpha_2 9t^2 + \alpha_1 3t + \alpha_0$$
$$e^{3t} = \alpha_2 6t + \alpha_1$$
$$e^{3t} = 2\alpha_2$$

The solution to this set of equations is

$$\alpha_2 = \frac{1}{2}e^{3t} \qquad \alpha_1 = (1-3t)e^{3t} \qquad \alpha_0 = \left(1 - 3t + \frac{9}{2}t^2\right)e^{3t}$$

Substituting these values into (1) and simplifying, we obtain

$$e^{\mathbf{A}t} = e^{3t}\begin{bmatrix} 1 & t & t^2/2 \\ 0 & 1 & t \\ 0 & 0 & 1 \end{bmatrix}$$

29.5. Find $e^{\mathbf{A}t}$ for $\mathbf{A} = \begin{bmatrix} 0 & 0 & 0 \\ 1 & 0 & 0 \\ 1 & 0 & 1 \end{bmatrix}$.

Here $n = 3$, so

$$e^{\mathbf{A}t} = \alpha_2\mathbf{A}^2t^2 + \alpha_1\mathbf{A}t + \alpha_0\mathbf{I}$$

$$= \begin{bmatrix} \alpha_0 & 0 & 0 \\ \alpha_1t & \alpha_0 & 0 \\ \alpha_2t^2 + \alpha_1t & 0 & \alpha_2t^2 + \alpha_1t + \alpha_0 \end{bmatrix} \qquad (1)$$

and $r(\lambda) = \alpha_2\lambda^2 + \alpha_1\lambda + \alpha_0$. The eigenvalues of $\mathbf{A}t$ are $\lambda_1 = \lambda_2 = 0$, $\lambda_3 = t$; hence, $\lambda = 0$ is an eigenvalue of multiplicity two, while $\lambda = t$ is an eigenvalue of multiplicity one. It then follows from Theorem 29.2 that $e^0 = r(0)$, $e^0 = r'(0)$, and $e^t = r(t)$. Since $r'(\lambda) = 2\alpha_2\lambda + \alpha_1$, these equations become

$$\begin{aligned} e^0 &= \alpha_2(0)^2 + \alpha_1(0) + \alpha_0 & & & 1 &= \alpha_0 \\ e^0 &= 2\alpha_2(0) + \alpha_1 & &\text{or} & 1 &= \alpha_1 \\ e^t &= \alpha_2(t)^2 + \alpha_1(t) + \alpha_0 & & & e^t &= \alpha_2t^2 + \alpha_1t + \alpha_0 \end{aligned}$$

Thus, $\alpha_2 = (e^t - t - 1)/t^2$, $\alpha_1 = 1$, and $\alpha_0 = 1$. Substituting these results into (1) and simplifying, we obtain

$$e^{\mathbf{A}t} = \begin{bmatrix} 1 & 0 & 0 \\ t & 1 & 0 \\ e^t - 1 & 0 & e^t \end{bmatrix}$$

29.6. Establish the necessary equations to find $e^{\mathbf{A}t}$ if

$$\mathbf{A} = \begin{bmatrix} 1 & 2 & 3 & 4 & 5 & 6 \\ 0 & 1 & 2 & 3 & 4 & 5 \\ 0 & 0 & 2 & 3 & 4 & 5 \\ 0 & 0 & 0 & 2 & 3 & 4 \\ 0 & 0 & 0 & 0 & 0 & 0 \\ 0 & 0 & 0 & 0 & 0 & 1 \end{bmatrix}$$

Here $n = 6$, so

$$e^{\mathbf{A}t} = \alpha_5\mathbf{A}^5t^5 + \alpha_4\mathbf{A}^4t^4 + \alpha_3\mathbf{A}^3t^3 + \alpha_2\mathbf{A}^2t^2 + \alpha_1\mathbf{A}t + \alpha_0\mathbf{I}$$

and

$$r(\lambda) = \alpha_5\lambda^5 + \alpha_4\lambda^4 + \alpha_3\lambda^3 + \alpha_2\lambda^2 + \alpha_1\lambda + \alpha_0$$

$$r'(\lambda) = 5\alpha_5\lambda^4 + 4\alpha_4\lambda^3 + 3\alpha_3\lambda^2 + 2\alpha_2\lambda + \alpha_1$$

$$r''(\lambda) = 20\alpha_5\lambda^3 + 12\alpha_4\lambda^2 + 6\alpha_3\lambda + 2\alpha_2$$

The eigenvalues of $\mathbf{A}t$ are $\lambda_1 = \lambda_2 = \lambda_3 = t$, $\lambda_4 = \lambda_5 = 2t$, and $\lambda_6 = 0$. (Recall that the determinant of a triangular matrix is equal to the product of the elements on the main diagonal.) Hence, $\lambda = t$ is an eigenvalue of multiplicity three, $\lambda = 2t$ is an eigenvalue of multiplicity two, and $\lambda = 0$ is an eigenvalue of multiplicity one. It now follows from Theorem 29.2 that

$$e^{2t} = r(2t) = \alpha_5(2t)^5 + \alpha_4(2t)^4 + \alpha_3(2t)^3 + \alpha_2(2t)^2 + \alpha_1(2t) + \alpha_0$$

$$e^{2t} = r'(2t) = 5\alpha_5(2t)^4 + 4\alpha_4(2t)^3 + 3\alpha_3(2t)^2 + 2\alpha_2(2t) + \alpha_1$$

$$e^{2t} = r''(2t) = 20\alpha_5(2t)^3 + 12\alpha_4(2t)^2 + 6\alpha_3(2t) + 2\alpha_2$$

$$e^{t} = r(t) = \alpha_5(t)^5 + \alpha_4(t)^4 + \alpha_3(t)^3 + \alpha_2(t)^2 + \alpha_1(t) + \alpha_0$$

$$e^{t} = r'(t) = 5\alpha_5(t)^4 + 4\alpha_4(t)^3 + 3\alpha_3(t)^2 + 2\alpha_2(t) + \alpha_1$$

$$e^{0} = r(0) = \alpha_5(0)^5 + \alpha_4(0)^4 + \alpha_3(0)^3 + \alpha_2(0)^2 + \alpha_1(0) + \alpha_0$$

or, more simply,

$$\begin{aligned} e^{2t} &= 32t^5\alpha_5 + 16t^4\alpha_4 + 8t^3\alpha_3 + 4t^2\alpha_2 + 2t\alpha_1 + \alpha_0 \\ e^{2t} &= 80t^4\alpha_5 + 32t^3\alpha_4 + 12t^2\alpha_3 + 4t\alpha_2 + \alpha_1 \\ e^{2t} &= 160t^3\alpha_5 + 48t^2\alpha_4 + 12t\alpha_3 + 2\alpha_2 \\ e^{t} &= t^5\alpha_5 + t^4\alpha_4 + t^3\alpha_3 + t^2\alpha_2 + t\alpha_1 + \alpha_0 \\ e^{t} &= 5t^4\alpha_5 + 4t^3\alpha_4 + 3t^2\alpha_3 + 2t\alpha_2 + \alpha_1 \\ 1 &= \alpha_0 \end{aligned}$$

29.7. Find $e^{\mathbf{A}t}e^{\mathbf{B}t}$ and $e^{(\mathbf{A}+\mathbf{B})t}$ for

$$\mathbf{A} = \begin{bmatrix} 0 & 1 \\ 0 & 0 \end{bmatrix} \quad \text{and} \quad \mathbf{B} = \begin{bmatrix} 0 & 0 \\ -1 & 0 \end{bmatrix}$$

and verify that, for these matrices, $e^{\mathbf{A}t}e^{\mathbf{B}t} \neq e^{(\mathbf{A}+\mathbf{B})t}$.

Here, $\mathbf{A} + \mathbf{B} = \begin{bmatrix} 0 & 1 \\ -1 & 0 \end{bmatrix}$. Using Theorem 29.1 and the result of Problem 29.3, we find that

$$e^{\mathbf{A}t} = \begin{bmatrix} 1 & t \\ 0 & 1 \end{bmatrix} \qquad e^{\mathbf{B}t} = \begin{bmatrix} 1 & 0 \\ -t & 1 \end{bmatrix} \qquad e^{(\mathbf{A}+\mathbf{B})t} = \begin{bmatrix} \cos t & \sin t \\ -\sin t & \cos t \end{bmatrix}$$

Thus, $$e^{\mathbf{A}t}e^{\mathbf{B}t} = \begin{bmatrix} 1 & t \\ 0 & 1 \end{bmatrix}\begin{bmatrix} 1 & 0 \\ -t & 1 \end{bmatrix} = \begin{bmatrix} 1-t^2 & t \\ -t & 1 \end{bmatrix} \neq e^{(\mathbf{A}+\mathbf{B})t}$$

29.8. Prove that $e^{\mathbf{A}t}e^{\mathbf{B}t} = e^{(\mathbf{A}+\mathbf{B})t}$ if and only if the matrices A and B commute.

If $\mathbf{AB} = \mathbf{BA}$, and only then, we have

$$(\mathbf{A}+\mathbf{B})^2 = (\mathbf{A}+\mathbf{B})(\mathbf{A}+\mathbf{B}) = \mathbf{A}^2 + \mathbf{AB} + \mathbf{BA} + \mathbf{B}^2 = \mathbf{A}^2 + 2\mathbf{AB} + \mathbf{B}^2$$
$$= \sum_{k=0}^{2} \binom{2}{k} \mathbf{A}^{n-k}\mathbf{B}^k$$

and, in general,
$$(\mathbf{A}+\mathbf{B})^n = \sum_{k=0}^{n} \binom{n}{k} \mathbf{A}^{n-k}\mathbf{B}^k \tag{1}$$

where $\binom{n}{k} = \dfrac{n!}{k!\,(n-k)!}$ is the binomial coefficient ("n things taken k at a time").

Now, according to the defining equation (*29.1*), we have for any A and B:

$$e^{\mathbf{A}t}e^{\mathbf{B}t} = \left(\sum_{n=0}^{\infty} \frac{1}{n!}\mathbf{A}^n t^n\right)\left(\sum_{n=0}^{\infty} \frac{1}{n!}\mathbf{B}^n t^n\right) = \sum_{n=0}^{\infty}\sum_{k=0}^{n} \frac{\mathbf{A}^{n-k}t^{n-k}}{(n-k)!}\frac{\mathbf{B}^k t^k}{k!}$$
$$= \sum_{n=0}^{\infty}\left[\sum_{k=0}^{n} \frac{\mathbf{A}^{n-k}\mathbf{B}^k}{(n-k)!\,k!}\right]t^n = \sum_{n=0}^{\infty}\left[\sum_{k=0}^{n}\binom{n}{k}\mathbf{A}^{n-k}\mathbf{B}^k\right]\frac{t^n}{n!} \tag{2}$$

and also
$$e^{(\mathbf{A}+\mathbf{B})t} = \sum_{n=0}^{\infty} \frac{1}{n!}(\mathbf{A}+\mathbf{B})^n t^n = \sum_{n=0}^{\infty} (\mathbf{A}+\mathbf{B})^n \frac{t^n}{n!} \tag{3}$$

We can equate the last series in (*3*) to the last series in (*2*) if and only if (*1*) holds; that is, if and only if A and B commute.

29.9. Prove that $e^{\mathbf{A}t}e^{-\mathbf{A}s} = e^{\mathbf{A}(t-s)}$.

Setting $t = 1$ in Problem 29.8, we conclude that $e^{\mathbf{A}}e^{\mathbf{B}} = e^{(\mathbf{A}+\mathbf{B})}$ if A and B commute. But the matrices $\mathbf{A}t$ and $-\mathbf{A}s$ commute, since

$$(\mathbf{A}t)(-\mathbf{A}s) = (\mathbf{AA})(-ts) = (\mathbf{AA})(-st) = (-\mathbf{A}s)(\mathbf{A}t)$$

Consequently, $e^{\mathbf{A}t}e^{-\mathbf{A}s} = e^{(\mathbf{A}t-\mathbf{A}s)} = e^{\mathbf{A}(t-s)}$.

29.10. Prove that $e^{\mathbf{0}} = \mathbf{I}$.

From the definition of matrix multiplication, $\mathbf{0}^n = \mathbf{0}$ for $n \geqq 1$. Hence,

$$e^{\mathbf{0}} = e^{\mathbf{0}t} = \sum_{n=0}^{\infty} \frac{1}{n!}\mathbf{0}^n t^n = \mathbf{I} + \sum_{n=1}^{\infty} \frac{1}{n!}\mathbf{0}^n t^n = \mathbf{I} + \mathbf{0} = \mathbf{I}$$

29.11. Prove that the eigenvalues of $\mathbf{A}t$ are t-times the eigenvalues of A.

Let λ be an eigenvalue of the n-rowed square matrix A; we must show that λt is an eigenvalue of $\mathbf{A}t$. Since λ is an eigenvalue of A, it follows from Section 28.7 that $\det(\mathbf{A} - \lambda\mathbf{I}) = 0$. But

$$\det(\mathbf{A}t - \lambda t\mathbf{I}) = \det[t(\mathbf{A} - \lambda\mathbf{I})] = t^n \det(\mathbf{A} - \lambda\mathbf{I}) = t^n(0) = 0$$

and we conclude, again from Section 28.7, that λt is an eigenvalue of $\mathbf{A}t$.

Supplementary Problems

Find $e^{\mathbf{A}t}$ for the following matrices $\mathbf{A}$.

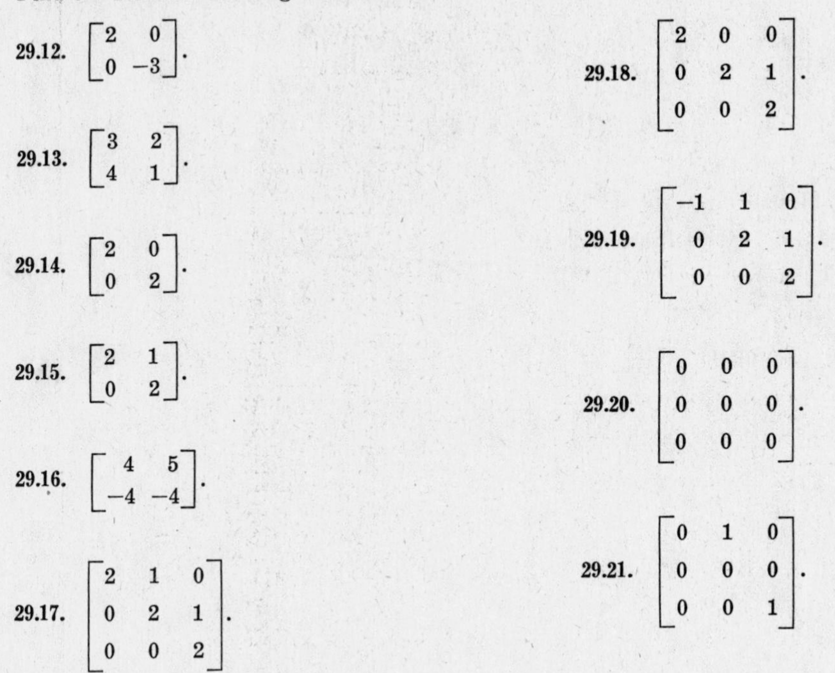

29.12. $\begin{bmatrix} 2 & 0 \\ 0 & -3 \end{bmatrix}$.

29.13. $\begin{bmatrix} 3 & 2 \\ 4 & 1 \end{bmatrix}$.

29.14. $\begin{bmatrix} 2 & 0 \\ 0 & 2 \end{bmatrix}$.

29.15. $\begin{bmatrix} 2 & 1 \\ 0 & 2 \end{bmatrix}$.

29.16. $\begin{bmatrix} 4 & 5 \\ -4 & -4 \end{bmatrix}$.

29.17. $\begin{bmatrix} 2 & 1 & 0 \\ 0 & 2 & 1 \\ 0 & 0 & 2 \end{bmatrix}$.

29.18. $\begin{bmatrix} 2 & 0 & 0 \\ 0 & 2 & 1 \\ 0 & 0 & 2 \end{bmatrix}$.

29.19. $\begin{bmatrix} -1 & 1 & 0 \\ 0 & 2 & 1 \\ 0 & 0 & 2 \end{bmatrix}$.

29.20. $\begin{bmatrix} 0 & 0 & 0 \\ 0 & 0 & 0 \\ 0 & 0 & 0 \end{bmatrix}$.

29.21. $\begin{bmatrix} 0 & 1 & 0 \\ 0 & 0 & 0 \\ 0 & 0 & 1 \end{bmatrix}$.

Answers to Supplementary Problems

29.12. $\lambda_1 = 2t,\ \lambda_2 = -3t;\quad \begin{bmatrix} e^{2t} & 0 \\ 0 & e^{-3t} \end{bmatrix}$

29.13. $\lambda_1 = -t,\ \lambda_2 = 5t;\quad \dfrac{1}{6}\begin{bmatrix} 4e^{5t} + 2e^{-t} & 2e^{5t} - 2e^{-t} \\ 4e^{5t} - 4e^{-t} & 2e^{5t} + 4e^{-t} \end{bmatrix}$

29.14. $\lambda_1 = \lambda_2 = 2t;\quad e^{2t}\begin{bmatrix} 1 & 0 \\ 0 & 1 \end{bmatrix}$

29.15. $\lambda_1 = \lambda_2 = 2t;\quad e^{2t}\begin{bmatrix} 1 & t \\ 0 & 1 \end{bmatrix}$

29.16. $\lambda_1 = 2ti,\ \lambda_2 = -2ti;\quad \begin{bmatrix} \cos 2t + 2\sin 2t & (5/2)\sin 2t \\ -2\sin 2t & \cos 2t - 2\sin 2t \end{bmatrix}$

29.17. $\lambda_1 = \lambda_2 = \lambda_3 = 2t;\quad e^{2t}\begin{bmatrix} 1 & t & t^2/2 \\ 0 & 1 & t \\ 0 & 0 & 1 \end{bmatrix}$

29.18. $\lambda_1 = \lambda_2 = \lambda_3 = 2t;\quad e^{2t}\begin{bmatrix} 1 & 0 & 0 \\ 0 & 1 & t \\ 0 & 0 & 1 \end{bmatrix}$

29.19. $\lambda_1 = -t,\ \lambda_2 = \lambda_3 = 2t;\quad \dfrac{1}{9}\begin{bmatrix} 9e^{-t} & -3e^{-t} + 3e^{2t} & e^{-t} - e^{2t} + 3te^{2t} \\ 0 & 9e^{2t} & 9te^{2t} \\ 0 & 0 & 9e^{2t} \end{bmatrix}$

29.20. $\lambda_1 = \lambda_2 = \lambda_3 = 0;\quad \begin{bmatrix} 1 & 0 & 0 \\ 0 & 1 & 0 \\ 0 & 0 & 1 \end{bmatrix}$ (see Problem 29.10)

29.21. $\lambda_1 = \lambda_2 = 0,\ \lambda_3 = t;\quad \begin{bmatrix} 1 & t & 0 \\ 0 & 1 & 0 \\ 0 & 0 & e^t \end{bmatrix}$

Chapter 30

Reduction of Linear Differential Equations to a First-Order System

Every initial-value problem of the form

$$b_n(t)\frac{d^nx}{dt^n} + b_{n-1}(t)\frac{d^{n-1}x}{dt^{n-1}} + \cdots + b_1(t)\dot{x} + b_0(t)x = g(t); \tag{30.1}$$

$$x(t_0) = c_0, \quad \dot{x}(t_0) = c_1, \quad \ldots, \quad x^{(n-1)}(t_0) = c_{n-1} \tag{30.2}$$

can be reduced to the first-order matrix system

$$\dot{\mathbf{x}}(t) = \mathbf{A}(t)\,\mathbf{x}(t) + \mathbf{f}(t)$$

$$\mathbf{x}(t_0) = \mathbf{c} \tag{30.3}$$

where $\mathbf{A}(t)$, $\mathbf{f}(t)$, $\mathbf{c}$, and the initial time t_0 are known. The method of reduction is as follows.

Step 1. Rewrite (*30.1*) so that $\frac{d^nx}{dt^n}$ appears by itself. Thus,

$$\frac{d^nx}{dt^n} = a_{n-1}(t)\frac{d^{n-1}x}{dt^{n-1}} + \cdots + a_1(t)\dot{x} + a_0(t)x + f(t) \tag{30.4}$$

where $a_j(t) = -b_j(t)/b_n(t)$ $(j = 0, 1, \ldots, n-1)$ and $f(t) = g(t)/b_n(t)$.

Step 2. Define n new variables (the same number as the order of the original differential equation), $x_1(t), x_2(t), \ldots, x_n(t)$, by the equations

$$x_1(t) = x(t), \quad x_2(t) = \frac{dx(t)}{dt}, \quad x_3(t) = \frac{d^2x(t)}{dt^2}, \quad \ldots, \quad x_n(t) = \frac{d^{n-1}x(t)}{dt^{n-1}} \tag{30.5}$$

These new variables are interrelated by the equations

$$\begin{aligned} \dot{x}_1(t) &= x_2(t) \\ \dot{x}_2(t) &= x_3(t) \\ \dot{x}_3(t) &= x_4(t) \\ &\cdots\cdots\cdots \\ \dot{x}_{n-1}(t) &= x_n(t) \end{aligned} \tag{30.6}$$

Step 3. Express $\frac{dx_n}{dt}$ in terms of the new variables. We proceed by first differentiating the last equation of (*30.5*) to obtain

$$\dot{x}_n(t) = \frac{d}{dt}\left[\frac{d^{n-1}x(t)}{dt^{n-1}}\right] = \frac{d^nx(t)}{dt^n}$$

Then, from (*30.4*) and (*30.5*),

$$\dot{x}_n(t) = a_{n-1}(t)\frac{d^{n-1}x(t)}{dt^{n-1}} + \cdots + a_1(t)\,\dot{x}(t) + a_0(t)\,x(t) + f(t)$$

$$= a_{n-1}(t)\,x_n(t) + \cdots + a_1(t)\,x_2(t) + a_0(t)\,x_1(t) + f(t)$$

For convenience, we rewrite this last equation so that $x_1(t)$ appears before $x_2(t)$, etc. Thus,

$$\dot{x}_n(t) = a_0(t)\,x_1(t) + a_1(t)\,x_2(t) + \cdots + a_{n-1}(t)\,x_n(t) + f(t) \tag{30.7}$$

Step 4. Equations (*30.6*) and (*30.7*) are a system of first-order linear differential equations in $x_1(t), x_2(t), \ldots, x_n(t)$. This system is equivalent to the single matrix equation $\dot{\mathbf{x}}(t) = \mathbf{A}(t)\,\mathbf{x}(t) + \mathbf{f}(t)$ if we define

$$\mathbf{x}(t) \equiv \begin{bmatrix} x_1(t) \\ x_2(t) \\ \vdots \\ x_n(t) \end{bmatrix} \tag{30.8}$$

$$\mathbf{f}(t) \equiv \begin{bmatrix} 0 \\ 0 \\ \vdots \\ 0 \\ f(t) \end{bmatrix} \tag{30.9}$$

$$\mathbf{A}(t) \equiv \begin{bmatrix} 0 & 1 & 0 & 0 & \cdots & 0 \\ 0 & 0 & 1 & 0 & \cdots & 0 \\ 0 & 0 & 0 & 1 & \cdots & 0 \\ \vdots & \vdots & \vdots & \vdots & & \vdots \\ 0 & 0 & 0 & 0 & \cdots & 1 \\ a_0(t) & a_1(t) & a_2(t) & a_3(t) & \cdots & a_{n-1}(t) \end{bmatrix} \tag{30.10}$$

Step 5. Define $\mathbf{c} \equiv \begin{bmatrix} c_0 \\ c_1 \\ \vdots \\ c_n \end{bmatrix}$

Then the initial conditions (*30.2*) can be given by the matrix (vector) equation $\mathbf{x}(t_0) = \mathbf{c}$. This last equation is an immediate consequence of (*30.8*), (*30.5*), and (*30.2*), since

$$\mathbf{x}(t_0) = \begin{bmatrix} x_1(t_0) \\ x_2(t_0) \\ \vdots \\ x_n(t_0) \end{bmatrix} = \begin{bmatrix} x(t_0) \\ \dot{x}(t_0) \\ \vdots \\ x^{(n-1)}(t_0) \end{bmatrix} = \begin{bmatrix} c_0 \\ c_1 \\ \vdots \\ c_{n-1} \end{bmatrix} \equiv \mathbf{c}$$

Observe that if no initial conditions are prescribed, Steps 1 through 4 by themselves reduce any linear differential equation (*30.1*) to the matrix equation $\dot{\mathbf{x}}(t) = \mathbf{A}(t)\,\mathbf{x}(t) + \mathbf{f}(t)$. (See Problem 30.4.)

When the original differential equation (*30.1*) has constant coefficients, the matrix system (*30.3*) can be solved by the general method of Chapter 31. Note that once $\mathbf{x}(t)$ is known, its first component, $x(t)$, gives the solution to the initial-value problem (*30.1*) and (*30.2*).

A *set* of linear differential equations with appropriate initial conditions also can be reduced to system (*30.3*). The procedure is completely analogous to the method just given for reducing *one* equation to matrix form; only, for a set of equations, Step 2 must be generalized so that new variables are defined for *each* of the unknown functions sought. (See Problems 30.5 through 30.7.)

Solved Problems

30.1. Put the initial-value problem

$$\ddot{x} + 2\dot{x} - 8x = e^t; \quad x(0) = 1, \quad \dot{x}(0) = -4$$

into form (*30.3*).

Following Step 1, we write $\ddot{x} = -2\dot{x} + 8x + e^t$; hence, $a_1(t) = -2$, $a_0(t) = 8$, and $f(t) = e^t$. Then, defining $x_1(t) = x$ and $x_2(t) = \dot{x}$ (the differential equation is second-order, so we need two new variables), we obtain $\dot{x}_1 = x_2$. Following Step 3, we find

$$\dot{x}_2 = \frac{d^2x}{dt^2} = -2\dot{x} + 8x + e^t = -2x_2 + 8x_1 + e^t$$

Thus,

$$\dot{x}_1 = 0x_1 + 1x_2 + 0$$
$$\dot{x}_2 = 8x_1 - 2x_2 + e^t$$

These equations are equivalent to the matrix equation $\dot{\mathbf{x}}(t) = \mathbf{A}(t)\,\mathbf{x}(t) + \mathbf{f}(t)$ if we define

$$\mathbf{x}(t) \equiv \begin{bmatrix} x_1(t) \\ x_2(t) \end{bmatrix} \qquad \mathbf{A}(t) \equiv \begin{bmatrix} 0 & 1 \\ 8 & -2 \end{bmatrix} \qquad \mathbf{f}(t) \equiv \begin{bmatrix} 0 \\ e^t \end{bmatrix}$$

Furthermore, if we also define $\mathbf{c} \equiv \begin{bmatrix} 1 \\ -4 \end{bmatrix}$, then the initial conditions can be given by $\mathbf{x}(t_0) = \mathbf{c}$, where $t_0 = 0$.

30.2. Put the initial-value problem

$$e^{-t}\frac{d^4x}{dt^4} - \frac{d^2x}{dt^2} + e^t t^2 \frac{dx}{dt} = 5e^{-t};$$

$$x(1) = 2, \quad \dot{x}(1) = 3, \quad \ddot{x}(1) = 4, \quad \dddot{x}(1) = 5$$

into form (*30.3*).

Following Step 1, we obtain

$$\frac{d^4x}{dt^4} = e^t\frac{d^2x}{dt^2} - t^2e^{2t}\frac{dx}{dt} + 5$$

Hence, $a_3(t) = 0$, $a_2(t) = e^t$, $a_1(t) = -t^2e^{2t}$, $a_0(t) = 0$, and $f(t) = 5$. If we define four new variables,

$$x_1(t) = x \qquad x_2(t) = \frac{dx}{dt} \qquad x_3(t) = \frac{d^2x}{dt^2} \qquad x_4(t) = \frac{d^3x}{dt^3}$$

we obtain $\dot{x}_1 = x_2$, $\dot{x}_2 = x_3$, $\dot{x}_3 = x_4$, and, upon following Step 3,

$$\dot{x}_4 = \frac{d^4x}{dt^4} = e^t\ddot{x} - t^2e^{2t}\dot{x} + 5 = e^tx_3 - t^2e^{2t}x_2 + 5$$

Thus,

$$\begin{aligned}\dot{x}_1 &= 0x_1 + 1x_2 + 0x_3 + 0x_4 + 0\\ \dot{x}_2 &= 0x_1 + 0x_2 + 1x_3 + 0x_4 + 0\\ \dot{x}_3 &= 0x_1 + 0x_2 + 0x_3 + 1x_4 + 0\\ \dot{x}_4 &= 0x_1 - t^2e^{2t}x_2 + e^tx_3 + 0x_4 + 5\end{aligned}$$

These equations are equivalent to the matrix equation $\dot{\mathbf{x}}(t) = \mathbf{A}(t)\,\mathbf{x}(t) + \mathbf{f}(t)$ if we define

$$\mathbf{x}(t) \equiv \begin{bmatrix} x_1(t)\\ x_2(t)\\ x_3(t)\\ x_4(t)\end{bmatrix} \qquad \mathbf{A}(t) \equiv \begin{bmatrix} 0 & 1 & 0 & 0\\ 0 & 0 & 1 & 0\\ 0 & 0 & 0 & 1\\ 0 & -t^2e^{2t} & e^t & 0\end{bmatrix} \qquad \mathbf{f}(t) \equiv \begin{bmatrix} 0\\ 0\\ 0\\ 5\end{bmatrix}$$

Furthermore, if we also define $\mathbf{c} \equiv \begin{bmatrix} 2\\ 3\\ 4\\ 5\end{bmatrix}$, then the initial conditions can be given by $\mathbf{x}(t_0) = \mathbf{c}$, where $t_0 = 1$.

30.3. Put the initial-value problem

$$\ddot{x} + 2\dot{x} - 8x = 0; \quad x(1) = 2, \quad \dot{x}(1) = 3$$

into form (*30.3*).

Proceeding as in Problem 30.1, with e^t replaced by zero, we define

$$\mathbf{x}(t) \equiv \begin{bmatrix} x_1(t)\\ x_2(t)\end{bmatrix} \qquad \mathbf{A}(t) \equiv \begin{bmatrix} 0 & 1\\ 8 & -2\end{bmatrix} \qquad \mathbf{f}(t) \equiv \begin{bmatrix} 0\\ 0\end{bmatrix}$$

The differential equation is then equivalent to the matrix equation $\dot{\mathbf{x}}(t) = \mathbf{A}(t)\,\mathbf{x}(t) + \mathbf{f}(t)$, or simply $\dot{\mathbf{x}}(t) = \mathbf{A}(t)\,\mathbf{x}(t)$, since $\mathbf{f}(t) = \mathbf{0}$. The initial conditions can be given by $\mathbf{x}(t_0) = \mathbf{c}$, if we define $t_0 = 1$ and $\mathbf{c} \equiv \begin{bmatrix} 2\\ 3\end{bmatrix}$.

30.4. Convert the differential equation $\ddot{x} - 6\dot{x} + 9x = t$ into the matrix equation

$$\dot{\mathbf{x}}(t) = \mathbf{A}(t)\,\mathbf{x}(t) + \mathbf{f}(t)$$

Here we omit Step 5, since the differential equation has no prescribed initial conditions. Following Step 1, we obtain

$$\ddot{x} = 6\dot{x} - 9x + t$$

Hence $a_1(t) = 6$, $a_0(t) = -9$, and $f(t) = t$. If we define two new variables, $x_1(t) = x$ and $x_2(t) = \dot{x}$, we have

$$\dot{x}_1 = x_2 \quad \text{and} \quad \dot{x}_2 = \ddot{x} = 6\dot{x} - 9x + t = 6x_2 - 9x_1 + t$$

Thus,

$$\begin{aligned}\dot{x}_1 &= 0x_1 + 1x_2 + 0\\ \dot{x}_2 &= -9x_1 + 6x_2 + t\end{aligned}$$

These equations are equivalent to the matrix equation $\dot{\mathbf{x}}(t) = \mathbf{A}(t)\,\mathbf{x}(t) + \mathbf{f}(t)$ if we define

$$\mathbf{x}(t) \equiv \begin{bmatrix} x_1(t)\\ x_2(t)\end{bmatrix} \qquad \mathbf{A}(t) \equiv \begin{bmatrix} 0 & 1\\ -9 & 6\end{bmatrix} \qquad \mathbf{f}(t) \equiv \begin{bmatrix} 0\\ t\end{bmatrix}$$

30.5. Put the following system into form *(30.1)*:

$$\dddot{x} = t\ddot{x} + x - \dot{y} + t + 1$$
$$\ddot{y} = (\sin t)\dot{x} + x - y + t^2;$$
$$x(1) = 2,\ \dot{x}(1) = 3,\ \ddot{x}(1) = 4,\ y(1) = 5,\ \dot{y}(1) = 6$$

Since this system contains a third-order differential equation in x *and* a second-order differential equation in y, we will need three new x-variables and two new y-variables. Generalizing Step 2, we define

$$x_1(t) = x \qquad x_2(t) = \frac{dx}{dt} \qquad x_3(t) = \frac{d^2x}{dt^2}$$

$$y_1(t) = y \qquad y_2(t) = \frac{dy}{dt}$$

Thus,

$$\dot{x}_1 = x_2$$
$$\dot{x}_2 = x_3$$
$$\dot{x}_3 = \frac{d^3x}{dt^3} = t\ddot{x} + x - \dot{y} + t + 1 = tx_3 + x_1 - y_2 + t + 1$$
$$\dot{y}_1 = y_2$$
$$\dot{y}_2 = \frac{d^2y}{dt^2} = (\sin t)\dot{x} + x - y + t^2 = (\sin t)x_2 + x_1 - y_1 + t^2$$

or,

$$\dot{x}_1 = 0x_1 + 1x_2 + 0x_3 + 0y_1 + 0y_2 + 0$$
$$\dot{x}_2 = 0x_1 + 0x_2 + 1x_3 + 0y_1 + 0y_2 + 0$$
$$\dot{x}_3 = 1x_1 + 0x_2 + tx_3 + 0y_1 - 1y_2 + (t+1)$$
$$\dot{y}_1 = 0x_1 + 0x_2 + 0x_3 + 0y_1 + 1y_2 + 0$$
$$\dot{y}_2 = 1x_1 + (\sin t)x_2 + 0x_3 - 1y_1 + 0y_2 + t^2$$

These equations are equivalent to the matrix equation $\dot{\mathbf{x}}(t) = \mathbf{A}(t)\mathbf{x}(t) + \mathbf{f}(t)$ if we define

$$\mathbf{x}(t) \equiv \begin{bmatrix} x_1(t) \\ x_2(t) \\ x_3(t) \\ y_1(t) \\ y_2(t) \end{bmatrix} \qquad \mathbf{A}(t) \equiv \begin{bmatrix} 0 & 1 & 0 & 0 & 0 \\ 0 & 0 & 1 & 0 & 0 \\ 1 & 0 & t & 0 & -1 \\ 0 & 0 & 0 & 0 & 1 \\ 1 & \sin t & 0 & -1 & 0 \end{bmatrix} \qquad \mathbf{f}(t) = \begin{bmatrix} 0 \\ 0 \\ t+1 \\ 0 \\ t^2 \end{bmatrix}$$

Furthermore, if we define $\mathbf{c} \equiv \begin{bmatrix} 2 \\ 3 \\ 4 \\ 5 \\ 6 \end{bmatrix}$ and $t_0 = 1$, then the initial condition can be given by $\mathbf{x}(t_0) = \mathbf{c}$.

30.6. Put the following system into form *(30.1)*:

$$\ddot{x} = -2\dot{x} - 5y + 3$$
$$\dot{y} = \dot{x} + 2y;$$
$$x(0) = 0,\ \dot{x}(0) = 0,\ y(0) = 1$$

Since the system contains a second-order differential equation in x and a first-order differential in y, we define the three new variables

$$x_1(t) = x \qquad x_2(t) = \frac{dx}{dt} \qquad y_1(t) = y$$

Then,

$$\dot{x}_1 = x_2$$
$$\dot{x}_2 = \ddot{x} = -2\dot{x} - 5y + 3 = -2x_2 - 5y_1 + 3$$
$$\dot{y}_1 = \dot{y} = \dot{x} + 2y = x_2 + 2y_1$$

or,

$$\dot{x}_1 = 0x_1 + 1x_2 + 0y_1 + 0$$
$$\dot{x}_2 = 0x_1 - 2x_2 - 5y_1 + 3$$
$$\dot{y}_1 = 0x_1 + 1x_2 + 2y_1 + 0$$

These equations are equivalent to the matrix equation $\dot{\mathbf{x}}(t) = \mathbf{A}(t)\,\mathbf{x}(t) + \mathbf{f}(t)$ if we define

$$\mathbf{x}(t) \equiv \begin{bmatrix} x_1(t) \\ x_2(t) \\ y_1(t) \end{bmatrix} \qquad \mathbf{A}(t) = \begin{bmatrix} 0 & 1 & 0 \\ 0 & -2 & -5 \\ 0 & 1 & 2 \end{bmatrix} \qquad \mathbf{f}(t) = \begin{bmatrix} 0 \\ 3 \\ 0 \end{bmatrix}$$

If we also define $t_0 = 0$ and $\mathbf{c} \equiv \begin{bmatrix} 0 \\ 0 \\ 1 \end{bmatrix}$, then the initial conditions can be given by $\mathbf{x}(t_0) = \mathbf{c}$.

30.7. Put the following system into matrix form:

$$\dot{x} = x + y$$
$$\dot{y} = 9x + y$$

We proceed exactly as in Problems 30.5 and 30.6, except that now there are no initial conditions to consider. Since the system consists of two first-order differential equations, we define two new variables $x_1(t) = x$ and $y_1(t) = y$. Thus,

$$\dot{x}_1 = \dot{x} = x + y = x_1 + y_1 + 0$$
$$\dot{y}_1 = \dot{y} = 9x + y = 9x_1 + y_1 + 0$$

If we define

$$\mathbf{x}(t) \equiv \begin{bmatrix} x_1(t) \\ y_1(t) \end{bmatrix} \qquad \mathbf{A}(t) \equiv \begin{bmatrix} 1 & 1 \\ 9 & 1 \end{bmatrix} \qquad \mathbf{f}(t) \equiv \begin{bmatrix} 0 \\ 0 \end{bmatrix}$$

then this last set of equations is equivalent to the matrix equation $\dot{\mathbf{x}}(t) = \mathbf{A}(t)\,\mathbf{x}(t) + \mathbf{f}(t)$, or simply to $\dot{\mathbf{x}}(t) = \mathbf{A}(t)\,\mathbf{x}(t)$, since $\mathbf{f}(t) = \mathbf{0}$.

Supplementary Problems

Define $\mathbf{x}(t)$, $\mathbf{A}(t)$, $\mathbf{f}(t)$, $\mathbf{c}$ and t_0 so that the given system is equivalent to system (*30.1*).

30.8. $\ddot{x} - 2\dot{x} + x = t + 1; \quad x(1) = 1, \quad \dot{x}(1) = 2.$

30.9. $2\ddot{x} + x = 4e^t; \quad x(0) = 1, \quad \dot{x}(0) = 1.$

30.10. $e^t\dddot{x} - t\ddot{x} + \dot{x} - e^t x = 0;$

$x(-1) = 1, \quad \dot{x}(-1) = 0, \quad \ddot{x}(-1) = 1.$

30.11. $\dddot{x} = t; \quad x(0) = 0, \quad \dot{x}(0) = 0, \quad \ddot{x}(0) = 0.$

30.12. $\ddot{x} = \dot{x} + \dot{y} - z + t$

$\ddot{y} = tx + \dot{y} - 2y + t^2 + 1$

$\dot{z} = x - y + \dot{y} + z;$

$x(1) = 1, \quad \dot{x}(1) = 15, \quad y(1) = 0, \quad \dot{y}(1) = -7, \quad z(1) = 4.$

30.13. $\ddot{x} = 2\dot{x} + 5y + 3$

$\dot{y} = -\dot{x} - 2y;$

$x(0) = 0, \quad \dot{x}(0) = 0, \quad y(0) = 1.$

30.14. $\dot{x} = x + 2y$

$\dot{y} = 4x + 3y;$

$x(7) = 2, \quad y(7) = -3.$

Answers to Supplementary Problems

30.8. $\mathbf{x}(t) \equiv \begin{bmatrix} x_1(t) \\ x_2(t) \end{bmatrix} \quad \mathbf{A}(t) \equiv \begin{bmatrix} 0 & 1 \\ -1 & 2 \end{bmatrix} \quad \mathbf{f}(t) \equiv \begin{bmatrix} 0 \\ t+1 \end{bmatrix} \quad \mathbf{c} \equiv \begin{bmatrix} 1 \\ 2 \end{bmatrix} \quad t_0 = 1$

30.9. $\mathbf{x}(t) \equiv \begin{bmatrix} x_1(t) \\ x_2(t) \end{bmatrix} \quad \mathbf{A}(t) \equiv \begin{bmatrix} 0 & 1 \\ -\frac{1}{2} & 0 \end{bmatrix} \quad \mathbf{f}(t) \equiv \begin{bmatrix} 0 \\ 2e^t \end{bmatrix} \quad \mathbf{c} \equiv \begin{bmatrix} 1 \\ 1 \end{bmatrix} \quad t_0 = 0$

30.10. $\mathbf{x}(t) \equiv \begin{bmatrix} x_1(t) \\ x_2(t) \\ x_3(t) \end{bmatrix} \quad \mathbf{A}(t) \equiv \begin{bmatrix} 0 & 1 & 0 \\ 0 & 0 & 1 \\ 1 & -e^{-t} & te^{-t} \end{bmatrix} \quad \mathbf{f}(t) \equiv \begin{bmatrix} 0 \\ 0 \\ 0 \end{bmatrix} \quad \mathbf{c} \equiv \begin{bmatrix} 1 \\ 0 \\ 1 \end{bmatrix} \quad t_0 = -1$

30.11. $\mathbf{x}(t) \equiv \begin{bmatrix} x_1(t) \\ x_2(t) \\ x_3(t) \end{bmatrix} \quad \mathbf{A}(t) \equiv \begin{bmatrix} 0 & 1 & 0 \\ 0 & 0 & 1 \\ 0 & 0 & 0 \end{bmatrix} \quad \mathbf{f}(t) \equiv \begin{bmatrix} 0 \\ 0 \\ t \end{bmatrix} \quad \mathbf{c} \equiv \begin{bmatrix} 0 \\ 0 \\ 0 \end{bmatrix} \quad t_0 = 0$

30.12. $\mathbf{x}(t) \equiv \begin{bmatrix} x_1(t) \\ x_2(t) \\ y_1(t) \\ y_2(t) \\ z_1(t) \end{bmatrix}$ $\mathbf{A}(t) \equiv \begin{bmatrix} 0 & 1 & 0 & 0 & 0 \\ 0 & 1 & 0 & 1 & -1 \\ 0 & 0 & 0 & 1 & 0 \\ t & 0 & -2 & 1 & 0 \\ 1 & 0 & -1 & 1 & 1 \end{bmatrix}$

$\mathbf{f}(t) \equiv \begin{bmatrix} 0 \\ t \\ 0 \\ t^2+1 \\ 0 \end{bmatrix}$ $\mathbf{c} \equiv \begin{bmatrix} 1 \\ 15 \\ 0 \\ -7 \\ 4 \end{bmatrix}$ $t_0 = 1$

30.13. $\mathbf{x}(t) \equiv \begin{bmatrix} x_1(t) \\ x_2(t) \\ y_1(t) \end{bmatrix}$ $\mathbf{A}(t) \equiv \begin{bmatrix} 0 & 1 & 0 \\ 0 & 2 & 5 \\ 0 & -1 & -2 \end{bmatrix}$ $\mathbf{f}(t) \equiv \begin{bmatrix} 0 \\ 3 \\ 0 \end{bmatrix}$ $\mathbf{c} \equiv \begin{bmatrix} 0 \\ 0 \\ 1 \end{bmatrix}$ $t_0 = 0$

30.14. $\mathbf{x}(t) \equiv \begin{bmatrix} x_1(t) \\ y_1(t) \end{bmatrix}$ $\mathbf{A}(t) \equiv \begin{bmatrix} 1 & 2 \\ 4 & 3 \end{bmatrix}$ $\mathbf{f}(t) \equiv \begin{bmatrix} 0 \\ 0 \end{bmatrix}$ $\mathbf{c} \equiv \begin{bmatrix} 2 \\ -3 \end{bmatrix}$ $t_0 = 7$

Chapter 31

Solutions of Linear Systems with Constant Coefficients

31.1 INTRODUCTION

By the procedure of Chapter 30, any linear system with constant coefficients can be reduced to a single matrix differential equation $\dot{\mathbf{x}}(t) = \mathbf{A}\mathbf{x}(t) + \mathbf{f}(t)$, where $\mathbf{A}$ is a constant matrix. Such an equation is formally equivalent to (*8.1*) with $p(x) = -a$, a constant, and its solution, given below, should be compared with the result of Problem 8.9.

31.2 SOLUTION OF THE INITIAL-VALUE PROBLEM

The matrix system

$$\dot{\mathbf{x}}(t) = \mathbf{A}\mathbf{x}(t) + \mathbf{f}(t); \quad \mathbf{x}(t_0) = \mathbf{c} \tag{31.1}$$

has the solution

$$\mathbf{x}(t) = e^{\mathbf{A}(t-t_0)}\mathbf{c} + e^{\mathbf{A}t}\int_{t_0}^{t} e^{-\mathbf{A}s}\mathbf{f}(s)\,ds \tag{31.2}$$

or equivalently (see Problem 29.9)

$$\mathbf{x}(t) = e^{\mathbf{A}(t-t_0)}\mathbf{c} + \int_{t_0}^{t} e^{\mathbf{A}(t-s)}\mathbf{f}(s)\,ds \tag{31.3}$$

In particular, if the initial-value problem is *homogeneous* (i.e., $\mathbf{f}(t) = \mathbf{0}$), then both equations (*31.2*) and (*31.3*) reduce to

$$\mathbf{x}(t) = e^{\mathbf{A}(t-t_0)}\mathbf{c} \tag{31.4}$$

In the above solutions, the matrices $e^{\mathbf{A}(t-t_0)}$, $e^{-\mathbf{A}s}$, and $e^{\mathbf{A}(t-s)}$ are easily computed from $e^{\mathbf{A}t}$ by replacing the variable t by $t-t_0$, $-s$, and $t-s$, respectively. Usually, $\mathbf{x}(t)$ is obtained quicker from (*31.3*) than from (*31.2*), since the former equation involves one less matrix multiplication. However, the integrals arising in (*31.3*) are generally more difficult to evaluate than those in (*31.2*). (See Problems 31.2 and 31.3.)

If no initial conditions are prescribed, the solution of $\dot{\mathbf{x}}(t) = \mathbf{A}\mathbf{x}(t) + \mathbf{f}(t)$ is

$$\mathbf{x}(t) = e^{\mathbf{A}t}\mathbf{k} + e^{\mathbf{A}t}\int e^{-\mathbf{A}t}\mathbf{f}(t)\,dt \tag{31.5}$$

or, when $\mathbf{f}(t) = \mathbf{0}$,

$$\mathbf{x}(t) = e^{\mathbf{A}t}\mathbf{k} \tag{31.6}$$

where $\mathbf{k}$ is an arbitrary constant vector. All constants of integration can be disregarded when computing the integral in (*31.5*), since they are already included in $\mathbf{k}$.

31.3 COMPARISON OF SOLUTION METHODS

We have developed three ways to solve linear differential equations with constant coefficients: (1) the characteristic equation method, coupled with either variation of parameters or the method of undetermined coefficients; (2) Laplace transforms; and (3) the matrix method. A comparison is therefore in order.

All three methods involve root finding and, usually, integration, two procedures which can be impossible to perform practically. Computing the roots of a polynomial equation is an obvious step in the characteristic equation method. It is also a part of the matrix method (where one must compute the eigenvalues of $\mathbf{A}$ in order to find $e^{\mathbf{A}t}$) and Laplace transforms (where one factors to quadratic terms in order to compute the inverse transform). Integration is a necessary step in both variation of parameters and the matrix method. With Laplace transforms, complex integration by residues is the usual means for obtaining $\mathcal{L}^{-1}$ when the inverse transform is not tabulated.

When the method of undetermined coefficients is applicable, then either the characteristic equation method or Laplace transforms is the most efficient technique. For more complicated differential equations, and also for systems of differential equations where the characteristic equation method is inappropriate, either Laplace transforms or the matrix method is the preferred technique. If $\mathcal{L}^{-1}$ is tabulated, then Laplace transforms is generally quicker than the matrix method; the same is true if $\mathcal{L}^{-1}$ is not tabulated but can be obtained from the theory of residues (a procedure which is beyond the scope of this book). When $\mathcal{L}^{-1}$ cannot be obtained easily, then the matrix method is more efficient if the integrations required by this method can be performed. In general, however, there is little to choose between Laplace transforms and the matrix method. In those cases where both methods prove intractable, numerical methods (Chapters 32-36) must be used.

One distinct advantage of the matrix method is that an expression for the solution, (*31.2*), is readily available. On occasion, this representation can be used to obtain information about the solution without explicitly calculating the solution.

Solved Problems

31.1. Solve $\ddot{x} + 2\dot{x} - 8x = 0;\; x(1) = 2,\; \dot{x}(1) = 3.$

From Problem 30.3, this initial-value problem is equivalent to (*31.1*) with

$$\mathbf{x}(t) = \begin{bmatrix} x_1(t) \\ x_2(t) \end{bmatrix} \qquad \mathbf{A} = \begin{bmatrix} 0 & 1 \\ 8 & -2 \end{bmatrix} \qquad \mathbf{f}(t) = \mathbf{0} \qquad \mathbf{c} = \begin{bmatrix} 2 \\ 3 \end{bmatrix} \qquad t_0 = 1$$

The solution to this system is given by (*31.4*). For this $\mathbf{A}$, $e^{\mathbf{A}t}$ is given in Problem 29.2; hence,

$$e^{\mathbf{A}(t-t_0)} = e^{\mathbf{A}(t-1)} = \frac{1}{6}\begin{bmatrix} 4e^{2(t-1)} + 2e^{-4(t-1)} & e^{2(t-1)} - e^{-4(t-1)} \\ 8e^{2(t-1)} - 8e^{-4(t-1)} & 2e^{2(t-1)} + 4e^{-4(t-1)} \end{bmatrix}$$

Therefore,

$$\begin{aligned} \mathbf{x}(t) &= e^{\mathbf{A}(t-1)}\mathbf{c} \\ &= \frac{1}{6}\begin{bmatrix} 4e^{2(t-1)} + 2e^{-4(t-1)} & e^{2(t-1)} - e^{-4(t-1)} \\ 8e^{2(t-1)} - 8e^{-4(t-1)} & 2e^{2(t-1)} + 4e^{-4(t-1)} \end{bmatrix}\begin{bmatrix} 2 \\ 3 \end{bmatrix} \\ &= \frac{1}{6}\begin{bmatrix} 2(4e^{2(t-1)} + 2e^{-4(t-1)}) + 3(e^{2(t-1)} - e^{-4(t-1)}) \\ 2(8e^{2(t-1)} - 8e^{-4(t-1)}) + 3(2e^{2(t-1)} + 4e^{-4(t-1)}) \end{bmatrix} \\ &= \begin{bmatrix} \frac{11}{6}e^{2(t-1)} + \frac{1}{6}e^{-4(t-1)} \\ \frac{22}{6}e^{2(t-1)} - \frac{4}{6}e^{-4(t-1)} \end{bmatrix} \end{aligned}$$

and the solution to the original initial-value problem is

$$x(t) = x_1(t) = \frac{11}{6}e^{2(t-1)} + \frac{1}{6}e^{-4(t-1)}$$

31.2. Solve $\ddot{x} + 2\dot{x} - 8x = e^t;\ x(0) = 1,\ \dot{x}(0) = -4.$

From Problem 30.1, this initial-value problem is equivalent to *(31.1)* with

$$\mathbf{x}(t) = \begin{bmatrix} x_1(t) \\ x_2(t) \end{bmatrix} \qquad \mathbf{A} = \begin{bmatrix} 0 & 1 \\ 8 & -2 \end{bmatrix} \qquad \mathbf{f}(t) = \begin{bmatrix} 0 \\ e^t \end{bmatrix} \qquad \mathbf{c} = \begin{bmatrix} 1 \\ -4 \end{bmatrix}$$

and $t_0 = 0$. The solution is given by either *(31.2)* or *(31.3)*. Here, we use *(31.2)*; the solution using *(31.3)* is found in Problem 31.3. For this $\mathbf{A}$, $e^{\mathbf{A}t}$ has already been calculated in Problem 29.2. Therefore,

$$e^{\mathbf{A}(t-t_0)}\mathbf{c} = e^{\mathbf{A}t}\mathbf{c} = \frac{1}{6}\begin{bmatrix} 4e^{2t} + 2e^{-4t} & e^{2t} - e^{-4t} \\ 8e^{2t} - 8e^{-4t} & 2e^{2t} + 4e^{-4t} \end{bmatrix}\begin{bmatrix} 1 \\ -4 \end{bmatrix} = \begin{bmatrix} e^{-4t} \\ -4e^{-4t} \end{bmatrix}$$

$$e^{-\mathbf{A}s}\mathbf{f}(s) = \frac{1}{6}\begin{bmatrix} 4e^{-2s} + 2e^{4s} & e^{-2s} - e^{4s} \\ 8e^{-2s} - 8e^{4s} & 2e^{-2s} + 4e^{4s} \end{bmatrix}\begin{bmatrix} 0 \\ e^s \end{bmatrix} = \begin{bmatrix} \frac{1}{6}e^{-s} - \frac{1}{6}e^{5s} \\ \frac{2}{6}e^{-s} + \frac{4}{6}e^{5s} \end{bmatrix}$$

$$\int_{t_0}^{t} e^{-\mathbf{A}s}\mathbf{f}(s)\,ds = \begin{bmatrix} \int_0^t \left(\frac{1}{6}e^{-s} - \frac{1}{6}e^{5s}\right) ds \\ \int_0^t \left(\frac{1}{3}e^{-s} + \frac{2}{3}e^{5s}\right) ds \end{bmatrix} = \frac{1}{30}\begin{bmatrix} -5e^{-t} - e^{5t} + 6 \\ -10e^{-t} + 4e^{5t} + 6 \end{bmatrix}$$

$$e^{\mathbf{A}t}\int_{t_0}^{t} e^{-\mathbf{A}s}\mathbf{f}(s)\,ds = \left(\frac{1}{6}\right)\left(\frac{1}{30}\right)\begin{bmatrix} 4e^{2t} + 2e^{-4t} & e^{2t} - e^{-4t} \\ 8e^{2t} - 8e^{-4t} & 2e^{2t} + 4e^{-4t} \end{bmatrix}\begin{bmatrix} -5e^{-t} - e^{5t} + 6 \\ -10e^{-t} + 4e^{5t} + 6 \end{bmatrix}$$

$$= \frac{1}{180}\begin{bmatrix} (4e^{2t} + 2e^{-4t})(-5e^{-t} - e^{5t} + 6) + (e^{2t} - e^{-4t})(-10e^{-t} + 4e^{5t} + 6) \\ (8e^{2t} - 8e^{-4t})(-5e^{-t} - e^{5t} + 6) + (2e^{2t} + 4e^{-4t})(-10e^{-t} + 4e^{5t} + 6) \end{bmatrix}$$

$$= \frac{1}{30}\begin{bmatrix} -6e^t + 5e^{2t} + e^{-4t} \\ -6e^t + 10e^{2t} - 4e^{-4t} \end{bmatrix}$$

Thus,

$$\mathbf{x}(t) = e^{\mathbf{A}(t-t_0)}\mathbf{c} + e^{\mathbf{A}t}\int_{t_0}^{t} e^{-\mathbf{A}s}\mathbf{f}(s)\,ds$$

$$= \begin{bmatrix} e^{-4t} \\ -4e^{-4t} \end{bmatrix} + \frac{1}{30}\begin{bmatrix} -6e^t + 5e^{2t} + e^{-4t} \\ -6e^t + 10e^{2t} - 4e^{-4t} \end{bmatrix} = \begin{bmatrix} \frac{31}{30}e^{-4t} + \frac{1}{6}e^{2t} - \frac{1}{5}e^t \\ -\frac{62}{15}e^{-4t} + \frac{1}{3}e^{2t} - \frac{1}{5}e^t \end{bmatrix}$$

and $x(t) = x_1(t) = \frac{31}{30}e^{-4t} + \frac{1}{6}e^{2t} - \frac{1}{5}e^t.$

31.3. Use *(31.3)* to solve the initial-value problem of Problem 31.2.

The vector $e^{\mathbf{A}(t-t_0)}\mathbf{c}$ remains $\begin{bmatrix} e^{-4t} \\ -4e^{-4t} \end{bmatrix}$. Furthermore,

$$e^{\mathbf{A}(t-s)}\mathbf{f}(s) = \frac{1}{6}\begin{bmatrix} 4e^{2(t-s)} + 2e^{-4(t-s)} & e^{2(t-s)} - e^{-4(t-s)} \\ 8e^{2(t-s)} - 8e^{-4(t-s)} & 2e^{2(t-s)} + 4e^{-4(t-s)} \end{bmatrix}\begin{bmatrix} 0 \\ e^s \end{bmatrix}$$

$$= \frac{1}{6}\begin{bmatrix} e^{(2t-s)} - e^{(-4t+5s)} \\ 2e^{(2t-s)} + 4e^{(-4t+5s)} \end{bmatrix}$$

$$\int_{t_0}^{t} e^{\mathbf{A}(t-s)}\mathbf{f}(s)\,ds = \frac{1}{6}\begin{bmatrix} \int_0^t [e^{(2t-s)} - e^{(-4t+5s)}]\,ds \\ \int_0^t [2e^{(2t-s)} + 4e^{(-4t+5s)}]\,ds \end{bmatrix}$$

$$= \frac{1}{6}\begin{bmatrix} \left[-e^{(2t-s)} - \frac{1}{5}e^{(-4t+5s)}\right]_{s=0}^{s=t} \\ \left[-2e^{(2t-s)} + \frac{4}{5}e^{(-4t+5s)}\right]_{s=0}^{s=t} \end{bmatrix} = \frac{1}{6}\begin{bmatrix} -\frac{6}{5}e^{t} + e^{2t} + \frac{1}{5}e^{-4t} \\ -\frac{6}{5}e^{t} + 2e^{2t} - \frac{4}{5}e^{-4t} \end{bmatrix}$$

Thus,

$$\mathbf{x}(t) = e^{\mathbf{A}(t-t_0)}\mathbf{c} + \int_{t_0}^{t} e^{\mathbf{A}(t-s)}\mathbf{f}(s)\,ds$$

$$= \begin{bmatrix} e^{-4t} \\ -4e^{-4t} \end{bmatrix} + \frac{1}{6}\begin{bmatrix} -\frac{6}{5}e^{t} + e^{2t} + \frac{1}{5}e^{-4t} \\ -\frac{6}{5}e^{t} + 2e^{2t} - \frac{4}{5}e^{-4t} \end{bmatrix} = \begin{bmatrix} \frac{31}{30}e^{-4t} + \frac{1}{6}e^{2t} - \frac{1}{5}e^{t} \\ -\frac{62}{15}e^{-4t} + \frac{1}{3}e^{2t} - \frac{1}{5}e^{t} \end{bmatrix}$$

as before.

31.4. Solve $\ddot{x} + x = 3;\ x(\pi) = 1,\ \dot{x}(\pi) = 2.$

This initial-value problem is equivalent to *(31.1)* with

$$\mathbf{x}(t) = \begin{bmatrix} x_1(t) \\ x_2(t) \end{bmatrix} \qquad \mathbf{A} = \begin{bmatrix} 0 & 1 \\ -1 & 0 \end{bmatrix} \qquad \mathbf{f}(t) = \begin{bmatrix} 0 \\ 3 \end{bmatrix} \qquad \mathbf{c} = \begin{bmatrix} 1 \\ 2 \end{bmatrix}$$

and $t_0 = \pi$. Then, using *(31.3)* and the results of Problem 29.3, we find that

$$e^{\mathbf{A}(t-t_0)}\mathbf{c} = \begin{bmatrix} \cos(t-\pi) & \sin(t-\pi) \\ -\sin(t-\pi) & \cos(t-\pi) \end{bmatrix}\begin{bmatrix} 1 \\ 2 \end{bmatrix} = \begin{bmatrix} \cos(t-\pi) + 2\sin(t-\pi) \\ -\sin(t-\pi) + 2\cos(t-\pi) \end{bmatrix}$$

$$e^{\mathbf{A}(t-s)}\mathbf{f}(s) = \begin{bmatrix} \cos(t-s) & \sin(t-s) \\ -\sin(t-s) & \cos(t-s) \end{bmatrix}\begin{bmatrix} 0 \\ 3 \end{bmatrix} = \begin{bmatrix} 3\sin(t-s) \\ 3\cos(t-s) \end{bmatrix}$$

$$\int_{t_0}^{t} e^{\mathbf{A}(t-s)}\mathbf{f}(s)\,ds = \begin{bmatrix} \int_{\pi}^{t} 3\sin(t-s)\,ds \\ \int_{\pi}^{t} 3\cos(t-s)\,ds \end{bmatrix}$$

$$= \begin{bmatrix} 3\cos(t-s)\Big|_{s=\pi}^{s=t} \\ -3\sin(t-s)\Big|_{s=\pi}^{s=t} \end{bmatrix} = \begin{bmatrix} 3 - 3\cos(t-\pi) \\ 3\sin(t-\pi) \end{bmatrix}$$

Thus,

$$\mathbf{x}(t) = e^{\mathbf{A}(t-t_0)}\mathbf{c} + \int_{t_0}^{t} e^{\mathbf{A}(t-s)}\mathbf{f}(s)\,ds$$

$$= \begin{bmatrix} \cos(t-\pi) + 2\sin(t-\pi) \\ -\sin(t-\pi) + 2\cos(t-\pi) \end{bmatrix} + \begin{bmatrix} 3 - 3\cos(t-\pi) \\ 3\sin(t-\pi) \end{bmatrix}$$

$$= \begin{bmatrix} 3 - 2\cos(t-\pi) + 2\sin(t-\pi) \\ 2\cos(t-\pi) + 2\sin(t-\pi) \end{bmatrix}$$

and $x(t) = x_1(t) = 3 - 2\cos(t-\pi) + 2\sin(t-\pi)$.

Noting that $\cos(t-\pi) = -\cos t$ and $\sin(t-\pi) = -\sin t$, we also obtain

$$x(t) = 3 + 2\cos t - 2\sin t$$

31.5. Solve the differential equation $\ddot{x} - 6\dot{x} + 9x = t$.

This differential equation is equivalent to the standard matrix differential equation with

$$\mathbf{x}(t) = \begin{bmatrix} x_1(t) \\ x_2(t) \end{bmatrix} \qquad \mathbf{A} = \begin{bmatrix} 0 & 1 \\ -9 & 6 \end{bmatrix} \qquad \mathbf{f}(t) = \begin{bmatrix} 0 \\ t \end{bmatrix}$$

(see Problem 30.4). For this $\mathbf{A}$, we compute

$$e^{\mathbf{A}t} = \begin{bmatrix} (1-3t)e^{3t} & te^{3t} \\ -9te^{3t} & (1+3t)e^{3t} \end{bmatrix} \qquad e^{-\mathbf{A}t} = \begin{bmatrix} (1+3t)e^{-3t} & -te^{-3t} \\ 9te^{-3t} & (1-3t)e^{-3t} \end{bmatrix}$$

Then, using (*31.5*), we obtain

$$e^{\mathbf{A}t}\mathbf{k} = \begin{bmatrix} (1-3t)e^{3t} & te^{3t} \\ -9te^{3t} & (1+3t)e^{3t} \end{bmatrix}\begin{bmatrix} k_1 \\ k_2 \end{bmatrix} = \begin{bmatrix} [(-3k_1+k_2)t + k_1]e^{3t} \\ [(-9k_1+3k_2)t + k_2]e^{3t} \end{bmatrix}$$

$$e^{-\mathbf{A}t}\mathbf{f}(t) = \begin{bmatrix} (1+3t)e^{-3t} & -te^{-3t} \\ 9te^{-3t} & (1-3t)e^{-3t} \end{bmatrix}\begin{bmatrix} 0 \\ t \end{bmatrix} = \begin{bmatrix} -t^2e^{-3t} \\ (t-3t^2)e^{-3t} \end{bmatrix}$$

$$\int e^{-\mathbf{A}t}\mathbf{f}(t)\,dt = \begin{bmatrix} -\int t^2e^{-3t}\,dt \\ \int (t-3t^2)e^{-3t}\,dt \end{bmatrix} = \begin{bmatrix} \left(\frac{1}{3}t^2 + \frac{2}{9}t + \frac{2}{27}\right)e^{-3t} \\ \left(t^2 + \frac{1}{3}t + \frac{1}{9}\right)e^{-3t} \end{bmatrix}$$

$$e^{\mathbf{A}t}\int e^{-\mathbf{A}t}\mathbf{f}(t)\,dt = \begin{bmatrix} (1-3t)e^{3t} & te^{3t} \\ -9te^{3t} & (1+3t)e^{3t} \end{bmatrix}\begin{bmatrix} \left(\frac{1}{3}t^2 + \frac{2}{9}t + \frac{2}{27}\right)e^{-3t} \\ \left(t^2 + \frac{1}{3}t + \frac{1}{9}\right)e^{-3t} \end{bmatrix} = \begin{bmatrix} \frac{1}{9}t + \frac{2}{27} \\ \frac{1}{9} \end{bmatrix}$$

and

$$\mathbf{x}(t) = e^{\mathbf{A}t}\mathbf{k} + e^{\mathbf{A}t}\int e^{-\mathbf{A}t}\mathbf{f}(t)\,dt$$

$$= \begin{bmatrix} [(-3k_1+k_2)t + k_1]e^{3t} + \frac{1}{9}t + \frac{2}{27} \\ [(-9k_1+3k_2)t + k_2]e^{3t} + \frac{1}{9} \end{bmatrix}$$

Thus, $x(t) = x_1(t) = [(-3k_1+k_2)t + k_1]e^{3t} + \frac{1}{9}t + \frac{2}{27} = (k_1 + k_3t)e^{3t} + \frac{1}{9}t + \frac{2}{27}$ where $k_3 = -3k_1 + k_2$.

31.6. Solve the system

$$\ddot{x} = -2\dot{x} - 5y + 3$$
$$\dot{y} = \dot{x} + 2y;$$
$$x(0) = 0, \quad \dot{x}(0) = 0, \quad y(0) = 1$$

This initial-value problem is equivalent to (31.1) with (see Problem 30.6)

$$\mathbf{x}(t) = \begin{bmatrix} x_1(t) \\ x_2(t) \\ y_1(t) \end{bmatrix} \qquad \mathbf{A} = \begin{bmatrix} 0 & 1 & 0 \\ 0 & -2 & -5 \\ 0 & 1 & 2 \end{bmatrix} \qquad \mathbf{f}(t) = \begin{bmatrix} 0 \\ 3 \\ 0 \end{bmatrix} \qquad \mathbf{c} = \begin{bmatrix} 0 \\ 0 \\ 1 \end{bmatrix}$$

and $t_0 = 0$. For this **A**, we compute

$$e^{\mathbf{A}t} = \begin{bmatrix} 1 & -2 + 2\cos t + \sin t & -5 + 5\cos t \\ 0 & \cos t - 2\sin t & -5\sin t \\ 0 & \sin t & \cos t + 2\sin t \end{bmatrix}$$

Then, using (31.3), we find that

$$e^{\mathbf{A}(t-t_0)}\mathbf{c} = \begin{bmatrix} 1 & -2 + 2\cos t + \sin t & -5 + 5\cos t \\ 0 & \cos t - 2\sin t & -5\sin t \\ 0 & \sin t & \cos t + 2\sin t \end{bmatrix} \begin{bmatrix} 0 \\ 0 \\ 1 \end{bmatrix} = \begin{bmatrix} -5 + 5\cos t \\ -5\sin t \\ \cos t + 2\sin t \end{bmatrix}$$

$$e^{\mathbf{A}(t-s)}\mathbf{f}(s) = \begin{bmatrix} 1 & -2 + 2\cos(t-s) + \sin(t-s) & -5 + 5\cos(t-s) \\ 0 & \cos(t-s) - 2\sin(t-s) & -5\sin(t-s) \\ 0 & \sin(t-s) & \cos(t-s) + 2\sin(t-s) \end{bmatrix} \begin{bmatrix} 0 \\ 3 \\ 0 \end{bmatrix}$$

$$= \begin{bmatrix} -6 + 6\cos(t-s) + 3\sin(t-s) \\ 3\cos(t-s) - 6\sin(t-s) \\ 3\sin(t-s) \end{bmatrix}$$

and

$$\int_{t_0}^{t} e^{\mathbf{A}(t-s)}\mathbf{f}(s)\,ds = \begin{bmatrix} \int_0^t [-6 + 6\cos(t-s) + 3\sin(t-s)]\,ds \\ \int_0^t [3\cos(t-s) - 6\sin(t-s)]\,ds \\ \int_0^t 3\sin(t-s)\,ds \end{bmatrix}$$

$$= \begin{bmatrix} \left[-6s - 6\sin(t-s) + 3\cos(t-s)\right]_{s=0}^{s=t} \\ \left[-3\sin(t-s) - 6\cos(t-s)\right]_{s=0}^{s=t} \\ 3\cos(t-s)\Big|_{s=0}^{s=t} \end{bmatrix}$$

$$= \begin{bmatrix} -6t + 3 + 6\sin t - 3\cos t \\ -6 + 3\sin t + 6\cos t \\ 3 - 3\cos t \end{bmatrix}$$

Therefore,
$$\mathbf{x}(t) = e^{\mathbf{A}(t-t_0)}\mathbf{c} + \int_{t_0}^{t} e^{\mathbf{A}(t-s)}\mathbf{f}(s)\,ds$$

$$= \begin{bmatrix} -5 + 5\cos t \\ -5\sin t \\ \cos t + 2\sin t \end{bmatrix} + \begin{bmatrix} -6t + 3 + 6\sin t - 3\cos t \\ -6 + 3\sin t + 6\cos t \\ 3 - 3\cos t \end{bmatrix}$$

$$= \begin{bmatrix} -2 - 6t + 2\cos t + 6\sin t \\ -6 + 6\cos t - 2\sin t \\ 3 - 2\cos t + 2\sin t \end{bmatrix}$$

Finally,
$$x(t) = x_1(t) = 2\cos t + 6\sin t - 2 - 6t$$
$$y(t) = y_1(t) = -2\cos t + 2\sin t + 3$$

31.7. Solve the system of differential equations
$$\dot{x} = x + y$$
$$\dot{y} = 9x + y$$

This set of equations is equivalent to the matrix system $\dot{\mathbf{x}}(t) = \mathbf{A}\mathbf{x}(t)$ with (See Problem 30.7)

$$\mathbf{x}(t) = \begin{bmatrix} x_1(t) \\ y_1(t) \end{bmatrix} \qquad \mathbf{A} = \begin{bmatrix} 1 & 1 \\ 9 & 1 \end{bmatrix}$$

The solution is given by (*31.6*). For this **A**, we compute

$$e^{\mathbf{A}t} = \frac{1}{6}\begin{bmatrix} 3e^{4t} + 3e^{-2t} & e^{4t} - e^{-2t} \\ 9e^{4t} - 9e^{-2t} & 3e^{4t} + 3e^{-2t} \end{bmatrix}$$

hence,
$$\mathbf{x}(t) = e^{\mathbf{A}t}\mathbf{k} = \frac{1}{6}\begin{bmatrix} 3e^{4t} + 3e^{-2t} & e^{4t} - e^{-2t} \\ 9e^{4t} - 9e^{-2t} & 3e^{4t} + 3e^{-2t} \end{bmatrix}\begin{bmatrix} k_1 \\ k_2 \end{bmatrix}$$

$$= \begin{bmatrix} \frac{1}{6}(3k_1 + k_2)e^{4t} + \frac{1}{6}(3k_1 - k_2)e^{-2t} \\ \frac{3}{6}(3k_1 + k_2)e^{4t} - \frac{3}{6}(3k_1 - k_2)e^{-2t} \end{bmatrix}$$

Thus,
$$x(t) = x_1(t) = \frac{1}{6}(3k_1 + k_2)e^{4t} + \frac{1}{6}(3k_1 - k_2)e^{-2t}$$
$$y(t) = y_1(t) = \frac{3}{6}(3k_1 + k_2)e^{4t} - \frac{3}{6}(3k_1 - k_2)e^{-2t}$$

If we define two new arbitrary constants $k_3 = (3k_1 + k_2)/6$ and $k_4 = (3k_1 - k_2)/6$, then

$$x(t) = k_3e^{4t} + k_4e^{-2t} \qquad y(t) = 3k_3e^{4t} - 3k_4e^{-2t}$$

Supplementary Problems

Solve each of the following systems by the matrix method. Note that e^{At} for Problems 31.8 through 31.12 has been computed in Problem 29.2.

31.8. $\ddot{x} + 2\dot{x} - 8x = 0; \quad x(1) = 1, \quad \dot{x}(1) = 0.$

31.9. $\ddot{x} + 2\dot{x} - 8x = 4; \quad x(0) = 0, \quad \dot{x}(0) = 0.$

31.10. $\ddot{x} + 2\dot{x} - 8x = 4; \quad x(1) = 0, \quad \dot{x}(1) = 0.$

31.11. $\ddot{x} + 2\dot{x} - 8x = 4; \quad x(0) = 1, \quad \dot{x}(0) = 2.$

31.12. $\ddot{x} + 2\dot{x} - 8x = 9e^{-t}; \quad x(0) = 0, \quad \dot{x}(0) = 0.$

31.13. The system of Problem 31.4, using (*31.2*).

31.14. $\ddot{x} = 2\dot{x} + 5y + 3,$

$\dot{y} = -\dot{x} - 2y;$

$x(0) = 0, \quad \dot{x}(0) = 0, \quad y(0) = 1.$

31.15. $\dot{x} = x + 2y$

$\dot{y} = 4x + 3y.$ (*Hint.* See Problem 29.1.)

31.16. $\dddot{x} = 6t; \quad x(0) = 0, \quad \dot{x}(0) = 0, \quad \ddot{x}(0) = 12.$

Answers to Supplementary Problems

31.8. $x = \frac{1}{3}e^{-4(t-1)} + \frac{2}{3}e^{2(t-1)}$

31.9. $x = \frac{1}{6}e^{-4t} + \frac{1}{3}e^{2t} - \frac{1}{2}$

31.10. $x = \frac{1}{6}e^{-4(t-1)} + \frac{1}{3}e^{2(t-1)} - \frac{1}{2}$

31.11. $x = \frac{1}{6}e^{-4t} + \frac{4}{3}e^{2t} - \frac{1}{2}$

31.12. $x = \frac{1}{2}e^{-4t} + \frac{1}{2}e^{2t} - e^{-t}$

31.14. $x = -8\cos t - 6\sin t + 8 + 6t$

$y = 4\cos t - 2\sin t - 3$

31.15. $x = k_3e^{5t} + k_4e^{-t} \quad y = 2k_3e^{5t} - k_4e^{-t}$

where $k_3 = \frac{1}{3}(k_1 + k_2)$ and $k_4 = \frac{1}{3}(2k_1 - k_2)$

31.16. $x = \frac{1}{4}t^4 + 6t^2$

Chapter 32

Simple Numerical Methods

32.1 GENERAL REMARKS

A *numerical method* for solving an initial-value problem is a procedure which produces approximate solutions at particular points using only the operations of addition, subtraction, multiplication, and division, and functional evaluations.

In Chapters 32 through 35, we restrict ourselves to first-order initial-value problems of the form [see (*3.1*)]

$$y' = f(x, y); \quad y(x_0) = y_0 \tag{32.1}$$

Generalizations to higher-order problems are given in Chapter 36.

Example 32.1. (*a*) For the initial-value problem $y' = -y + x + 2$; $y(0) = 2$, we have $f(x, y) = -y + x + 2$, $x_0 = 0$, and $y_0 = 2$. (*b*) For the problem $y' = y^2 + 1$; $y(1) = 0$, we have $f(x, y) = y^2 + 1$, $x_0 = 1$, and $y_0 = 0$. (*c*) For the problem $y' = 3$; $y(0) = 0$, we have $f(x, y) \equiv 3$, $x_0 = 0$, and $y_0 = 0$.

Observe that, in a particular problem, $f(x, y)$ may be independent of x, of y, or of x and y.

All numerical methods will involve finding approximate solutions at $x_0, x_1, x_2, \ldots$, where the difference between any two successive x-values is a constant, h; that is, $x_{n+1} - x_n = h$ $(n = 0, 1, 2, \ldots)$. The value h is chosen arbitrarily; in general, the smaller h is, the more accurate the approximate solution becomes.

The approximate solution at x_n will be designated by $y(x_n)$, or simply y_n. The true solution at x_n will be denoted by either $Y(x_n)$ or Y_n. Note that once y_n is known, (*32.1*) can be used to obtain y'_n as

$$y'_n = f(x_n, y_n) \tag{32.2}$$

32.2 EULER'S METHOD

$$y_{n+1} = y_n + hy'_n \tag{32.3}$$

or, by (*32.2*),

$$y_{n+1} = y_n + hf(x_n, y_n) \tag{32.4}$$

(See Problems 32.1 through 32.8.)

32.3 HEUN'S METHOD

This is an improvement on Euler's method, and is given by

$$y_{n+1} = y_n + \frac{h}{2}[y'_n + f(x_n + h,\, y_n + hy'_n)] \tag{32.5}$$

or, from (*32.2*),

$$y_{n+1} = y_n + \frac{h}{2}[f(x_n, y_n) + f(x_n + h,\, y_n + hy'_n)] \tag{32.6}$$

(See Problems 32.9 through 32.11.)

32.4 THREE-TERM TAYLOR SERIES METHOD

$$y_{n+1} = y_n + hy'_n + \frac{h^2}{2}y''_n \tag{32.7}$$

(See Problems 32.12 and 32.13.) Here, y''_n is obtained by differentiating the given first-order equation with respect to x, and then evaluating at $x = x_n$.

32.5 NYSTROM'S METHOD

$$y_{n+1} = y_{n-1} + 2hy'_n \tag{32.8}$$

(See Problems 32.14 through 32.16.) Nystrom's method is not self-starting; this difficulty is explained and resolved in Problem 32.14.

32.6 ORDER OF A NUMERICAL METHOD

A numerical method is of *order n*, where n is a positive integer, if the method is exact for polynomials of degree n or less. In other words, if the true solution of an initial-value problem is a polynomial of degree n or less, then the approximate solution and the true solution will be identical for a method of order n.

In general, the higher the order, the more accurate the method. Euler's method is of order one, whereas each of the other three methods is of order two.

Taylor series methods are generally avoided, since they require the operation of differentiation, which can be tedious and involved. Since both Heun's method and Nystrom's method are of the same order as the three-term Taylor series method and do not involve differentiation, they are to be preferred.

Solved Problems

32.1. Find $y(1)$ for $y' = y - x$; $y(0) = 2$, using Euler's method with $h = \frac{1}{4}$.

For this problem, $x_0 = 0$, $y_0 = 2$, and $f(x,y) = y - x$; so (32.2) becomes $y'_n = y_n - x_n$. Because $h = \frac{1}{4}$,

$$x_1 = x_0 + h = \frac{1}{4} \qquad x_2 = x_1 + h = \frac{1}{2} \qquad x_3 = x_2 + h = \frac{3}{4} \qquad x_4 = x_3 + h = 1$$

Using (32.3) with $n = 0, 1, 2, 3$ successively, we now compute the corresponding y-values.

n = 0: $y_1 = y_0 + hy'_0$

But $y'_0 = f(x_0, y_0) = y_0 - x_0 = 2 - 0 = 2$

Hence, $y_1 = 2 + \frac{1}{4}(2) = \frac{5}{2}$

n = 1: $y_2 = y_1 + hy'_1$

But $y'_1 = f(x_1, y_1) = y_1 - x_1 = \frac{5}{2} - \frac{1}{4} = \frac{9}{4}$

Hence, $y_2 = \frac{5}{2} + \frac{1}{4}\left(\frac{9}{4}\right) = \frac{49}{16}$

$n = 2$: $y_3 = y_2 + hy'_2$

But $y'_2 = f(x_2, y_2) = y_2 - x_2 = \frac{49}{16} - \frac{1}{2} = \frac{41}{16}$

Hence, $y_3 = \frac{49}{16} + \frac{1}{4}\left(\frac{41}{16}\right) = \frac{237}{64}$

$n = 3$: $y_4 = y_3 + hy'_3$

But $y'_3 = f(x_3, y_3) = y_3 - x_3 = \frac{237}{64} - \frac{3}{4} = \frac{189}{64}$

Hence, $y_4 = \frac{237}{64} + \frac{1}{4}\left(\frac{189}{64}\right) = \frac{1137}{256}$

Thus, $y(1) = y_4 = \frac{1137}{256} = 4.441$. Note that the true solution is $Y(x) = e^x + x + 1$, so that $Y(1) = 4.718$.

32.2. Solve Problem 32.1 with $h = 0.1$.

With $h = 0.1$, $y(1) = y_{10}$. As before, $y'_n = y_n - x_n$. Then, using (*32.3*) with $n = 0, 1, \ldots, 9$ successively, we obtain

$n = 0$: $x_0 = 0$, $y_0 = 2$, $y'_0 = y_0 - x_0 = 2 - 0 = 2$

$y_1 = y_0 + hy'_0 = 2 + (0.1)(2) = 2.2$

$n = 1$: $x_1 = 0.1$, $y_1 = 2.2$, $y'_1 = y_1 - x_1 = 2.2 - 0.1 = 2.1$

$y_2 = y_1 + hy'_1 = 2.2 + (0.1)(2.1) = 2.41$

$n = 2$: $x_2 = 0.2$, $y_2 = 2.41$, $y'_2 = y_2 - x_2 = 2.41 - 0.2 = 2.21$

$y_3 = y_2 + hy'_2 = 2.41 + (0.1)(2.21) = 2.631$

$n = 3$: $x_3 = 0.3$, $y_3 = 2.631$, $y'_3 = y_3 - x_3 = 2.631 - 0.3 = 2.331$

$y_4 = y_3 + hy'_3 = 2.631 + (0.1)(2.331) = 2.864$

$n = 4$: $x_4 = 0.4$, $y_4 = 2.864$, $y'_4 = y_4 - x_4 = 2.864 - 0.4 = 2.464$

$y_5 = y_4 + hy'_4 = 2.864 + (0.1)(2.464) = 3.110$

$n = 5$: $x_5 = 0.5$, $y_5 = 3.110$, $y'_5 = y_5 - x_5 = 3.110 - 0.5 = 2.610$

$y_6 = y_5 + hy'_5 = 3.110 + (0.1)(2.610) = 3.371$

$n = 6$: $x_6 = 0.6$, $y_6 = 3.371$, $y'_6 = y_6 - x_6 = 3.371 - 0.6 = 2.771$

$y_7 = y_6 + hy'_6 = 3.371 + (0.1)(2.771) = 3.648$

$n = 7$: $x_7 = 0.7$, $y_7 = 3.648$, $y'_7 = y_7 - x_7 = 3.648 - 0.7 = 2.948$

$y_8 = y_7 + hy'_7 = 3.648 + (0.1)(2.948) = 3.943$

$n = 8$: $x_8 = 0.8$, $y_8 = 3.943$, $y'_8 = y_8 - x_8 = 3.943 - 0.8 = 3.143$

$y_9 = y_8 + hy'_8 = 3.943 + (0.1)(3.143) = 4.257$

$n = 9$: $x_9 = 0.9$, $y_9 = 4.257$, $y'_9 = y_9 - x_9 = 4.257 - 0.9 = 3.357$

$y_{10} = y_9 + hy'_9 = 4.257 + (0.1)(3.357) = 4.593$

The above results are displayed in Table 32-1. For comparison, Table 32-1 also contains results for $h = 0.05$, $h = 0.01$, and $h = 0.005$, with all computations rounded to four decimal places. Note that more accurate results are obtained when smaller values of h are used.

Table 32-1

Method: EULER'S METHOD					
Problem: $y' = y - x;\ y(0) = 2$					
x_n	y_n				True solution
	$h = 0.1$	$h = 0.05$	$h = 0.01$	$h = 0.005$	$Y(x) = e^x + x + 1$
0.0	2.0000	2.0000	2.0000	2.0000	2.0000
0.1	2.2000	2.2025	2.2046	2.2049	2.2052
0.2	2.4100	2.4155	2.4202	2.4208	2.4214
0.3	2.6310	2.6401	2.6478	2.6489	2.6499
0.4	2.8641	2.8775	2.8889	2.8903	2.8918
0.5	3.1105	3.1289	3.1446	3.1467	3.1487
0.6	3.3716	3.3959	3.4167	3.4194	3.4221
0.7	3.6487	3.6799	3.7068	3.7102	3.7138
0.8	3.9436	3.9829	4.0167	4.0211	4.0255
0.9	4.2579	4.3066	4.3486	4.3541	4.3596
1.0	4.5937	4.6533	4.7048	4.7115	4.7183

Table 32-2

Method: EULER'S METHOD					
Problem: $y' = y;\ y(0) = 1$					
x_n	y_n				True solution
	$h = 0.1$	$h = 0.05$	$h = 0.01$	$h = 0.005$	$Y(x) = e^x$
0.0	1.0000	1.0000	1.0000	1.0000	1.0000
0.1	1.1000	1.1025	1.1046	1.1049	1.1052
0.2	1.2100	1.2155	1.2202	1.2208	1.2214
0.3	1.3310	1.3401	1.3478	1.3489	1.3499
0.4	1.4641	1.4775	1.4889	1.4903	1.4918
0.5	1.6105	1.6289	1.6446	1.6467	1.6487
0.6	1.7716	1.7959	1.8167	1.8194	1.8221
0.7	1.9487	1.9799	2.0068	2.0102	2.0138
0.8	2.1436	2.1829	2.2167	2.2211	2.2255
0.9	2.3579	2.4066	2.4486	2.4541	2.4596
1.0	2.5937	2.6533	2.7048	2.7115	2.7183

32.3. **Find $y(0.5)$ for $y' = y$; $y(0) = 1$, using Euler's method with $h = 0.1$.**

For this problem, $f(x,y) = y$, $x_0 = 0$, and $y_0 = 1$; hence, from (32.2), $y'_n = f(x_n, y_n) = y_n$. With $h = 0.1$, $y(0.5) = y_5$. Then, using (32.3) with $n = 0, 1, 2, 3, 4$ successively, we obtain

$$n = 0: \quad x_0 = 0, \quad y_0 = 1, \quad y'_0 = y_0 = 1$$
$$y_1 = y_0 + hy'_0 = 1 + (0.1)(1) = 1.1$$

$$n = 1: \quad x_1 = 0.1, \quad y_1 = 1.1, \quad y'_1 = y_1 = 1.1$$
$$y_2 = y_1 + hy'_1 = 1.1 + (0.1)(1.1) = 1.21$$

$$n = 2: \quad x_2 = 0.2, \quad y_2 = 1.21, \quad y'_2 = y_2 = 1.21$$
$$y_3 = y_2 + hy'_2 = 1.21 + (0.1)(1.21) = 1.331$$

$$n = 3: \quad x_3 = 0.3, \quad y_3 = 1.331, \quad y'_3 = y_3 = 1.331$$
$$y_4 = y_3 + hy'_3 = 1.331 + (0.1)(1.331) = 1.464$$

$$n = 4: \quad x_4 = 0.4, \quad y_4 = 1.464, \quad y'_4 = y_4 = 1.464$$
$$y_5 = y_4 + hy'_4 = 1.464 + (0.1)(1.464) = 1.610$$

Thus, $y(0.5) = y_5 = 1.610$. Note that since the true solution is $Y(x) = e^x$, $Y(0.5) = e^{0.5} = 1.649$.

32.4. **Find $y(1)$ for $y' = y$; $y(0) = 1$, using Euler's method with $h = 0.1$.**

We proceed exactly as in Problem 32.3, except that we now calculate through $n = 9$. The results of these computations are given in Table 32-2. For comparison, Table 32-2 also contains results for $h = 0.05$, $h = 0.01$, and $h = 0.005$, with all calculations rounded to four decimal places.

32.5. **Find $y(1)$ for $y' = y^2 + 1$; $y(0) = 0$, using Euler's method with $h = 0.1$.**

Here, $f(x,y) = y^2 + 1$, $x_0 = 0$, and $y_0 = 0$; hence, from (32.3), $y'_n = f(x_n, y_n) = (y_n)^2 + 1$. With $h = 0.1$, $y(1) = y_{10}$. Then, using (32.3) with $n = 0, 1, \ldots, 9$ successively, we obtain

$$n = 0: \quad x_0 = 0, \quad y_0 = 0, \quad y'_0 = (y_0)^2 + 1 = (0)^2 + 1 = 1$$
$$y_1 = y_0 + hy'_0 = 0 + (0.1)(1) = 0.1$$

$$n = 1: \quad x_1 = 0.1, \quad y_1 = 0.1, \quad y'_1 = (y_1)^2 + 1 = (0.1)^2 + 1 = 1.01$$
$$y_2 = y_1 + hy'_1 = 0.1 + (0.1)(1.01) = 0.201$$

$$n = 2: \quad x_2 = 0.2, \quad y_2 = 0.201$$
$$y'_2 = (y_2)^2 + 1 = (0.201)^2 + 1 = 1.040$$
$$y_3 = y_2 + hy'_2 = 0.201 + (0.1)(1.040) = 0.305$$

$$n = 3: \quad x_3 = 0.3, \quad y_3 = 0.305$$
$$y'_3 = (y_3)^2 + 1 = (0.305)^2 + 1 = 1.093$$
$$y_4 = y_3 + hy'_3 = 0.305 + (0.1)(1.093) = 0.414$$

$$n = 4: \quad x_4 = 0.4, \quad y_4 = 0.414$$
$$y'_4 = (y_4)^2 + 1 = (0.414)^2 + 1 = 1.171$$
$$y_5 = y_4 + hy'_4 = 0.414 + (0.1)(1.171) = 0.531$$

Continuing in this manner, we find that $y_{10} = 1.396$.

The calculations are found in Table 32-3. For comparison, Table 32-3 also contains results for $h = 0.05$, $h = 0.01$, and $h = 0.005$, with all computations rounded to four decimal places. The true solution to this problem is $Y(x) = \tan x$, hence $Y(1) = 1.557$.

Table 32-3

Method: EULER'S METHOD					
Problem: $y' = y^2 + 1;\ y(0) = 0$					
x_n	y_n				True solution
	$h = 0.1$	$h = 0.05$	$h = 0.01$	$h = 0.005$	$Y(x) = \tan x$
0.0	0.0000	0.0000	0.0000	0.0000	0.0000
0.1	0.1000	0.1001	0.1003	0.1003	0.1003
0.2	0.2010	0.2018	0.2025	0.2026	0.2027
0.3	0.3050	0.3070	0.3088	0.3091	0.3093
0.4	0.4143	0.4183	0.4218	0.4223	0.4228
0.5	0.5315	0.5384	0.5446	0.5455	0.5463
0.6	0.6598	0.6711	0.6814	0.6827	0.6841
0.7	0.8033	0.8212	0.8378	0.8400	0.8423
0.8	0.9678	0.9959	1.0223	1.0260	1.0296
0.9	1.1615	1.2055	1.2482	1.2541	1.2602
1.0	1.3964	1.4663	1.5370	1.5470	1.5574

32.6. Find $y(9)$ for $y' = 3;\ y(1) = 6$, using Euler's method with $h = 2$.

Here, $f(x,y) \equiv 3$, $x_0 = 1$, and $y_0 = 6$; hence, $y'_n = f(x_n, y_n) \equiv 3$. With $h = 2$, we have $y(9) = y(x_0 + 4h) = y(x_4) = y_4$. Then,

$n = 0$: $\quad x_0 = 1, \quad y_0 = 6, \quad y'_0 = 3$

$$y_1 = y_0 + hy'_0 = 6 + (2)(3) = 12$$

$n = 1$: $\quad x_1 = 3, \quad y_1 = 12, \quad y'_1 = 3$

$$y_2 = y_1 + hy'_1 = 12 + (2)(3) = 18$$

$n = 2$: $\quad x_2 = 5, \quad y_2 = 18, \quad y'_2 = 3$

$$y_3 = y_2 + hy'_2 = 18 + (2)(3) = 24$$

$n = 3$: $\quad x_3 = 7, \quad y_3 = 24, \quad y'_3 = 3$

$$y_4 = y_3 + hy'_3 = 24 + (2)(3) = 30$$

Thus, $y(9) = y_4 = 30$.

The true solution of this problem is $Y(x) = 3x + 3$; hence, $Y(9) = 30$, which agrees exactly with $y(9)$. Because $Y(x)$ is a first-degree polynomial and Euler's method is a first-order method (Section 32.6), this agreement is expected even with a very large h.

32.7. Give an analytic derivation of Euler's method.

Let $Y(x)$ represent the true solution. Then, using the definition of the derivative, we have

$$Y'(x_n) = \lim_{\Delta x \to 0} \frac{Y(x_n + \Delta x) - Y(x_n)}{\Delta x}$$

If Δx is small, then

$$Y'(x_n) \simeq \frac{Y(x_n + \Delta x) - Y(x_n)}{\Delta x}$$

Setting $\Delta x = h$ and solving for $Y(x_n + \Delta x) = Y(x_{n+1})$, we obtain

$$Y(x_{n+1}) \simeq Y(x_n) + hY'(x_n) \tag{1}$$

Finally, if we use y_n and y'_n to approximate $Y(x_n)$ and $Y'(x_n)$, respectively, the right side of (1) can be used to approximate $Y(x_{n+1})$. Thus,

$$y_{n+1} = y_n + hy'_n$$

which is Euler's method.

32.8. Give a geometric derivation of Euler's method.

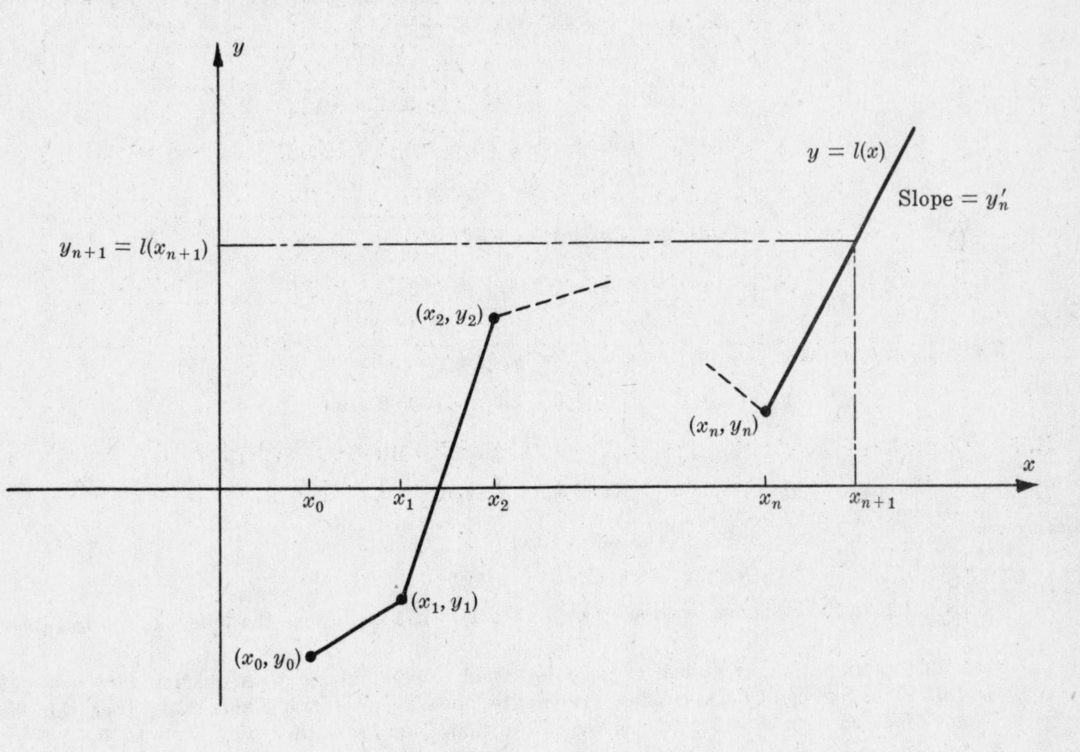

Fig. 32-1

Assume that $y_n = y(x_n)$ has already been computed, so that y'_n is also known, via (32.2). Draw a straight line $l(x)$ emanating from (x_n, y_n) and having slope y'_n, and use $l(x)$ to approximate $Y(x)$ on the interval $[x_n, x_{n+1}]$ (see Fig. 32-1). The value $l(x_{n+1})$ is taken to be y_{n+1}. Thus

$$l(x) = (y'_n)x + [y_n - (y'_n)x_n]$$

and

$$l(x_{n+1}) = (y'_n)x_{n+1} + [y_n - (y'_n)x_n]$$

$$= y_n + (y'_n)(x_{n+1} - x_n) = y_n + hy'_n$$

Hence, $y_{n+1} = y_n + hy'_n$, which is Euler's method.

32.9. Find $y(1)$ for $y' = y - x;\ y(0) = 2$, using Heun's method with $h = 0.1$.

Here $x_0 = 0$, $y_0 = 2$, and $f(x, y) = y - x$; hence, $y'_n = f(x_n, y_n) = y_n - x_n$. Furthermore, since $h = 0.1$, $y(1) = y_{10}$. Then, using (32.5) with $n = 0, 1, \ldots, 9$ successively, we obtain

$n = 0$: $x_0 = 0, \quad y_0 = 2, \quad y'_0 = y_0 - x_0 = 2$

$$f(x_0 + h, y_0 + hy'_0) = f[0 + 0.1, 2 + (0.1)(2)] = f(0.1, 2.2) = 2.2 - 0.1 = 2.1$$

$$y_1 = y_0 + (h/2)[y'_0 + f(x_0 + h, y_0 + hy'_0)] = 2 + \frac{1}{2}(0.1)(2 + 2.1) = 2.205$$

$n = 1$: $x_1 = 0.1, \quad y_1 = 2.205, \quad y'_1 = y_1 - x_1 = 2.205 - 0.1 = 2.105$

$$f(x_1 + h, y_1 + hy'_1) = f[0.1 + 0.1, 2.205 + (0.1)(2.105)] = f(0.2, 2.416) = 2.416 - 0.2 = 2.216$$

$$y_2 = y_1 + (h/2)[y'_1 + f(x_1 + h, y_1 + hy'_1)] = 2.205 + \frac{1}{2}(0.1)(2.105 + 2.216) = 2.421$$

$n = 2$: $x_2 = 0.2, \quad y_2 = 2.421, \quad y'_2 = y_2 - x_2 = 2.421 - 0.2 = 2.221$

$$f(x_2 + h, y_2 + hy'_2) = f[0.2 + 0.1, 2.421 + (0.1)(2.221)] = f(0.3, 2.643) = 2.643 - 0.3 = 2.343$$

$$y_3 = y_2 + (h/2)[y'_2 + f(x_2 + h, y_2 + hy'_2)] = 2.421 + \frac{1}{2}(0.1)(2.221 + 2.343) = 2.649$$

$n = 3$: $x_3 = 0.3, \quad y_3 = 2.649, \quad y'_3 = y_3 - x_3 = 2.649 - 0.3 = 2.349$

$$f(x_3 + h, y_3 + hy'_3) = f[0.3 + 0.1, 2.649 + (0.1)(2.349)] = f(0.4, 2.884) = 2.884 - 0.4 = 2.484$$

$$y_4 = y_3 + (h/2)[y'_3 + f(x_3 + h, y_3 + hy'_3)] = 2.649 + \frac{1}{2}(0.1)(2.349 + 2.484) = 2.891$$

Continuing in this manner, we find that $y_{10} = y(1) = 4.714$, whereas the true solution is $Y(1) = 4.718$.

The computations rounded to seven decimal places, along with results for $h = 0.05$ and $h = 0.01$, are given in Table 32-4. Compare these results with Table 32.1. For this problem, Heun's method with $h = 0.1$ is more accurate than Euler's method with $h = 0.005$.

Table 32-4

Method: HEUN'S METHOD				
Problem: $y' = y - x;\ y(0) = 2$				
x_n	y_n			True solution
	$h = 0.1$	$h = 0.05$	$h = 0.01$	$Y(x) = e^x + x + 1$
0.0	2.0000000	2.0000000	2.0000000	2.0000000
0.1	2.2050000	2.2051266	2.2051691	2.2051709
0.2	2.4210250	2.4213047	2.4213987	2.4214028
0.3	2.6492326	2.6496963	2.6498521	2.6498588
0.4	2.8909021	2.8915852	2.8918148	2.8918247
0.5	3.1474468	3.1483904	3.1487076	3.1487213
0.6	3.4204287	3.4216801	3.4221007	3.4221188
0.7	3.7115737	3.7131870	3.7137294	3.7137527
0.8	4.0227889	4.0248265	4.0255115	4.0255409
0.9	4.3561818	4.3587148	4.3595665	4.3596031
1.0	4.7140809	4.7171911	4.7182369	4.7182818

Table 32-5

Method: HEUN'S METHOD				
Problem: $y' = y;\ y(0) = 1$				
x_n	y_n			True solution
	$h = 0.1$	$h = 0.05$	$h = 0.01$	$Y(x) = e^x$
0.0	1.0000000	1.0000000	1.0000000	1.0000000
0.1	1.1050000	1.1051266	1.1051691	1.1051709
0.2	1.2210250	1.2213047	1.2213987	1.2214028
0.3	1.3492326	1.3496963	1.3498521	1.3498588
0.4	1.4909021	1.4915852	1.4918148	1.4918247
0.5	1.6474468	1.6483904	1.6487076	1.6487213
0.6	1.8204287	1.8216801	1.8221007	1.8221188
0.7	2.0115737	2.0131873	2.0137294	2.0137527
0.8	2.2227889	2.2248265	2.2255115	2.2255409
0.9	2.4561818	2.4587148	2.4595665	2.4596031
1.0	2.7140809	2.7171911	2.7182369	2.7182818

32.10. Find $y(1)$ for $y' = y$; $y(0) = 1$, using Heun's method with $h = 0.1$.

Here $x_0 = 0$, $y_0 = 1$, and $f(x,y) = y$; hence, $y'_n = f(x_n, y_n) = y_n$. Then, using (*32.5*) with $n = 0, 1, \ldots, 9$ successively, we obtain

n = 0: $x_0 = 0, \quad y_0 = 1, \quad y'_0 = y_0 = 1$

$$f(x_0 + h, y_0 + hy'_0) = f[0 + 0.1, 1 + (0.1)(1)] = f(0.1, 1.1) = 1.1$$

$$y_1 = y_0 + (h/2)[y'_0 + f(x_0 + h, y_0 + hy'_0)] = 1 + \frac{1}{2}(0.1)(1 + 1.1) = 1.105$$

n = 1: $x_1 = 0.1, \quad y_1 = 1.105, \quad y'_1 = y_1 = 1.105$

$$f(x_1 + h, y_1 + hy'_1) = f[0.1 + 0.1, 1.105 + (0.1)(1.105)] = f(0.2, 1.216) = 1.216$$

$$y_2 = y_1 + (h/2)[y'_1 + f(x_1 + h, y_1 + hy'_1)] = 1.105 + \frac{1}{2}(0.1)(1.105 + 1.216) = 1.221$$

n = 2: $x_2 = 0.2, \quad y_2 = 1.221, \quad y'_2 = y_2 = 1.221$

$$f(x_2 + h, y_2 + hy'_2) = f[0.2 + 0.1, 1.221 + (0.1)(1.221)] = f(0.3, 1.343) = 1.343$$

$$y_3 = y_2 + (h/2)[y'_2 + f(x_2 + h, y_2 + hy'_2)] = 1.221 + \frac{1}{2}(0.1)(1.221 + 1.343) = 1.349$$

n = 3: $x_3 = 0.3, \quad y_3 = 1.349, \quad y'_3 = y_3 = 1.349$

$$f(x_3 + h, y_3 + hy'_3) = f[0.3 + 0.1, 1.349 + (0.1)(1.349)] = f(0.4, 1.484) = 1.484$$

$$y_4 = y_3 + (h/2)[y'_3 + f(x_3 + h, y_3 + hy'_3)] = 1.349 + \frac{1}{2}(0.1)(1.349 + 1.484) = 1.491$$

Continuing in this manner, we find $y_{10} = y(1) = 2.714$, whereas the true solution is $Y(1) = 2.718$.

The computations, rounded to seven decimal places, along with results for $h = 0.05$ and $h = 0.01$, are given in Table 32-5. Compare these results with Table 32-2.

32.11. Find $y(1.6)$ for $y' = 2x$; $y(1) = 1$, using Heun's method with $h = 0.2$.

Here $x_0 = 1$, $y_0 = 1$, and $f(x,y) = 2x$; hence, $y'_n = f(x_n, y_n) = 2x_n$. Then, using (*32.5*) with $n = 0, 1, 2$ successively, we obtain

n = 0: $x_0 = 1, \quad y_0 = 1, \quad y'_0 = 2x_0 = 2(1) = 2$

$$f(x_0 + h, y_0 + hy'_0) = f[1 + 0.2, 1 + (0.2)(2)] = f(1.2, 1.4) = 2(1.2) = 2.4$$

$$y_1 = y_0 + (h/2)[y'_0 + f(x_0 + h, y_0 + hy'_0)] = 1 + \frac{1}{2}(0.2)(2 + 2.4) = 1.44$$

n = 1: $x_1 = x_0 + h = 1 + 0.2 = 1.2, \quad y_1 = 1.44$

$$y'_1 = 2x_1 = 2(1.2) = 2.4$$

$$f(x_1+h,\ y_1+hy_1') \;=\; f[1.2+0.2,\ 1.44+(0.2)(2.4)]$$
$$= \; f(1.4,\ 1.92) \;=\; 2(1.4) \;=\; 2.8$$

$$y_2 \;=\; y_1 \;+\; (h/2)[y_1' + f(x_1+h,\ y_1+hy_1')]$$
$$= \; 1.44 \;+\; \frac{1}{2}(0.2)(2.4+2.8) \;=\; 1.96$$

n = 2: $x_2 \;=\; 1.4, \quad y_2 \;=\; 1.96, \quad y_2' \;=\; 2x_2 \;=\; 2(1.4) \;=\; 2.8$

$$f(x_2+h,\ y_2+hy_2') \;=\; f[1.4+0.2,\ 1.96+(0.2)(2.8)]$$
$$= \; f(1.6, 2.52) \;=\; 2(1.6) \;=\; 3.2$$

$$y_3 \;=\; y_2 \;+\; (h/2)[y_2' + f(x_2+h,\ y_2+hy_2')]$$
$$= \; 1.96 \;+\; \frac{1}{2}(0.2)(2.8+3.2) \;=\; 2.56$$

Thus, $y(1.6) = y_3 = 2.56$.

The true solution is $Y(x) = x^2$, hence $Y(1.6) = y(1.6) = 2.56$. Since the true solution is a second-degree polynomial and Heun's method is a second-order method, this agreement is expected.

32.12. Find $y(1)$ for $y' = y - x;\ y(0) = 2$, using the three-term Taylor series method with $h = 0.1$.

Here $x_0 = 0$, $y_0 = 2$, and $f(x, y) = y - x$. Since $y' = y - x$, it follows upon differentiating this equation that $y'' = y' - 1$. Thus, $y_n' = y_n - x_n$ and $y'' = y_n' - 1$. Then, using (*32.7*) with $n = 0, 1, \ldots, 9$ successively, we obtain

n = 0: $x_0 \;=\; 0, \quad y_0 \;=\; 2, \quad y_0' \;=\; y_0 - x_0 \;=\; 2 - 0 \;=\; 2$

$$y_0'' \;=\; y_0' - 1 \;=\; 2 - 1 \;=\; 1$$

$$y_1 \;=\; y_0 + hy_0' + \frac{h^2}{2}y_0'' \;=\; 2 + (0.1)2 + \frac{0.01}{2}(1) \;=\; 2.205$$

n = 1: $x_1 \;=\; 0.1, \quad y_1 \;=\; 2.205, \quad y_1' \;=\; y_1 - x_1 \;=\; 2.205 - 0.1 \;=\; 2.105$

$$y_1'' \;=\; y_1' - 1 \;=\; 2.105 - 1 \;=\; 1.105$$

$$y_2 \;=\; y_1 + hy_1' + \frac{h^2}{2}y_1''$$
$$= \; 2.205 + (0.1)(2.105) + \frac{0.01}{2}(1.105) \;=\; 2.421$$

n = 2: $x_2 \;=\; 0.2, \quad y_2 \;=\; 2.421, \quad y_2' \;=\; y_2 - x_2 \;=\; 2.421 - 0.2 \;=\; 2.221$

$$y_2'' \;=\; y_2' - 1 \;=\; 2.221 - 1 \;=\; 1.221$$

$$y_3 \;=\; y_2 + hy_2' + \frac{h^2}{2}y_2''$$
$$= \; 2.421 + (0.1)(2.221) + \frac{0.01}{2}(1.221) \;=\; 2.649$$

n = 3: $x_3 \;=\; 0.3, \quad y_3 \;=\; 2.649, \quad y_3' \;=\; y_3 - x_3 \;=\; 2.649 - 0.3 \;=\; 2.349$

$$y_3'' \;=\; y_3' - 1 \;=\; 2.349 - 1 \;=\; 1.349$$

$$y_4 \;=\; y_3 + hy_3' + \frac{h^2}{2}y_3''$$
$$= \; 2.649 + (0.1)(2.349) + \frac{0.01}{2}(1.349) \;=\; 2.891$$

Continuing in this manner (see Table 32-6), we find that $y_{10} = y(1) = 4.714$. Compare the results with both Table 32-1 and Table 32-4.

Table 32-6

Method: THREE-TERM TAYLOR SERIES METHOD

Problem: $y' = y - x; \quad y(0) = 2$

x_n	y_n			True solution
	$h = 0.1$	$h = 0.05$	$h = 0.01$	$Y(x) = e^x + x + 1$
0.0	2.0000000	2.0000000	2.0000000	2.0000000
0.1	2.2050000	2.2051266	2.2051691	2.2051709
0.2	2.4210250	2.4213047	2.4213987	2.4214028
0.3	2.6492327	2.6496963	2.6498521	2.6498588
0.4	2.8909021	2.8915852	2.8918148	2.8918247
0.5	3.1474468	3.1483904	3.1487076	3.1487213
0.6	3.4204287	3.4216801	3.4221007	3.4221188
0.7	3.7115737	3.7131870	3.7137294	3.7137527
0.8	4.0227889	4.0248265	4.0255115	4.0255409
0.9	4.3561818	4.3587148	4.3595665	4.3596031
1.0	4.7140809	4.7171911	4.7182369	4.7182818

Table 32-7

Method: THREE-TERM TAYLOR SERIES METHOD

Problem: $y' = y^2 + 1; \quad y(0) = 0$

x_n	y_n			True solution
	$h = 0.1$	$h = 0.05$	$h = 0.01$	$Y(x) = \tan x$
0.0	0.0000000	0.0000000	0.0000000	0.0000000
0.1	0.1000000	0.1002503	0.1003313	0.1003347
0.2	0.2020100	0.2025322	0.2027028	0.2027100
0.3	0.3081933	0.3090436	0.3093243	0.3093363
0.4	0.4210663	0.4223474	0.4227749	0.4227932
0.5	0.5437532	0.5456384	0.5462751	0.5463025
0.6	0.6803652	0.6831445	0.6840955	0.6841368
0.7	0.8366079	0.8407771	0.8422249	0.8422884
0.8	1.0208208	1.0272615	1.0295378	1.0296386
0.9	1.2458742	1.2562453	1.2599903	1.2601582
1.0	1.5328917	1.5505515	1.5571088	1.5574077

32.13. Find $y(1)$ for $y' = y^2 + 1$; $y(0) = 0$, using the three-term Taylor series method with $h = 0.1$.

Here $x_0 = 0$, $y_0 = 0$, and $f(x, y) = y^2 + 1$. Since $y' = y^2 + 1$, it follows that $y'' = 2yy'$; hence $y'_n = y_n^2 + 1$ and $y''_n = 2y_n y'_n$. Then, using (*32.7*) repeatedly, we obtain

$n = 0$: $x_0 = 0,\quad y_0 = 0,\quad y'_0 = y_0^2 + 1 = (0)^2 + 1 = 1$

$$y''_0 = 2y_0 y'_0 = 2(0)(1) = 0$$

$$y_1 = y_0 + hy'_0 + \frac{h^2}{2} y''_0 = 0 + (0.1)(1) + \frac{0.01}{2}(0) = 0.1$$

$n = 1$: $x_1 = 0.1,\quad y_1 = 0.1,\quad y'_1 = y_1^2 + 1 = (0.1)^2 + 1 = 1.01$

$$y''_1 = 2y_1 y'_1 = 2(0.1)(1.01) = 0.202$$

$$y_2 = y_1 + hy'_1 + \frac{h^2}{2} y''_1 = 0.1 + (0.1)(1.01) + \frac{0.01}{2}(0.202) = 0.202$$

$n = 2$: $x_2 = 0.2,\quad y_2 = 0.202,\quad y'_2 = y_2^2 + 1 = (0.202)^2 + 1 = 1.041$

$$y''_2 = 2y_2 y'_2 = 2(0.202)(1.041) = 0.421$$

$$y_3 = y_2 + hy'_2 + \frac{h^2}{2} y''_2 = 0.202 + (0.1)(1.041) + \frac{0.01}{2}(0.421) = 0.308$$

$n = 3$: $x_3 = 0.3,\quad y_3 = 0.308,\quad y'_3 = y_3^2 + 1 = (0.308)^2 + 1 = 1.095$

$$y''_3 = 2y_3 y'_3 = 2(0.308)(1.095) = 0.675$$

$$y_4 = y_3 + hy'_3 + \frac{h^2}{2} y''_3 = 0.308 + (0.1)(1.095) + \frac{0.01}{2}(0.675) = 0.421$$

Continuing in this manner (see Table 32-7), we find that $y_{10} = y(1) = 1.533$. Compare the results with Table 32-3.

32.14. Find $y(1)$ for $y' = y - x$; $y(0) = 2$, using Nystrom's method with $h = 0.1$.

Here $x_0 = 0$, $y_0 = 2$, and $y'_n = y_n - x_n$. Substituting $n = 0$ into (*32.8*), we have $y_1 = y_{-1} + 2hy'_0$, which is unusable since y_{-1} is not defined. Starting Nystrom's method with $n = 1$, we obtain $y_2 = y_0 + 2hy'_1$. But to compute y'_1, we need y_1, a value that is still unknown.

STARTING VALUES. Nystrom's method cannot be used until additional starting values are generated; in particular, y_1 must be known. This value can be obtained by either Heun's method or the three-term Taylor series method. From Section 32.6, Heun's method is preferred. Note that since Euler's method is of lower order than Nystrom's method, it is less accurate and, therefore, would not be used to start Nystrom's method.

Using Heun's method (Problem 32.9) on the present problem, we find $y_1 = 2.205$. Then, using Nystrom's method with $n = 1, 2, \ldots, 9$ successively, we obtain

$n = 1$: $x_1 = 0.1,\quad y_1 = 2.205,\quad y'_1 = y_1 - x_1 = 2.205 - 0.1 = 2.105$

$$y_2 = y_0 + 2hy'_1 = 2 + 2(0.1)(2.105) = 2.421$$

$n = 2$: $x_2 = 0.2,\quad y_2 = 2.421,\quad y'_2 = y_2 - x_2 = 2.421 - 0.2 = 2.221$

$$y_3 = y_1 + 2hy'_2 = 2.205 + 2(0.1)(2.221) = 2.649$$

$n = 3$: $x_3 = 0.3,\quad y_3 = 2.649,\quad y'_3 = y_3 - x_3 = 2.649 - 0.3 = 2.349$

$$y_4 = y_2 + 2hy'_3 = 2.421 + 2(0.1)(2.349) = 2.891$$

$n = 4$: $x_4 = 0.4,\quad y_4 = 2.891,\quad y'_4 = y_4 - x_4 = 2.891 - 0.4 = 2.491$

$$y_5 = y_3 + 2hy'_4 = 2.649 + 2(0.1)(2.491) = 3.147$$

Continuing in this manner (see Table 32-8), we find that $y(1) = y_{10} = 4.714$. Compare the result with Tables 32-1, 32-4, and 32-6.

Table 32-8

Method: NYSTROM'S METHOD				
Problem: $y' = y - x;\ y(0) = 2$				
x_n	y_n			True solution
	$h = 0.1$	$h = 0.05$	$h = 0.01$	$Y(x) = e^x + x + 1$
0.0	2.0000000	2.0000000	2.0000000	2.0000000
0.1	2.2050000	2.2051250	2.2051691	2.2051709
0.2	2.4210000	2.4213013	2.4213987	2.4214028
0.3	2.6492000	2.6496905	2.6498521	2.6498588
0.4	2.8908400	2.8915767	2.8918148	2.8918247
0.5	3.1473680	3.1483786	3.1487075	3.1487213
0.6	3.4203136	3.4216643	3.4221006	3.4221188
0.7	3.7114307	3.7131667	3.7137292	3.7137527
0.8	4.0225997	4.0248007	4.0255113	4.0255409
0.9	4.3559507	4.3586828	4.3595662	4.3596031
1.0	4.7137899	4.7171516	4.7182365	4.7182818

Table 32-9

Method: NYSTROM'S METHOD				
Problem: $y' = y;\ y(0) = 1$				
x_n	y_n			True solution
	$h = 0.1$	$h = 0.05$	$h = 0.01$	$Y(x) = e^x$
0.0	1.0000000	1.0000000	1.0000000	1.0000000
0.1	1.1050000	1.1051250	1.1051691	1.1051709
0.2	1.2210000	1.2213013	1.2213987	1.2214028
0.3	1.3492000	1.3496905	1.3498521	1.3498588
0.4	1.4908400	1.4915767	1.4918148	1.4918247
0.5	1.6473680	1.6483786	1.6487075	1.6487213
0.6	1.8203136	1.8216643	1.8221006	1.8221188
0.7	2.0114307	2.0131667	2.0137292	2.0137527
0.8	2.2225997	2.2248007	2.2255113	2.2255409
0.9	2.4559507	2.4586828	2.4595662	2.4596031
1.0	2.7137899	2.7171516	2.7182365	2.7182818

32.15. Find $y(1)$ for $y' = y$; $y(0) = 1$, using Nystrom's method with $h = 0.1$.

Here $x_0 = 0$, $y_0 = 1$, and $f(x,y) = y$. Using Heun's method (see Problem 32.10), we compute $y_1 = 1.105$. Then, using (*32.8*), we obtain

$n = 1$: $\quad x_1 = 0.1, \quad y_1 = 1.105, \quad y_1' = y_1 = 1.105$

$$y_2 = y_0 + 2hy_1' = 1 + 2(0.1)(1.105) = 1.221$$

$n = 2$: $\quad x_2 = 0.2, \quad y_2 = 1.221, \quad y_2' = y_2 = 1.221$

$$y_3 = y_1 + 2hy_2' = 1.105 + 2(0.1)(1.221) = 1.349$$

$n = 3$: $\quad x_3 = 0.3, \quad y_3 = 1.349, \quad y_3' = y_3 = 1.349$

$$y_4 = y_2 + 2hy_3' = 1.221 + 2(0.1)(1.349) = 1.491$$

$n = 4$: $\quad x_4 = 0.4, \quad y_4 = 1.491, \quad y_4' = y_4 = 1.491$

$$y_5 = y_3 + 2hy_4' = 1.349 + 2(0.1)(1.491) = 1.647$$

Continuing in this manner (see Table 32-9), we find that $y(1) = y_{10} = 2.714$. Compare the result with Table 32-5 and Table 32-2.

32.16. Find $y(1)$ for $y' = y^2 + 1$; $y(0) = 0$, using Nystrom's method with $h = 0.1$.

Here $x_0 = 0$, $y_0 = 0$, and $f(x,y) = y^2 + 1$. Using Heun's method, we calculate $y_1 = 0.101$. Then, using (*32.8*), we obtain

$n = 1$: $\quad x_1 = 0.1, \quad y_1 = 0.101, \quad y_1' = (y_1)^2 + 1 = (0.101)^2 + 1 = 1.01$

$$y_2 = y_0 + 2hy_1' = 0 + 2(0.1)(1.01) = 0.202$$

$n = 2$: $\quad x_2 = 0.2, \quad y_2 = 0.202, \quad y_2' = (y_2)^2 + 1 = (0.202)^2 + 1 = 1.041$

$$y_3 = y_1 + 2hy_2' = 0.101 + 2(0.1)(1.041) = 0.309$$

$n = 3$: $\quad x_3 = 0.3, \quad y_3 = 0.309, \quad y_3' = (y_3)^2 + 1 = (0.309)^2 + 1 = 1.095$

$$y_4 = y_2 + 2hy_3' = (0.202) + 2(0.1)(1.095) = 0.421$$

$n = 4$: $\quad x_4 = 0.4, \quad y_4 = 0.421, \quad y_4' = (y_4)^2 + 1 = (0.421)^2 + 1 = 1.177$

$$y_5 = y_3 + 2hy_4' = 0.309 + 2(0.1)(1.177) = 0.544$$

Continuing in this manner (see Table 32-10), we find that $y(1) = y_{10} = 1.530$. Compare the result with Table 32-7 and Table 32-3.

Table 32-10

Method: NYSTROM'S METHOD				
Problem: $y' = y^2 + 1;\ y(0) = 0$				
x_n	y_n			True solution
	$h = 0.1$	$h = 0.05$	$h = 0.01$	$Y(x) = \tan x$
0.0	0.0000000	0.0000000	0.0000000	0.0000000
0.1	0.1005000	0.1002506	0.1003313	0.1003347
0.2	0.2020201	0.2025328	0.2027028	0.2027100
0.3	0.3086624	0.3090439	0.3093243	0.3093363
0.4	0.4210746	0.4223460	0.4227749	0.4227932
0.5	0.5441232	0.5456322	0.5462750	0.5463025
0.6	0.6802886	0.6831269	0.6840953	0.6841368
0.7	0.8366817	0.8407344	0.8422245	0.8422884
0.8	1.0202958	1.0271647	1.0295368	1.0296386
0.9	1.2448824	1.2560293	1.2599880	1.2601582
1.0	1.5302422	1.5500594	1.5571037	1.5574077

Supplementary Problems

Carry all computations to three decimal places.

32.17. Find $y(1.0)$ for $y' = -y;\ y(0) = 1$, using Euler's method with $h = 0.1$.

32.18. Find $y(0.5)$ for $y' = 2x;\ y(0) = 0$, using Euler's method with $h = 0.1$.

32.19. Find $y(0.5)$ for $y' = -y + x + 2;\ y(0) = 2$, using Euler's method with $h = 0.1$.

32.20. Find $y(0.5)$ for $y' = 4x^3;\ y(0) = 0$, using Euler's method with $h = 0.1$.

32.21. Redo Problem 32.17 using Heun's method.

32.22. Redo Problem 32.18 using Heun's method.

32.23. Redo Problem 32.19 using Heun's method.

32.24. Redo Problem 32.20 using Heun's method.

32.25. Redo Problem 32.17 using the three-term Taylor series method.

32.26. Redo Problem 32.19 using the three-term Taylor series method.

32.27. Redo Problem 32.17 using Nystrom's method. Find y_1 by Heun's method.

32.28. Redo Problem 32.19 using Nystrom's method. Find y_1 by Heun's method.

32.29. Redo Problem 32.20 using Nystrom's method. (*Hint.* Since y' does not depend on y, y'_1 can be found directly. Heun's method, therefore, need not be used to generate starting values.)

Answers to Supplementary Problems

For comparison with other methods to be presented in subsequent chapters, answers are rounded off to four decimal places (for Euler's method) or seven decimal places (for the second-order methods). Answers are carried through $x = 1.0$, and are given for additional values of h.

32.17.

Method: EULER'S METHOD				
Problem: $y' = -y;\ y(0) = 1$				
x_n	y_n			True solution
	$h = 0.1$	$h = 0.05$	$h = 0.01$	$Y(x) = e^{-x}$
0.0	1.0000	1.0000	1.0000	1.0000
0.1	0.9000	0.9025	0.9044	0.9048
0.2	0.8100	0.8145	0.8179	0.8187
0.3	0.7290	0.7351	0.7397	0.7408
0.4	0.6561	0.6634	0.6690	0.6703
0.5	0.5905	0.5987	0.6050	0.6065
0.6	0.5314	0.5404	0.5472	0.5488
0.7	0.4783	0.4877	0.4948	0.4966
0.8	0.4305	0.4401	0.4475	0.4493
0.9	0.3874	0.3972	0.4047	0.4066
1.0	0.3487	0.3585	0.3660	0.3679

32.18.

Method: EULER'S METHOD				
Problem: $y' = 2x;\ y(0) = 0$				
x_n	y_n			True solution
	$h = 0.1$	$h = 0.05$	$h = 0.01$	$Y(x) = x^2$
0.0	0.0000	0.0000	0.0000	0.0000
0.1	0.0000	0.0050	0.0090	0.0100
0.2	0.0200	0.0300	0.0380	0.0400
0.3	0.0600	0.0750	0.0870	0.0900
0.4	0.1200	0.1400	0.1560	0.1600
0.5	0.2000	0.2250	0.2450	0.2500
0.6	0.3000	0.3300	0.3540	0.3600
0.7	0.4200	0.4550	0.4830	0.4900
0.8	0.5600	0.6000	0.6320	0.6400
0.9	0.7200	0.7650	0.8010	0.8100
1.0	0.9000	0.9500	0.9900	1.0000

32.19.

Method: EULER'S METHOD				
Problem: $y' = -y + x + 2;\ y(0) = 2$				
x_n	y_n			True solution
	$h = 0.1$	$h = 0.05$	$h = 0.01$	$Y(x) = e^{-x} + x + 1$
0.0	2.0000	2.0000	2.0000	2.0000
0.1	2.0000	2.0025	2.0044	2.0048
0.2	2.0100	2.0145	2.0179	2.0187
0.3	2.0290	2.0351	2.0397	2.0408
0.4	2.0561	2.0634	2.0690	2.0703
0.5	2.0905	2.0987	2.1050	2.1065
0.6	2.1314	2.1404	2.1472	2.1488
0.7	2.1783	2.1877	2.1948	2.1966
0.8	2.2305	2.2401	2.2475	2.2493
0.9	2.2874	2.2972	2.3047	2.3066
1.0	2.3487	2.3585	2.3660	2.3679

32.20.

Method: EULER'S METHOD				
Problem: $y' = 4x^3;\ y(0) = 0$				
x_n	y_n			True solution
	$h = 0.1$	$h = 0.05$	$h = 0.01$	$Y(x) = x^4$
0.0	0.0000	0.0000	0.0000	0.0000
0.1	0.0000	0.0000	0.0001	0.0001
0.2	0.0004	0.0009	0.0014	0.0016
0.3	0.0036	0.0056	0.0076	0.0081
0.4	0.0144	0.0196	0.0243	0.0256
0.5	0.0400	0.0506	0.0600	0.0625
0.6	0.0900	0.1089	0.1253	0.1296
0.7	0.1764	0.2070	0.2333	0.2401
0.8	0.3136	0.3600	0.3994	0.4096
0.9	0.5184	0.5852	0.6416	0.6561
1.0	0.8100	0.9025	0.9801	1.0000

32.21.

Method: HEUN'S METHOD				
Problem: $y' = -y;\ y(0) = 1$				
x_n	y_n			True solution
	$h = 0.1$	$h = 0.05$	$h = 0.01$	$Y(x) = e^{-x}$
0.0	1.0000000	1.0000000	1.0000000	1.0000000
0.1	0.9050000	0.9048766	0.9048389	0.9048374
0.2	0.8190250	0.8188016	0.8187335	0.8187308
0.3	0.7412176	0.7409144	0.7408220	0.7408182
0.4	0.6708020	0.6704361	0.6703246	0.6703201
0.5	0.6070758	0.6066619	0.6065358	0.6065307
0.6	0.5494036	0.5489541	0.5488172	0.5488116
0.7	0.4972102	0.4967357	0.4965911	0.4965853
0.8	0.4499753	0.4494845	0.4493350	0.4493290
0.9	0.4072276	0.4067280	0.4065758	0.4065697
1.0	0.3685410	0.3680386	0.3678856	0.3678794

32.22. Since $Y(x) = x^2$ is a second-degree polynomial, Heun's method is exact and $y(0.5) = Y(0.5) = 0.25$.

32.23.

Method: HEUN'S METHOD				
Problem: $y' = -y + x + 2;\ y(0) = 2$				
x_n	y_n			True solution
	$h = 0.1$	$h = 0.05$	$h = 0.01$	$Y(x) = e^{-x} + x + 1$
0.0	2.000000	2.000000	2.000000	2.000000
0.1	2.005000	2.004877	2.004839	2.004837
0.2	2.019025	2.018802	2.018734	2.018731
0.3	2.041218	2.040914	2.040822	2.040818
0.4	2.070802	2.070436	2.070325	2.070320
0.5	2.107076	2.106662	2.106536	2.106531
0.6	2.149404	2.148954	2.148817	2.148812
0.7	2.197210	2.196736	2.196591	2.196585
0.8	2.249975	2.249485	2.249335	2.249329
0.9	2.307228	2.306728	2.306576	2.306570
1.0	2.368541	2.368039	2.367886	2.367879

32.24.

Method: HEUN'S METHOD

Problem: $y' = 4x^3;\ y(0) = 0$

x_n	y_n			True solution
	$h = 0.1$	$h = 0.05$	$h = 0.01$	$Y(x) = x^4$
0.0	0.0000000	0.0000000	0.0000000	0.0000000
0.1	0.0002000	0.0001250	0.0001010	0.0001000
0.2	0.0020000	0.0017000	0.0016040	0.0016000
0.3	0.0090000	0.0083250	0.0081090	0.0081000
0.4	0.0272000	0.0260000	0.0256160	0.0256000
0.5	0.0650000	0.0631250	0.0625250	0.0625000
0.6	0.1332000	0.1305000	0.1296360	0.1296000
0.7	0.2450000	0.2413250	0.2401490	0.2401000
0.8	0.4160000	0.4112000	0.4096640	0.4096000
0.9	0.6642000	0.6581250	0.6561810	0.6561000
1.0	1.0100000	1.0025000	1.0001000	1.0000000

32.25.

Method: THREE-TERM TAYLOR SERIES METHOD

Problem: $y' = -y;\ y(0) = 1$

x_n	y_n			True solution
	$h = 0.1$	$h = 0.05$	$h = 0.01$	$Y(x) = e^{-x}$
0.0	1.0000000	1.0000000	1.0000000	1.0000000
0.1	0.9050000	0.9048766	0.9048389	0.9048374
0.2	0.8190250	0.8188016	0.8187335	0.8187308
0.3	0.7412176	0.7409144	0.7408220	0.7408182
0.4	0.6708020	0.6704361	0.6703246	0.6703201
0.5	0.6070758	0.6066619	0.6065358	0.6065306
0.6	0.5494036	0.5489541	0.5488172	0.5488116
0.7	0.4972102	0.4967357	0.4965911	0.4965853
0.8	0.4499753	0.4494845	0.4493350	0.4493290
0.9	0.4072276	0.4067280	0.4065758	0.4065697
1.0	0.3685410	0.3680386	0.3678856	0.3678794

32.26.

Method: THREE-TERM TAYLOR SERIES METHOD				
Problem: $y' = -y + x + 2;\ y(0) = 2$				
x_n	y_n			True solution
	$h = 0.1$	$h = 0.05$	$h = 0.01$	$Y(x) = e^{-x} + x + 1$
0.0	2.000000	2.000000	2.000000	2.000000
0.1	2.005000	2.004877	2.004839	2.004837
0.2	2.019025	2.018802	2.018734	2.018731
0.3	2.041218	2.040914	2.040822	2.040818
0.4	2.070802	2.070436	2.070325	2.070320
0.5	2.107076	2.106662	2.106536	2.106531
0.6	2.149404	2.148954	2.148817	2.148812
0.7	2.197210	2.196736	2.196591	2.196585
0.8	2.249975	2.249485	2.249335	2.249329
0.9	2.307228	2.306728	2.306576	2.306570
1.0	2.368541	2.368039	2.367886	2.367879

32.27.

Method: NYSTROM'S METHOD				
Problem: $y' = -y;\ y(0) = 1$				
x_n	y_n			True solution
	$h = 0.1$	$h = 0.05$	$h = 0.01$	$Y(x) = e^{-x}$
0.0	1.0000000	1.0000000	1.0000000	1.0000000
0.1	0.9050000	0.9048750	0.9048389	0.9048374
0.2	0.8190000	0.8187988	0.8187335	0.8187308
0.3	0.7412000	0.7409105	0.7408219	0.7408182
0.4	0.6707600	0.6704313	0.6703245	0.6703201
0.5	0.6070480	0.6066565	0.6065357	0.6065306
0.6	0.5493504	0.5489482	0.5488171	0.5488116
0.7	0.4971779	0.4967294	0.4965911	0.4965853
0.8	0.4499145	0.4494779	0.4493350	0.4493290
0.9	0.4071950	0.4067212	0.4065758	0.4065697
1.0	0.3684758	0.3680317	0.3678856	0.3678794

32.28.

Method: NYSTROM'S METHOD				
Problem: $y' = -y + x + 2;\ y(0) = 2$				
x_n	y_n			True solution
	$h = 0.1$	$h = 0.05$	$h = 0.01$	$Y(x) = e^{-x} + x + 1$
0.0	2.000000	2.000000	2.000000	2.000000
0.1	2.005000	2.004875	2.004839	2.004837
0.2	2.019000	2.018799	2.018734	2.018731
0.3	2.041200	2.040911	2.040822	2.040818
0.4	2.070760	2.070431	2.070325	2.070320
0.5	2.107048	2.106657	2.106536	2.106531
0.6	2.149350	2.148948	2.148817	2.148817
0.7	2.197178	2.196729	2.196591	2.196585
0.8	2.249915	2.249478	2.249335	2.249329
0.9	2.307195	2.306721	2.306576	2.306570
1.0	2.368476	2.368032	2.367886	2.367879

32.29.

Method: NYSTROM'S METHOD				
Problem: $y' = 4x^3;\ y(0) = 0$				
x_n	y_n			True solution
	$h = 0.1$	$h = 0.05$	$h = 0.01$	$Y(x) = x^4$
0.0	0.0000000	0.0000000	0.0000000	0.0000000
0.1	0.0000000	0.0000500	0.0000980	0.0001000
0.2	0.0008000	0.0014000	0.0015920	0.0016000
0.3	0.0064000	0.0076500	0.0080820	0.0081000
0.4	0.0224000	0.0248000	0.0255680	0.0256000
0.5	0.0576000	0.0612500	0.0624500	0.0625000
0.6	0.1224000	0.1278000	0.1295280	0.1296000
0.7	0.2304000	0.2376500	0.2400020	0.2401000
0.8	0.3968000	0.4064000	0.4094720	0.4096000
0.9	0.6400000	0.6520500	0.6559380	0.6561000
1.0	0.9800000	0.9950000	0.9998000	1.0000000

Chapter 33

Runge-Kutta Methods

33.1 INTRODUCTION

Runge-Kutta methods are a particular set of self-starting numerical methods. Euler's method, (*32.3*), is a first-order Runge-Kutta method; Heun's method, (*32.5*), is a Runge-Kutta method of order two.

Runge-Kutta methods can be used to generate an entire solution, but these methods are slower and more involved than the predictor-corrector methods presented in Chapter 34. The latter methods, however, require starting values (see Problem 32.14), which are best obtained by Runge-Kutta methods.

33.2 A THIRD-ORDER RUNGE-KUTTA METHOD

$$y_{n+1} = y_n + \tfrac{1}{6}(k_1 + 4k_2 + k_3) \tag{33.1}$$

where

$$k_1 = hf(x_n, y_n)$$

$$k_2 = hf(x_n + \tfrac{1}{2}h,\ y_n + \tfrac{1}{2}k_1)$$

$$k_3 = hf(x_n + h,\ y_n - k_1 + 2k_2)$$

In using (*33.1*) [or (*33.2*) below], one first computes the various k's and then determines y_{n+1}. Note that since each k depends on x_n and y_n, it must be calculated anew for each n.

33.3 A FOURTH-ORDER RUNGE-KUTTA METHOD

$$y_{n+1} = y_n + \tfrac{1}{6}(k_1 + 2k_2 + 2k_3 + k_4) \tag{33.2}$$

where

$$k_1 = hf(x_n, y_n)$$

$$k_2 = hf(x_n + \tfrac{1}{2}h,\ y_n + \tfrac{1}{2}k_1)$$

$$k_3 = hf(x_n + \tfrac{1}{2}h,\ y_n + \tfrac{1}{2}k_2)$$

$$k_4 = hf(x_n + h,\ y_n + k_3)$$

Solved Problems

33.1. Find $y(1)$ for $y' = y - x$; $y(0) = 2$, using a third-order Runge-Kutta method with $h = 0.1$.

Here $f(x,y) = y - x$. Using (33.1) with $n = 0, 1, 2, \ldots, 9$, we compute

n = 0: $x_0 = 0, \quad y_0 = 2$

$$k_1 = hf(x_0, y_0) = hf(0,2) = (0.1)(2-0) = 0.2$$

$$k_2 = hf(x_0 + \tfrac{1}{2}h,\ y_0 + \tfrac{1}{2}k_1) = hf[0 + \tfrac{1}{2}(0.1),\ 2 + \tfrac{1}{2}(0.2)] = hf(0.05, 2.1) = (0.1)(2.1 - 0.05) = 0.205$$

$$k_3 = hf(x_0 + h,\ y_0 - k_1 + 2k_2) = hf[0 + 0.1,\ 2 - 0.2 + 2(0.205)] = hf(0.1, 2.21) = (0.1)(2.21 - 0.1) = 0.211$$

$$y_1 = y_0 + \tfrac{1}{6}(k_1 + 4k_2 + k_3) = 2 + \tfrac{1}{6}[0.2 + 4(0.205) + 0.211] = 2.205$$

n = 1: $x_1 = 0.1, \quad y_1 = 2.205$

$$k_1 = hf(x_1, y_1) = hf(0.1, 2.205) = (0.1)(2.205 - 0.1) = 0.211$$

$$k_2 = hf(x_1 + \tfrac{1}{2}h,\ y_1 + \tfrac{1}{2}k_1) = hf[0.1 + \tfrac{1}{2}(0.1),\ 2.205 + \tfrac{1}{2}(0.211)] = hf(0.15, 2.311) = (0.1)(2.311 - 0.15) = 0.216$$

$$k_3 = hf(x_1 + h,\ y_1 - k_1 + 2k_2) = hf[0.1 + 0.1,\ 2.205 - 0.211 + 2(0.216)] = hf(0.2, 2.426) = (0.1)(2.426 - 0.2) = 0.223$$

$$y_2 = y_1 + \tfrac{1}{6}(k_1 + 4k_2 + k_3) = 2.205 + \tfrac{1}{6}[0.211 + 4(0.216) + 0.223] = 2.421$$

n = 2: $x_2 = 0.2, \quad y_2 = 2.421$

$$k_1 = hf(x_2, y_2) = hf(0.2, 2.421) = (0.1)(2.421 - 0.2) = 0.222$$

$$k_2 = hf(x_2 + \tfrac{1}{2}h,\ y_2 + \tfrac{1}{2}k_1) = hf[0.2 + \tfrac{1}{2}(0.1),\ 2.421 + \tfrac{1}{2}(0.222)] = hf(0.25, 2.532) = (0.1)(2.532 - 0.25) = 0.228$$

$$k_3 = hf(x_2 + h,\ y_2 - k_1 + 2k_2) = hf[0.2 + 0.1,\ 2.421 - 0.222 + 2(0.228)] = hf(0.3, 2.655) = (0.1)(2.655 - 0.3) = 0.236$$

$$y_3 = y_2 + \tfrac{1}{6}(k_1 + 4k_2 + k_3) = 2.421 + \tfrac{1}{6}[0.222 + 4(0.228) + 0.236] = 2.649$$

Continuing in this manner (see Table 33-1), we obtain $y(1) = y_{10} = 4.718$.

Note from Table 33-1 that this third-order Runge-Kutta method with $h = 0.1$ is more accurate than Euler's method with $h = 0.005$ (Table 32-1, page 204) or Heun's method with $h = 0.05$ (Table 32-4, page 209). Also compare Table 33-1 with Tables 32-6, page 212, and 32-8, page 214.

33.2. Find $y(1)$ for $y' = y - x$; $y(0) = 2$, using a fourth-order Runge-Kutta method with $h = 0.1$.

Again $f(x,y) = y - x$. Using (33.2) with $n = 0, 1, \ldots, 9$, we compute

n = 0: $x_0 = 0, \quad y_0 = 2$

$$k_1 = hf(x_0, y_0) = hf(0,2) = (0.1)(2-0) = 0.2$$

$$k_2 = hf(x_0 + \tfrac{1}{2}h,\ y_0 + \tfrac{1}{2}k_1) = hf[0 + \tfrac{1}{2}(0.1),\ 2 + \tfrac{1}{2}(0.2)] = hf(0.05, 2.1) = (0.1)(2.1 - 0.05) = 0.205$$

Table 33-1

Method: THIRD-ORDER RUNGE-KUTTA METHOD

Problem: $y' = y - x; \quad y(0) = 2$

x_n	y_n		True solution
	$h = 0.1$	$h = 0.05$	$Y(x) = e^x + x + 1$
0.0	2.0000000	2.0000000	2.0000000
0.1	2.2051667	2.2051704	2.2051709
0.2	2.4213934	2.4214015	2.4214028
0.3	2.6498432	2.6498568	2.6498588
0.4	2.8918017	2.8918217	2.8918247
0.5	3.1486896	3.1487172	3.1487213
0.6	3.4220767	3.4221133	3.4221188
0.7	3.7136985	3.7137457	3.7137527
0.8	4.0254724	4.0255320	4.0255409
0.9	4.3595180	4.3595920	4.3596031
1.0	4.7181773	4.7182682	4.7182818

Table 33-2

Method: FOURTH-ORDER RUNGE-KUTTA METHOD

Problem: $y' = y - x; \quad y(0) = 2$

x_n	$h = 0.1$ y_n	True solution $Y(x) = e^x + x + 1$
0.0	2.0000000	2.0000000
0.1	2.2051708	2.2051709
0.2	2.4214026	2.4214028
0.3	2.6498585	2.6498588
0.4	2.8918242	2.8918247
0.5	3.1487206	3.1487213
0.6	3.4221180	3.4221188
0.7	3.7137516	3.7137527
0.8	4.0255396	4.0255409
0.9	4.3596014	4.3596031
1.0	4.7182797	4.7182818

$$k_3 = hf(x_0 + \tfrac{1}{2}h,\ y_0 + \tfrac{1}{2}k_2) = hf[0 + \tfrac{1}{2}(0.1),\ 2 + \tfrac{1}{2}(0.205)]$$
$$= hf(0.05, 2.103) = (0.1)(2.103 - 0.05) = 0.205$$

$$k_4 = hf(x_0 + h,\ y_0 + k_3) = hf(0 + 0.1,\ 2 + 0.205)$$
$$= hf(0.1, 2.205) = (0.1)(2.205 - 0.1) = 0.211$$

$$y_1 = y_0 + \tfrac{1}{6}(k_1 + 2k_2 + 2k_3 + k_4)$$
$$= 2 + \tfrac{1}{6}[0.2 + 2(0.205) + 2(0.205) + 0.211] = 2.205$$

$n = 1$: $x_1 = 0.1, \quad y_1 = 2.205$

$$k_1 = hf(x_1, y_1) = hf(0.1, 2.205) = (0.1)(2.205 - 0.1) = 0.211$$

$$k_2 = hf(x_1 + \tfrac{1}{2}h,\ y_1 + \tfrac{1}{2}k_1) = hf[0.1 + \tfrac{1}{2}(0.1),\ 2.205 + \tfrac{1}{2}(0.211)]$$
$$= hf(0.15, 2.311) = (0.1)(2.311 - 0.15) = 0.216$$

$$k_3 = hf(x_1 + \tfrac{1}{2}h,\ y_1 + \tfrac{1}{2}k_2) = hf[0.1 + \tfrac{1}{2}(0.1),\ 2.205 + \tfrac{1}{2}(0.216)]$$
$$= hf(0.15, 2.313) = (0.1)(2.313 - 0.15) = 0.216$$

$$k_4 = hf(x_1 + h,\ y_1 + k_3) = hf(0.1 + 0.1,\ 2.205 + 0.216)$$
$$= hf(0.2, 2.421) = (0.1)(2.421 - 0.2) = 0.222$$

$$y_2 = y_1 + \tfrac{1}{6}(k_1 + 2k_2 + 2k_3 + k_4)$$
$$= 2.205 + \tfrac{1}{6}[0.211 + 2(0.216) + 2(0.216) + 0.222] = 2.421$$

$n = 2$: $x_2 = 0.2, \quad y_2 = 2.421$

$$k_1 = hf(x_2, y_2) = hf(0.2, 2.421) = (0.1)(2.421 - 0.2) = 0.222$$

$$k_2 = hf(x_2 + \tfrac{1}{2}h,\ y_2 + \tfrac{1}{2}k_1) = hf[0.2 + \tfrac{1}{2}(0.1),\ 2.421 + \tfrac{1}{2}(0.222)]$$
$$= hf(0.25, 2.532) = (0.1)(2.532 - 0.25) = 0.228$$

$$k_3 = hf(x_2 + \tfrac{1}{2}h,\ y_2 + \tfrac{1}{2}k_2) = hf[0.2 + \tfrac{1}{2}(0.1),\ 2.421 + \tfrac{1}{2}(0.228)]$$
$$= hf(0.25, 2.535) = (0.1)(2.535 - 0.25) = 0.229$$

$$k_4 = hf(x_2 + h,\ y_2 + k_3) = hf(0.2 + 0.1,\ 2.421 + 0.229)$$
$$= hf(0.3, 2.650) = (0.1)(2.650 - 0.3) = 0.235$$

$$y_3 = y_2 + \tfrac{1}{6}(k_1 + 2k_2 + 2k_3 + k_4)$$
$$= 2.421 + \tfrac{1}{6}[0.222 + 2(0.228) + 2(0.229) + 0.235] = 2.650$$

Continuing in this manner (Table 33-2), we obtain $y(1) = y_{10} = 4.718$. Note that these results are more accurate than those given in Table 33-1.

33.3. Find $y(1)$ for $y' = y$; $y(0) = 1$, using a fourth-order Runge-Kutta method with $h = 0.1$.

Here $f(x, y) = y$. Using *(33.2)*, we compute

$n = 0$: $x_0 = 0, \quad y_0 = 1$

$$k_1 = hf(x_0, y_0) = hf(0, 1) = (0.1)(1) = 0.1$$

$$k_2 = hf(x_0 + \tfrac{1}{2}h,\ y_0 + \tfrac{1}{2}k_1) = hf[0 + \tfrac{1}{2}(0.1),\ 1 + \tfrac{1}{2}(0.1)]$$
$$= hf(0.05, 1.05) = (0.1)(1.05) = 0.105$$

$$k_3 = hf(x_0 + \tfrac{1}{2}h,\ y_0 + \tfrac{1}{2}k_2) = hf[0 + \tfrac{1}{2}(0.1),\ 1 + \tfrac{1}{2}(0.105)]$$
$$= hf(0.05, 1.053) = (0.1)(1.053) = 0.105$$

$$k_4 = hf(x_0+h,\ y_0+k_3) = hf(0+0.1, 1+0.105)$$
$$= hf(0.1, 1.105) = (0.1)(1.105) = 0.111$$

$$y_1 = y_0 + \tfrac{1}{6}(k_1+2k_2+2k_3+k_4)$$
$$= 1 + \tfrac{1}{6}[0.1 + 2(0.105) + 2(0.105) + 0.111] = 1.105$$

n = 1: $x_1 = 0.1, \quad y_1 = 1.105$

$$k_1 = hf(x_1, y_1) = hf(0.1, 1.105) = (0.1)(1.105) = 0.111$$

$$k_2 = hf(x_1+\tfrac{1}{2}h,\ y_1+\tfrac{1}{2}k_1) = hf[0.1+\tfrac{1}{2}(0.1),\ 1.105+\tfrac{1}{2}(0.111)]$$
$$= hf(0.15, 1.161) = (0.1)(1.161) = 0.116$$

$$k_3 = hf(x_1+\tfrac{1}{2}h,\ y_1+\tfrac{1}{2}k_2) = hf[0.1+\tfrac{1}{2}(0.1),\ 1.105+\tfrac{1}{2}(0.116)]$$
$$= hf(0.15, 1.163) = (0.1)(1.163) = 0.116$$

$$k_4 = hf(x_1+h,\ y_1+k_3) = hf(0.1+0.1,\ 1.105+0.116)$$
$$= hf(0.2, 1.221) = (0.1)(1.221) = 0.122$$

$$y_2 = y_1 + \tfrac{1}{6}(k_1+2k_2+2k_3+k_4)$$
$$= 1.105 + \tfrac{1}{6}[0.111 + 2(0.116) + 2(0.116) + 0.122] = 1.221$$

n = 2: $x_2 = 0.2, \quad y_2 = 1.221$

$$k_1 = hf(x_2, y_2) = hf(0.2, 1.221) = (0.1)(1.221) = 0.122$$

$$k_2 = hf(x_2+\tfrac{1}{2}h,\ y_2+\tfrac{1}{2}k_1) = hf[0.2+\tfrac{1}{2}(0.1),\ 1.221+\tfrac{1}{2}(0.122)]$$
$$= hf(0.25, 1.282) = (0.1)(1.282) = 0.128$$

$$k_3 = hf(x_2+\tfrac{1}{2}h,\ y_2+\tfrac{1}{2}k_2) = hf[0.2+\tfrac{1}{2}(0.1),\ 1.221+\tfrac{1}{2}(0.128)]$$
$$= hf(0.25, 1.285) = (0.1)(1.285) = 0.129$$

$$k_4 = hf(x_2+h,\ y_2+k_3) = hf(0.2+0.1,\ 1.221+0.129)$$
$$= hf(0.3, 1.350) = (0.1)(1.350) = 0.135$$

$$y_3 = y_2 + \tfrac{1}{6}(k_1+2k_2+2k_3+k_4)$$
$$= 1.221 + \tfrac{1}{6}[0.122 + 2(0.128) + 2(0.129) + 0.135] = 1.350$$

Continuing in this manner (Table 33-3), we obtain $y(1) = y_{10} = 2.718$. Compare the results with Tables 32-2, 32-5, and 32-9 on pages 204, 209, and 214, respectively.

33.4. Find $y(1)$ for $y' = y^2+1;\ y(0) = 0$, using a fourth-order Runge-Kutta method with $h = 0.1$.

For this problem, $f(x,y) = y^2+1$. Using (*33.2*), we compute

n = 0: $x_0 = 0, \quad y_0 = 0$

$$k_1 = hf(x_0, y_0) = hf(0,0) = (0.1)[(0)^2+1] = 0.1$$

$$k_2 = hf(x_0+\tfrac{1}{2}h,\ y_0+\tfrac{1}{2}k_1) = hf[0+\tfrac{1}{2}(0.1),\ 0+\tfrac{1}{2}(0.1)]$$
$$= hf(0.05, 0.05) = (0.1)[(0.05)^2+1] = 0.1$$

$$k_3 = hf(x_0+\tfrac{1}{2}h,\ y_0+\tfrac{1}{2}k_2) = hf[0+\tfrac{1}{2}(0.1),\ 0+\tfrac{1}{2}(0.1)]$$
$$= hf(0.05, 0.05) = (0.1)[(0.05)^2+1] = 0.1$$

Table 33-3

Method: FOURTH-ORDER RUNGE-KUTTA METHOD		
Problem: $y' = y;\ y(0) = 1$		
x_n	$h = 0.1$ y_n	True solution $Y(x) = e^x$
0.0	1.0000000	1.0000000
0.1	1.1051708	1.1051709
0.2	1.2214026	1.2214028
0.3	1.3498585	1.3498588
0.4	1.4918242	1.4918247
0.5	1.6487206	1.6487213
0.6	1.8221180	1.8221188
0.7	2.0137516	2.0137527
0.8	2.2255396	2.2255409
0.9	2.4596014	2.4596031
1.0	2.7182797	2.7182818

Table 33-4

Method: FOURTH-ORDER RUNGE-KUTTA METHOD		
Problem: $y' = y^2 + 1;\ y(0) = 0$		
x_n	$h = 0.1$ y_n	True solution $Y(x) = \tan x$
0.0	0.0000000	0.0000000
0.1	0.1003346	0.1003347
0.2	0.2027099	0.2027100
0.3	0.3093360	0.3093363
0.4	0.4227930	0.4227932
0.5	0.5463023	0.5463025
0.6	0.6841368	0.6841368
0.7	0.8422886	0.8422884
0.8	1.0296391	1.0296386
0.9	1.2601588	1.2601582
1.0	1.5574064	1.5574077

$$\begin{aligned} k_4 &= hf(x_0+h,\, y_0+k_3) = hf[0+0.1, 0+0.1] \\ &= hf(0.1, 0.1) = (0.1)[(0.1)^2+1] = 0.101 \end{aligned}$$

$$\begin{aligned} y_1 &= y_0 + \tfrac{1}{6}(k_1+2k_2+2k_3+k_4) \\ &= 0 + \tfrac{1}{6}[0.1+2(0.1)+2(0.1)+0.101] = 0.1 \end{aligned}$$

n = 1: $x_1 = 0.1, \quad y_1 = 0.1$

$$k_1 = hf(x_1, y_1) = hf(0.1, 0.1) = (0.1)[(0.1)^2+1] = 0.101$$

$$\begin{aligned} k_2 &= hf(x_1+\tfrac{1}{2}h,\, y_1+\tfrac{1}{2}k_1) = hf[0.1+\tfrac{1}{2}(0.1),\, (0.1)+\tfrac{1}{2}(0.101)] \\ &= hf(0.15, 0.151) = (0.1)[(0.151)^2+1] = 0.102 \end{aligned}$$

$$\begin{aligned} k_3 &= hf(x_1+\tfrac{1}{2}h,\, y_1+\tfrac{1}{2}k_2) = hf[0.1+\tfrac{1}{2}(0.1),\, (0.1)+\tfrac{1}{2}(0.102)] \\ &= hf(0.15, 0.151) = (0.1)[(0.151)^2+1] = 0.102 \end{aligned}$$

$$\begin{aligned} k_4 &= hf(x_1+h,\, y_1+k_3) = hf(0.1+0.1,\, 0.1+0.102) \\ &= hf(0.2, 0.202) = (0.1)[(0.202)^2+1] = 0.104 \end{aligned}$$

$$\begin{aligned} y_2 &= y_1 + \tfrac{1}{6}(k_1+2k_2+2k_3+k_4) \\ &= 0.1 + \tfrac{1}{6}[0.101+2(0.102)+2(0.102)+0.104] = 0.202 \end{aligned}$$

n = 2: $x_2 = 0.2, \quad y_2 = 0.202$

$$k_1 = hf(x_2, y_2) = hf(0.2, 0.202) = (0.1)[(0.202)^2+1] = 0.104$$

$$\begin{aligned} k_2 &= hf(x_2+\tfrac{1}{2}h,\, y_2+\tfrac{1}{2}k_1) = hf[0.2+\tfrac{1}{2}(0.1),\, 0.202+\tfrac{1}{2}(0.104)] \\ &= hf(0.25, 0.254) = (0.1)[(0.254)^2+1] = 0.106 \end{aligned}$$

$$\begin{aligned} k_3 &= hf(x_2+\tfrac{1}{2}h,\, y_2+\tfrac{1}{2}k_2) = hf[0.2+\tfrac{1}{2}(0.1),\, 0.202+\tfrac{1}{2}(0.106)] \\ &= hf(0.25, 0.255) = (0.1)[(0.255)^2+1] = 0.107 \end{aligned}$$

$$\begin{aligned} k_4 &= hf(x_2+h,\, y_2+k_3) = hf(0.2+0.1,\, 0.202+0.107) \\ &= hf(0.3, 0.309) = (0.1)[(0.309)^2+1] = 0.110 \end{aligned}$$

$$\begin{aligned} y_3 &= y_2 + \tfrac{1}{6}(k_1+2k_2+2k_3+k_4) \\ &= 0.202 + \tfrac{1}{6}[0.104+2(0.106)+2(0.107)+0.110] = 0.309 \end{aligned}$$

Continuing in this manner (Table 33-4), we obtain $y(1) = y_{10} = 1.557$. Compare the results with Tables 32-3, 32-7, and 32-10 on pages 206, 212, and 216, respectively.

Supplementary Problems

33.5. Find $y(0.3)$ for $y' = -y$; $y(0) = 1$, using a third-order Runge-Kutta method with $h = 0.1$. Carry all computations to three decimal places.

In Problems 33.6 through 33.9, use a fourth-order Runge-Kutta method to find $y(0.3)$ with $h = 0.1$. Carry all computations to three decimal places.

33.6. $y' = -y; \quad y(0) = 1.$

33.7. $y' = -y + x + 2; \quad y(0) = 2.$

33.8. $y' = 4x^3; \quad y(0) = 0.$

33.9. $y' = 5x^4; \quad y(0) = 0.$

Answers to Supplementary Problems

For comparison with methods given in other chapters, all answers are rounded off to seven decimal places through $x = 1.0$.

33.5.

Method: THIRD-ORDER RUNGE-KUTTA METHOD			
Problem: $y' = -y;\ y(0) = 1$			
x_n	y_n		True solution
	$h = 0.1$	$h = 0.05$	$Y(x) = e^{-x}$
0.0	1.0000000	1.0000000	1.0000000
0.1	0.9048333	0.9048369	0.9048374
0.2	0.8187234	0.8187299	0.8187308
0.3	0.7408082	0.7408170	0.7408182
0.4	0.6703079	0.6703186	0.6703201
0.5	0.6065170	0.6065290	0.6065306
0.6	0.5487968	0.5488099	0.5488116
0.7	0.4965696	0.4965834	0.4965853
0.8	0.4493127	0.4493270	0.4493290
0.9	0.4065531	0.4065677	0.4065697
1.0	0.3678628	0.3678775	0.3678794

33.6.

Method: FOURTH-ORDER RUNGE-KUTTA METHOD		
Problem: $y' = -y;\ y(0) = 1$		
x_n	$h = 0.1$ y_n	True solution $Y(x) = e^{-x}$
0.0	1.0000000	1.0000000
0.1	0.9048375	0.9048374
0.2	0.8187309	0.8187308
0.3	0.7408184	0.7408182
0.4	0.6703203	0.6703201
0.5	0.6065309	0.6065307
0.6	0.5488119	0.5488116
0.7	0.4965856	0.4965853
0.8	0.4493293	0.4493290
0.9	0.4065700	0.4065697
1.0	0.3678798	0.3678794

33.7.

Method: FOURTH-ORDER RUNGE-KUTTA METHOD		
Problem: $y' = -y + x + 2; \quad y(0) = 2$		
x_n	$h = 0.1$ y_n	True solution $Y(x) = e^{-x} + x + 1$
0.0	2.000000	2.000000
0.1	2.004838	2.004837
0.2	2.018731	2.018731
0.3	2.040818	2.040818
0.4	2.070320	2.070320
0.5	2.106531	2.106531
0.6	2.148812	2.148812
0.7	2.196586	2.196585
0.8	2.249329	2.249329
0.9	2.306570	2.306570
1.0	2.367880	2.367879

33.8. Since $Y(x) = x^4$ is a fourth-degree polynomial, a fourth-order Runge-Kutta method is exact and $y(0.3) = Y(0.3) = 0.0081$.

33.9.

Method: FOURTH-ORDER RUNGE-KUTTA METHOD		
Problem: $y' = 5x^4; \quad y(0) = 0$		
x_n	$h = 0.1$ y_n	True solution $Y(x) = x^5$
0.0	0.0000000	0.0000000
0.1	0.0000104	0.0000100
0.2	0.0003208	0.0003200
0.3	0.0024313	0.0024300
0.4	0.0102417	0.0102400
0.5	0.0312521	0.0312500
0.6	0.0777625	0.0777600
0.7	0.1680729	0.1680700
0.8	0.3276833	0.3276800
0.9	0.5904938	0.5904900
1.0	1.0000042	1.0000000

Chapter 34

Predictor-Corrector Methods

34.1 INTRODUCTION

A *predictor-corrector* method is a set of two equations for y_{n+1}. The first equation, called the *predictor*, is used to predict (obtain a first approximation to) y_{n+1}; the second equation, called the *corrector*, is then used to obtain a corrected value (second approximation) to y_{n+1}. In general, the corrector depends on the predicted value.

34.2 A SECOND-ORDER METHOD

A simple second-order predictor-corrector method is obtained by combining Nystrom's method, (*32.8*), with the *trapezoidal method*

$$y_{n+1} = y_n + \frac{h}{2}(y'_n + y'_{n+1})$$

The resulting equations are

$$\text{predictor:} \quad y_{n+1} = y_{n-1} + 2hy'_n$$

$$\text{corrector:} \quad y_{n+1} = y_n + \frac{h}{2}(y'_n + y'_{n+1}) \tag{34.1}$$

For notational convenience, we designate the predicted values of y_n by py_n. It then follows from (*32.2*) that

$$py'_n = f(x_n, py_n) \tag{34.2}$$

and (*34.1*) becomes

$$\text{predictor:} \quad py_{n+1} = y_{n-1} + 2hy'_n$$

$$\text{corrector:} \quad y_{n+1} = y_n + \frac{h}{2}(y'_n + py'_{n+1}) \tag{34.3}$$

(See Problems 34.1 and 34.2.)

34.3 MILNE'S METHOD

This fourth-order method is given by:

$$\text{predictor:} \quad py_{n+1} = y_{n-3} + \frac{4h}{3}(2y'_n - y'_{n-1} + 2y'_{n-2})$$

$$\text{corrector:} \quad y_{n+1} = y_{n-1} + \frac{h}{3}(py'_{n+1} + 4y'_n + y'_{n-1}) \tag{34.4}$$

(See Problems 34.3, 34.5, and 34.7.)

34.4 HAMMING'S METHOD

This is another fourth-order method; it uses the same predictor as Milne's method.

$$\text{predictor:} \quad py_{n+1} = y_{n-3} + \frac{4h}{3}(2y'_n - y'_{n-1} + 2y'_{n-2})$$

$$\text{corrector:} \quad y_{n+1} = \frac{1}{8}(9y_n - y_{n-2}) + \frac{3h}{8}(py'_{n+1} + 2y'_n - y'_{n-1}) \tag{34.5}$$

(See Problems 34.4, 34.6, and 34.8.) In general, when many values of x_n are required (for example, if $y(30)$ is computed with $h=0.1$, thereby necessitating 301 different values of x_n), Hamming's method is preferred to Milne's method. In contrast, if only a few values of x_n are required, Milne's method is preferred. Note that for all problems in this chapter, only eleven different values of x_n $(x_0, x_1, \ldots, x_{10})$ are required; hence, Milne's method gives slightly better results than Hamming's method.

34.5 STARTING VALUES

Each of the three methods presented above requires more starting values (see Problem 32.14) than are available in a first-order initial-value problem. In particular, besides the initial condition y_0, method (*34.3*) requires y_1; and both Milne's method and Hamming's method require y_1, y_2, and y_3.

To generate starting values for the second-order method (*34.3*), a second-order Runge-Kutta method (e.g. Heun's method) is generally used. A fourth-order Runge-Kutta method, (*33.2*), is commonly used to generate starting values for both Milne's method and Hamming's method, which are themselves of order four. Generating starting values by Taylor series methods should be avoided (compare Section 32.6).

Solved Problems

34.1. Find $y(1)$ for $y' = y - x$; $y(0) = 2$, using a second-order predictor-corrector method with $h = 0.1$.

For this problem, $f(x,y) = y - x$, $x_0 = 0$, and $y_0 = 2$; hence, $y'_0 = y_0 - x_0 = 2 - 0 = 2$. From Heun's method, Problem 32.9, we obtain the additional required starting value $y_1 = 2.205$. Then, using (*34.3*) and (*34.2*), we compute

n = 1:
$$x_1 = 0.1, \quad y_1 = 2.205$$
$$y'_1 = y_1 - x_1 = 2.205 - 0.1 = 2.105$$
$$py_2 = y_0 + 2hy'_1 = 2 + 2(0.1)(2.105) = 2.421$$
$$py'_2 = py_2 - x_2 = 2.421 - 0.2 = 2.221$$
$$y_2 = y_1 + \frac{h}{2}(y'_1 + py'_2) = 2.205 + \frac{0.1}{2}(2.105 + 2.221) = 2.421$$

n = 2:
$$x_2 = 0.2, \quad y_2 = 2.421$$
$$y'_2 = y_2 - x_2 = 2.421 - 0.2 = 2.221$$
$$py_3 = y_1 + 2hy'_2 = 2.205 + 2(0.1)(2.221) = 2.649$$
$$py'_3 = py_3 - x_3 = 2.649 - 0.3 = 2.349$$
$$y_3 = y_2 + \frac{h}{2}(y'_2 + py'_3) = 2.421 + \frac{0.1}{2}(2.221 + 2.349) = 2.650$$

n = 3:
$$x_3 = 0.3, \quad y_3 = 2.650$$
$$y'_3 = y_3 - x_3 = 2.650 - 0.3 = 2.350$$
$$py_4 = y_2 + 2hy'_3 = 2.421 + 2(0.1)(2.350) = 2.891$$
$$py'_4 = py_4 - x_4 = 2.891 - 0.4 = 2.491$$
$$y_4 = y_3 + \frac{h}{2}(y'_3 + py'_4) = 2.650 + \frac{0.1}{2}(2.350 + 2.491) = 2.892$$

Continuing in this manner (Table 34-1), we find that $y(1) = 4.719$. Compare the results with Table 32-8, page 214.

Table 34-1

Method: SECOND-ORDER PREDICTOR-CORRECTOR METHOD			
Problem: $y' = y - x; \quad y(0) = 2$			
x_n	$h = 0.1$		True solution
	py_n	y_n	$Y(x) = e^x + x + 1$
0.0	–	2.0000000	2.0000000
0.1	–	2.2050000	2.2051709
0.2	2.4210000	2.4213000	2.4214028
0.3	2.6492600	2.6498280	2.6498588
0.4	2.8912656	2.8918827	2.8918247
0.5	3.1482045	3.1488870	3.1487213
0.6	3.4216601	3.4224144	3.4221188
0.7	3.7133699	3.7142036	3.7137527
0.8	4.0252551	4.0261766	4.0255409
0.9	4.3594389	4.3604573	4.3596031
1.0	4.7182680	4.7193936	4.7182818

Table 34-2

Method: SECOND-ORDER PREDICTOR-CORRECTOR METHOD			
Problem: $y' = y^2 + 1; \quad y(0) = 0$			
x_n	$h = 0.1$		True solution
	py_n	y_n	$Y(x) = \tan x$
0.0	–	0.0000000	0.0000000
0.1	–	0.1005000	0.1003347
0.2	0.2020201	0.2030456	0.2027100
0.3	0.3087455	0.3098732	0.3093363
0.4	0.4222500	0.4235890	0.4227932
0.5	0.5457587	0.5474530	0.5463025
0.6	0.6835300	0.6857989	0.6841368
0.7	0.8415170	0.8447225	0.8422884
0.8	1.0285101	1.0332919	1.0296386
0.9	1.2582609	1.2658376	1.2601582
1.0	1.5537608	1.5666634	1.5574077

34.2. Find $y(1)$ for $y' = y^2 + 1;\ y(0) = 0$, using a second-order predictor-corrector method with $h = 0.1$.

Here, $f(x, y) = y^2 + 1$, $x_0 = 0$, and $y_0 = 0$; hence $y_0' = y_0^2 + 1 = 1$. Using Heun's method, we calculate the additional starting value $y_1 = 0.101$. Then using (*34.3*) and (*34.2*), we obtain

$n = 1$:
$$x_1 = 0.1, \quad y_1 = 0.101$$
$$y_1' = (y_1)^2 + 1 = (0.101)^2 + 1 = 1.01$$
$$py_2 = y_0 + 2hy_1' = 0 + 2(0.1)(1.01) = 0.202$$
$$py_2' = (py_2)^2 + 1 = (0.202)^2 + 1 = 1.041$$
$$y_2 = y_1 + \frac{h}{2}(y_1' + py_2') = 0.101 + \frac{0.1}{2}(1.01 + 1.041) = 0.204$$

$n = 2$:
$$x_2 = 0.2, \quad y_2 = 0.204$$
$$y_2' = (y_2)^2 + 1 = (0.204)^2 + 1 = 1.042$$
$$py_3 = y_1 + 2hy_2' = 0.101 + 2(0.1)(1.042) = 0.309$$
$$py_3' = (py_3)^2 + 1 = (0.309)^2 + 1 = 1.095$$
$$y_3 = y_2 + \frac{h}{2}(y_2' + py_3') = 0.204 + \frac{0.1}{2}(1.042 + 1.095) = 0.311$$

$n = 3$:
$$x_3 = 0.3, \quad y_3 = 0.311$$
$$y_3' = (y_3)^2 + 1 = (0.311)^2 + 1 = 1.097$$
$$py_4 = y_2 + 2hy_3' = 0.204 + 2(0.1)(1.097) = 0.423$$
$$py_4' = (py_4)^2 + 1 = (0.423)^2 + 1 = 1.179$$
$$y_4 = y_3 + \frac{h}{2}(y_3' + py_4') = 0.311 + \frac{0.1}{2}(1.097 + 1.179) = 0.425$$

Continuing in this manner (Table 34-2), we obtain $y(1) = 1.566$. Compare the results with Table 32-10 on page 216.

34.3. Find $y(1)$ for $y' = y - x;\ y(0) = 2$, using Milne's method with $h = 0.1$.

Here $f(x, y) = y - x$, $x_0 = 0$, and $y_0 = 2$. Using Table 33-2, page 225, we find for the three additional required starting values: $y_1 = 2.2051708$, $y_2 = 2.4214026$, and $y_3 = 2.6498585$. Thus,

$$y_0' = y_0 - x_0 = 2 - 0 = 2 \qquad y_1' = y_1 - x_1 = 2.1051708$$
$$y_2' = y_2 - x_2 = 2.2214026 \qquad y_3' = y_3 - x_3 = 2.3498585$$

Then, using (*34.4*), we compute

$n = 3$:
$$py_4 = y_0 + \frac{4h}{3}(2y_3' - y_2' + 2y_1')$$
$$= 2 + \frac{4(0.1)}{3}[2(2.3498585) - 2.2214026 + 2(2.1051708)]$$
$$= 2.8918208$$
$$py_4' = py_4 - x_4 = 2.4918208$$
$$y_4 = y_2 + \frac{h}{3}(py_4' + 4y_3' + y_2')$$
$$= 2.4214026 + \frac{0.1}{3}[2.4918208 + 4(2.3498585) + 2.2214026]$$
$$= 2.8918245$$

$n = 4$: $\quad x_4 = 0.4, \quad y'_4 = y_4 - x_4 = 2.4918245$

$$py_5 = y_1 + \frac{4h}{3}(2y'_4 - y'_3 + 2y'_2)$$
$$= 2.2051708 + \frac{4(0.1)}{3}[2(2.4918245) - 2.3498585 + 2(2.2214026)]$$
$$= 3.1487169$$

$$py'_5 = py_5 - x_5 = 2.6487169$$

$$y_5 = y_3 + \frac{h}{3}(py'_5 + 4y'_4 + y'_3)$$
$$= 2.6498585 + \frac{0.1}{3}[2.6487169 + 4(2.4918245) + 2.3498585]$$
$$= 3.1487209$$

$n = 5$: $\quad x_5 = 0.5, \quad y'_5 = y_5 - x_5 = 2.6487209$

$$py_6 = y_2 + \frac{4h}{3}(2y'_5 - y'_4 + 2y'_3)$$
$$= 2.4214026 + \frac{4(0.1)}{3}[2(2.6487209) - 2.4918245 + 2(2.3498585)]$$
$$= 3.4221138$$

$$py'_6 = py_6 - x_6 = 2.8221138$$

$$y_6 = y_4 + \frac{h}{3}(py'_6 + 4y'_5 + y'_4)$$
$$= 2.8918245 + \frac{0.1}{3}[2.8221138 + 4(2.6487209) + 2.4918245]$$
$$= 3.4221186$$

Continuing in this manner (Table 34-3), we obtain $y(1) = 4.7182815$. Compare the results with Tables 33-2, 32-4, 32-6, and 32-8.

34.4. Redo Problem 34.3 by Hamming's method.

We again use Table 33-2 to find the additional three starting values. Therefore, y_0, y_1, y_2, y_3, and their derivatives are exactly as given in Problem 34.3. Then, using (*34.5*), we compute

$n = 3$:
$$py_4 = y_0 + \frac{4h}{3}(2y'_3 - y'_2 + 2y'_1)$$
$$= 2 + \frac{4(0.1)}{3}[2(2.3498585) - 2.2214026 + 2(2.1051708)]$$
$$= 2.8918208$$

$$py'_4 = py_4 - x_4 = 2.4918208$$

$$y_4 = \frac{1}{8}(9y_3 - y_1) + \frac{3h}{8}(py'_4 + 2y'_3 - y'_2)$$
$$= \frac{1}{8}[9(2.6498585) - 2.2051708] + \frac{3(0.1)}{8}[2.4918208 + 2(2.3498585) - 2.2214026]$$
$$= 2.8918245$$

$n = 4$: $\quad x_4 = 0.4, \quad y'_4 = y_4 - x_4 = 2.4918245$

$$py_5 = y_1 + \frac{4h}{3}(2y'_4 - y'_3 + 2y'_2)$$
$$= 2.2051708 + \frac{4(0.1)}{3}[2(2.4918245) - 2.3498585 + 2(2.2214026)]$$
$$= 3.1487169$$

Table 34-3

Method: MILNE'S METHOD			
Problem: $y' = y - x;\ y(0) = 2$			
x_n	$h = 0.1$		True solution
	py_n	y_n	$Y(x) = e^x + x + 1$
0.0	—	2.0000000	2.0000000
0.1	—	2.2051708	2.2051709
0.2	—	2.4214026	2.4214028
0.3	—	2.6498585	2.6498588
0.4	2.8918208	2.8918245	2.8918247
0.5	3.1487169	3.1487209	3.1487213
0.6	3.4221138	3.4221186	3.4221188
0.7	3.7137472	3.7137524	3.7137527
0.8	4.0255349	4.0255407	4.0255409
0.9	4.3595964	4.3596027	4.3596031
1.0	4.7182745	4.7182815	4.7182818

Table 34-4

Method: HAMMING'S METHOD			
Problem: $y' = y - x;\ y(0) = 2$			
x_n	$h = 0.1$		True solution
	py_n	y_n	$Y(x) = e^x + x + 1$
0.0	—	2.0000000	2.0000000
0.1	—	2.2051708	2.2051709
0.2	—	2.4214026	2.4214028
0.3	—	2.6498585	2.6498588
0.4	2.8918208	2.8918245	2.8918247
0.5	3.1487169	3.1487213	3.1487213
0.6	3.4221139	3.4221191	3.4221188
0.7	3.7137473	3.7137533	3.7137527
0.8	4.0255352	4.0255419	4.0255409
0.9	4.3595971	4.3596046	4.3596031
1.0	4.7182756	4.7182838	4.7182818

$$py_5' = py_5 - x_5 = 2.6487169$$

$$y_5 = \frac{1}{8}(9y_4 - y_2) + \frac{3h}{8}(py_5' + 2y_4' - y_3')$$

$$= \frac{1}{8}[9(2.8918245) - 2.4214026] + \frac{3(0.1)}{8}[2.6487169 + 2(2.4918245) - 2.3498585]$$

$$= 3.1487213$$

$n = 5$: $x_5 = 0.5, \quad y_5' = y_5 - x_5 = 2.6487213$

$$py_6 = y_2 + \frac{4h}{3}(2y_5' - y_4' + 2y_3')$$

$$= 2.4214026 + \frac{4(0.1)}{3}[2(2.6487213) - 2.4918245 + 2(2.3498585)]$$

$$= 3.4221139$$

$$py_6' = py_6 - x_6 = 2.8221139$$

$$y_6 = \frac{1}{8}(9y_5 - y_3) + \frac{3h}{8}(py_6' + 2y_5' - y_4')$$

$$= \frac{1}{8}[9(3.1487213) - 2.6498585] + \frac{3(0.1)}{8}[2.8221139 + 2(2.6487213) - 2.4918245]$$

$$= 3.4221191$$

Continuing in this manner (Table 34-4), we obtain $y(1) = 4.7182838$. Compare the results with Table 34-3.

34.5. Find $y(1)$ for $y' = y$; $y(0) = 1$, using Milne's method with $h = 0.1$.

Here $f(x, y) = y$, $x_0 = 0$, and $y_0 = 1$. From Table 33-3, page 228, we find as the three additional starting values $y_1 = 1.1051708$, $y_2 = 1.2214026$, and $y_3 = 1.3498585$. Note that $y_1' = y_1$, $y_2' = y_2$, and $y_3' = y_3$. Then, using (*34.4*), we compute

$n = 3$: $$py_4 = y_0 + \frac{4h}{3}(2y_3' - y_2' + 2y_1')$$

$$= 1 + \frac{4(0.1)}{3}[2(1.3498585) - 1.2214026 + 2(1.1051708)]$$

$$= 1.4918208$$

$$py_4' = py_4 = 1.4918208$$

$$y_4 = y_2 + \frac{h}{3}(py_4' + 4y_3' + y_2')$$

$$= 1.2214026 + \frac{0.1}{3}[1.4918208 + 4(1.3498585) + 1.2214026]$$

$$= 1.4918245$$

$n = 4$: $x_4 = 0.4, \quad y_4' = y_4 = 1.4918245$

$$py_5 = y_1 + \frac{4h}{3}(2y_4' - y_3' + 2y_2')$$

$$= 1.1051708 + \frac{4(0.1)}{3}[2(1.4918245) - 1.3498585 + 2(1.2214026)]$$

$$= 1.6487169$$

$$py_5' = py_5 = 1.6487169$$

$$y_5 = y_3 + \frac{h}{3}(py_5' + 4y_4' + y_3')$$

$$= 1.3498585 + \frac{0.1}{3}[1.6487169 + 4(1.4918245) + 1.3498585]$$

$$= 1.6487209$$

$n = 5$: $x_5 = 0.5, \quad y'_5 = y_5 = 1.6487209$

$$py_6 = y_2 + \frac{4h}{3}(2y'_5 - y'_4 + 2y'_3)$$
$$= 1.2214026 + \frac{4(0.1)}{3}[2(1.6487209) - 1.4918245 + 2(1.3498585)]$$
$$= 1.8221138$$
$$py'_6 = py_6 = 1.8221138$$
$$y_6 = y_4 + \frac{h}{3}(py'_6 + 4y'_5 + y'_4)$$
$$= 1.4918245 + \frac{0.1}{3}[1.8221138 + 4(1.6487209) + 1.4918245]$$
$$= 1.8221186$$

Continuing in this manner (Table 34-5), we obtain $y(1) = 2.7182815$. Compare these results to Tables 33-3, 32-2, 32-5, and 32-9.

34.6. Redo Problem 34.5 using Hamming's method.

The values of y_0, y_1, y_2, y_3, and their derivatives are exactly as given in Problem 34.5. Using (*34.5*), we compute

$n = 3$:
$$py_4 = y_0 + \frac{4h}{3}(2y'_3 - y'_2 + 2y'_1)$$
$$= 1 + \frac{4(0.1)}{3}[2(1.3498585) - 1.2214026 + 2(1.1051708)]$$
$$= 1.4918208$$
$$py'_4 = py_4 = 1.4918208$$
$$y_4 = \frac{1}{8}(9y_3 - y_1) + \frac{3h}{8}(py'_4 + 2y'_3 - y'_2)$$
$$= \frac{1}{8}[9(1.3498585) - 1.1051708] + \frac{3(0.1)}{8}[1.4918208 + 2(1.3498585) - 1.2214026]$$
$$= 1.4918245$$

$n = 4$: $x_4 = 0.4, \quad y_4 = y'_4 = 1.4918245$

$$py_5 = y_1 + \frac{4h}{3}(2y'_4 - y'_3 + 2y'_2)$$
$$= 1.1051708 + \frac{4(0.1)}{3}[2(1.4918245) - 1.3498585 + 2(1.2214026)]$$
$$= 1.6487169$$
$$py'_5 = py_5 = 1.6487169$$
$$y_5 = \frac{1}{8}(9y_4 - y_2) + \frac{3h}{8}(py'_5 + 2y'_4 - y'_3)$$
$$= \frac{1}{8}[9(1.4918245) - 1.2214026] + \frac{3(0.1)}{8}[1.6487169 + 2(1.4918245) - 1.3498585]$$
$$= 1.6487213$$

$n = 5$: $x_5 = 0.5, \quad y'_5 = y_5 = 1.6487213$

$$py_6 = y_2 + \frac{4h}{3}(2y'_5 - y'_4 + 2y'_3)$$
$$= 1.2214026 + \frac{4(0.1)}{3}[2(1.6487213) - 1.4918245 + 2(1.3498585)]$$
$$= 1.8221139$$

Table 34-5

Method: MILNE'S METHOD			
Problem: $y' = y;\ y(0) = 1$			
x_n	$h = 0.1$		True solution
	py_n	y_n	$Y(x) = e^x$
0.0	–	1.0000000	1.0000000
0.1	–	1.1051708	1.1051709
0.2	–	1.2214026	1.2214028
0.3	–	1.3498585	1.3498588
0.4	1.4918208	1.4918245	1.4918247
0.5	1.6487169	1.6487209	1.6487213
0.6	1.8221138	1.8221186	1.8221188
0.7	2.0137472	2.0137524	2.0137527
0.8	2.2255349	2.2255407	2.2255409
0.9	2.4595964	2.4596027	2.4596031
1.0	2.7182745	2.7182815	2.7182818

Table 34-6

Method: HAMMING'S METHOD			
Problem: $y' = y;\ y(0) = 1$			
x_n	$h = 0.1$		True solution
	py_n	y_n	$Y(x) = e^x$
0.0	–	1.0000000	1.0000000
0.1	–	1.1051708	1.1051709
0.2	–	1.2214026	1.2214028
0.3	–	1.3498585	1.3498588
0.4	1.4918208	1.4918245	1.4918247
0.5	1.6487169	1.6487213	1.6487213
0.6	1.8221139	1.8221191	1.8221188
0.7	2.0137473	2.0137533	2.0137527
0.8	2.2255352	2.2255419	2.2255409
0.9	2.4595971	2.4596046	2.4596031
1.0	2.7182756	2.7182838	2.7182818

$$py_6' = py_6 = 1.8221139$$

$$\begin{aligned} y_6 &= \frac{1}{8}(9y_5 - y_3) + \frac{3h}{8}(py_6' + 2y_5' - y_4') \\ &= \frac{1}{8}[9(1.6487213) - 1.3498585] + \frac{3(0.1)}{8}[1.8221139 + 2(1.6487213) - 1.4918245] \\ &= 1.8221191 \end{aligned}$$

Continuing in this manner (Table 34-6), we obtain $y(1) = 2.7182838$. Compare the results with Table 32-5.

34.7. Find $y(1)$ for $y' = y^2 + 1$; $y(0) = 0$, using Milne's method with $h = 0.1$.

Here $f(x,y) = y^2 + 1$, $x_0 = 0$, $y_0 = 0$. Using Table 33-4, page 228, we find as the three additional starting values $y_1 = 0.1003346$, $y_2 = 0.2027099$, and $y_3 = 0.3093360$. Thus,

$$\begin{aligned} y_0' &= (y_0)^2 + 1 = (0)^2 + 1 = 1 \\ y_1' &= (y_1)^2 + 1 = (0.1003346)^2 + 1 = 1.0100670 \\ y_2' &= (y_2)^2 + 1 = (0.2027099)^2 + 1 = 1.0410913 \\ y_3' &= (y_3)^2 + 1 = (0.3093360)^2 + 1 = 1.0956888 \end{aligned}$$

Then, using *(34.4)*, we compute

$n = 3$:

$$\begin{aligned} py_4 &= y_0 + \frac{4h}{3}(2y_3' - y_2' + 2y_1') \\ &= 0 + \frac{4(0.1)}{3}[2(1.0956888) - 1.0410913 + 2(1.0100670)] \\ &= 0.4227227 \end{aligned}$$

$$py_4' = (py_4)^2 + 1 = (0.4227227)^2 + 1 = 1.1786945$$

$$\begin{aligned} y_4 &= y_2 + \frac{h}{3}(py_4' + 4y_3' + y_2') \\ &= 0.2027099 + \frac{0.1}{3}[1.1786945 + 4(1.0956888) + 1.0410913] \\ &= 0.4227946 \end{aligned}$$

$n = 4$: $x_4 = 0.4$, $y_4' = (y_4)^2 + 1 = (0.4227946)^2 + 1 = 1.1787553$

$$\begin{aligned} py_5 &= y_1 + \frac{4h}{3}(2y_4' - y_3' + 2y_2') \\ &= 0.1003346 + \frac{4(0.1)}{3}[2(1.1787553) - 1.0956888 + 2(1.0410913)] \\ &= 0.5462019 \end{aligned}$$

$$py_5' = (py_5)^2 + 1 = (0.5462019)^2 + 1 = 1.2983365$$

$$\begin{aligned} y_5 &= y_3 + \frac{h}{3}(py_5' + 4y_4' + y_3') \\ &= 0.3093360 + \frac{0.1}{3}[1.2983365 + 4(1.1787553) + 1.0956888] \\ &= 0.5463042 \end{aligned}$$

$n = 5$: $x_5 = 0.5$, $y_5' = (y_5)^2 + 1 = (0.5463042)^2 + 1 = 1.2984484$

$$\begin{aligned} py_6 &= y_2 + \frac{4h}{3}(2y_5' - y_4' + 2y_3') \\ &= 0.2027099 + \frac{4(0.1)}{3}[2(1.2984484) - 1.1787553 + 2(1.0956888)] \\ &= 0.6839791 \end{aligned}$$

$$py'_6 = (py_6)^2 + 1 = (0.6839791)^2 + 1 = 1.4678274$$

$$\begin{aligned} y_6 &= y_4 + \frac{h}{3}(py'_6 + 4y'_5 + y'_4) \\ &= 0.4227946 + \frac{0.1}{3}[1.4678274 + 4(1.2984484) + 1.1787553] \\ &= 0.6841405 \end{aligned}$$

Continuing in this manner (Table 34-7), we obtain $y(1) = 1.5573578$. Compare the results to Tables 33-4, 32-3, 32-7, and 32-10.

34.8. Redo Problem 34.7 using Hamming's method.

The values of y_0, y_1, y_2, y_3, and their derivatives are exactly as given in Problem 34.7. Using (*34.5*), we compute

$$n = 3: \quad \begin{aligned} py_4 &= y_0 + \frac{4h}{3}(2y'_3 - y'_2 + 2y'_1) \\ &= 0 + \frac{4(0.1)}{3}[2(1.0956888) - 1.0410913 + 2(1.0100670)] \\ &= 0.4227227 \end{aligned}$$

$$py'_4 = (py_4)^2 + 1 = (0.4227227)^2 + 1 = 1.1786945$$

$$\begin{aligned} y_4 &= \frac{1}{8}(9y_3 - y_1) + \frac{3h}{8}(py'_4 + 2y'_3 - y'_2) \\ &= \frac{1}{8}[9(0.3093360) - 0.1003346] + \frac{3(0.1)}{8}[1.1786945 + 2(1.0956888) - 1.0410913] \\ &= 0.4227980 \end{aligned}$$

$$n = 4: \quad x_4 = 0.4, \quad y'_4 = (y_4)^2 + 1 = (0.4227980)^2 + 1 = 1.1787581$$

$$\begin{aligned} py_5 &= y_1 + \frac{4h}{3}(2y'_4 - y'_3 + 2y'_2) \\ &= 0.1003346 + \frac{4(0.1)}{3}[2(1.1787581) - 1.0956888 + 2(1.0410913)] \\ &= 0.5462026 \end{aligned}$$

$$py'_5 = (py_5)^2 + 1 = (0.5462026)^2 + 1 = 1.2983373$$

$$\begin{aligned} y_5 &= \frac{1}{8}(9y_4 - y_2) + \frac{3h}{8}(py'_5 + 2y'_4 - y'_3) \\ &= \frac{1}{8}[9(0.4227980) - 0.2027099] + \frac{3(0.1)}{8}[1.2983373 + 2(1.1787581) - 1.0956888] \\ &= 0.5463152 \end{aligned}$$

$$n = 5: \quad x_5 = 0.5, \quad y'_5 = (y_5)^2 + 1 = (0.5463152)^2 + 1 = 1.2984603$$

$$\begin{aligned} py_6 &= y_2 + \frac{4h}{3}(2y'_5 - y'_4 + 2y'_3) \\ &= 0.2027099 + \frac{4(0.1)}{3}[2(1.2984603) - 1.1787581 + 2(1.0956888)] \\ &= 0.6839819 \end{aligned}$$

$$py'_6 = (py_6)^2 + 1 = (0.6839819)^2 + 1 = 1.4678312$$

$$\begin{aligned} y_6 &= \frac{1}{8}(9y_5 - y_3) + \frac{3h}{8}(py'_6 + 2y'_5 - y'_4) \\ &= \frac{1}{8}[9(0.5463152) - 0.3093360] + \frac{3(0.1)}{8}[1.4678312 + 2(1.2984603) - 1.1787581] \\ &= 0.6841624 \end{aligned}$$

Continuing in this manner (Table 34-8), we obtain $y(1) = 1.5576487$. Compare the results with Table 34-7.

Table 34-7

Method: MILNE'S METHOD			
Problem: $y' = y^2 + 1;\ y(0) = 0$			
x_n	$h = 0.1$		True solution
	py_n	y_n	$Y(x) = \tan x$
0.0	–	0.0000000	0.0000000
0.1	–	0.1003346	0.1003347
0.2	–	0.2027099	0.2027100
0.3	–	0.3093360	0.3093363
0.4	0.4227227	0.4227946	0.4227932
0.5	0.5462019	0.5463042	0.5463025
0.6	0.6839791	0.6841405	0.6841368
0.7	0.8420238	0.8422924	0.8422884
0.8	1.0291628	1.0296421	1.0296386
0.9	1.2592330	1.2601516	1.2601582
1.0	1.5554357	1.5573578	1.5574077

Table 34-8

Method: HAMMING'S METHOD			
Problem: $y' = y^2 + 1;\ y(0) = 0$			
x_n	$h = 0.1$		True solution
	py_n	y_n	$Y(x) = \tan x$
0.0	–	0.0000000	0.0000000
0.1	–	0.1003346	0.1003347
0.2	–	0.2027099	0.2027100
0.3	–	0.3093360	0.3093363
0.4	0.4227227	0.4227980	0.4227932
0.5	0.5462026	0.5463152	0.5463025
0.6	0.6839819	0.6841624	0.6841368
0.7	0.8420310	0.8423346	0.8422884
0.8	1.0291844	1.0297194	1.0296386
0.9	1.2592849	1.2602984	1.2601582
1.0	1.5555540	1.5576487	1.5574077

Supplementary Problems

34.9. Find $y(0.5)$ for $y' = -y$; $y(0) = 1$, using a second-order predictor-corrector method with $h = 0.1$.

34.10. Find $y(0.5)$ for $y' = -y + x + 2$; $y(0) = 2$, using a second-order predictor-corrector method with $h = 0.1$.

34.11. Redo Problem 34.9 using Milne's method.

34.12. Redo Problem 34.10 using Milne's method.

34.13. Redo Problem 34.9 using Hamming's method.

34.14. Redo Problem 34.10 using Hamming's method.

34.15. Find $y(5)$ for $y' = 4x^3$; $y(0) = 0$, using Milne's method with $h = 1$.

Answers to Supplementary Problems

For comparison with methods given in other chapters, all answers are given through $x = 1.0$.

34.9.

Method: SECOND-ORDER PREDICTOR-CORRECTOR METHOD

Problem: $y' = -y$; $y(0) = 1$

x_n	$h = 0.1$		True solution
	py_n	y_n	$Y(x) = e^{-x}$
0.0	—	1.0000000	1.0000000
0.1	—	0.9050000	0.9048374
0.2	0.8190000	0.8188000	0.8187308
0.3	0.7412400	0.7407980	0.7408182
0.4	0.6706404	0.6702261	0.6703201
0.5	0.6067528	0.6063771	0.6065307
0.6	0.5489507	0.5486108	0.5488116
0.7	0.4966550	0.4963475	0.4965853
0.8	0.4493413	0.4490630	0.4493290
0.9	0.4065349	0.4062831	0.4065697
1.0	0.3678064	0.3675787	0.3678794

34.10.

Method: SECOND-ORDER PREDICTOR-CORRECTOR METHOD			
Problem: $y' = -y + x + 2;\ y(0) = 2$			
x_n	$h = 0.1$		True solution
	py_n	y_n	$Y(x) = e^{-x} + x + 1$
0.0	–	2.000000	2.000000
0.1	–	2.005000	2.004837
0.2	2.019000	2.018800	2.018731
0.3	2.041240	2.040798	2.040818
0.4	2.070640	2.070226	2.070320
0.5	2.106753	2.106377	2.106531
0.6	2.148951	2.148611	2.148812
0.7	2.196655	2.196348	2.196585
0.8	2.249341	2.249063	2.249329
0.9	2.306535	2.306283	2.306570
1.0	2.367806	2.367579	2.367879

34.11.

Method: MILNE'S METHOD			
Problem: $y' = -y;\ y(0) = 1$			
x_n	$h = 0.1$		True solution
	py_n	y_n	$Y(x) = e^{-x}$
0.0	–	1.0000000	1.0000000
0.1	–	0.9048375	0.9048374
0.2	–	0.8187309	0.8187308
0.3	–	0.7408184	0.7408182
0.4	0.6703225	0.6703200	0.6703201
0.5	0.6065331	0.6065307	0.6065307
0.6	0.5488138	0.5488114	0.5488116
0.7	0.4965875	0.4965852	0.4965853
0.8	0.4493306	0.4493287	0.4493290
0.9	0.4065714	0.4065695	0.4065697
1.0	0.3678807	0.3678791	0.3678794

34.12.

Method: MILNE'S METHOD			
Problem: $y' = -y + x + 2;\ \ y(0) = 2$			
x_n	$h = 0.1$		True solution
	py_n	y_n	$Y(x) = e^{-x} + x + 1$
0.0	—	2.000000	2.000000
0.1	—	2.004838	2.004837
0.2	—	2.018731	2.018731
0.3	—	2.040818	2.040818
0.4	2.070323	2.070320	2.070320
0.5	2.106533	2.106531	2.106531
0.6	2.148814	2.148811	2.148812
0.7	2.196588	2.196585	2.196585
0.8	2.249331	2.249329	2.249329
0.9	2.306571	2.306570	2.306570
1.0	2.367881	2.367879	2.367879

34.13.

Method: HAMMING'S METHOD			
Problem: $y' = -y;\ \ y(0) = 1$			
x_n	$h = 0.1$		True solution
	py_n	y_n	$Y(x) = e^{-x}$
0.0	—	1.0000000	1.0000000
0.1	—	0.9048375	0.9048374
0.2	—	0.8187309	0.8187308
0.3	—	0.7408184	0.7408182
0.4	0.6703225	0.6703200	0.6703201
0.5	0.6065331	0.6065303	0.6065307
0.6	0.5488139	0.5488110	0.5488116
0.7	0.4965875	0.4965844	0.4965853
0.8	0.4493309	0.4493278	0.4493290
0.9	0.4065712	0.4065684	0.4065697
1.0	0.3678806	0.3678780	0.3678794

34.14.

Method: HAMMING'S METHOD			
Problem: $y' = -y + x + 2;\ y(0) = 2$			
x_n	$h = 0.1$		True solution
	py_n	y_n	$Y(x) = e^{-x} + x + 1$
0.0	–	2.000000	2.000000
0.1	–	2.004838	2.004837
0.2	–	2.018731	2.018731
0.3	–	2.040818	2.040818
0.4	2.070323	2.070320	2.070320
0.5	2.106533	2.106530	2.106531
0.6	2.148814	2.148811	2.148812
0.7	2.196588	2.196584	2.196585
0.8	2.249331	2.249328	2.249329
0.9	2.306571	2.306568	2.306570
1.0	2.367881	2.367878	2.367879

34.15. Since $Y(x) = x^4$ is a fourth-degree polynomial, Milne's method is exact and $y(5) = Y(5) = 625$.

Chapter 35

Modified Predictor-Corrector Methods

35.1 INTRODUCTION

From an error analysis of predictor-corrector methods, one can modify (that is, improve) the predicted value of y_n. This modified value of y_n is then used in the corrector instead of the originally calculated predicted value.

For notational convenience we again let py_n represent the predicted value of y_n; in addition, we denote the modified value of y_n by my_n. It follows from *(32.2)* that

$$my'_n = f(x_n, my_n) \tag{35.1}$$

35.2 MODIFIED MILNE'S METHOD

$$\begin{aligned}
\text{predictor:} \quad & py_{n+1} = y_{n-3} + \frac{4h}{3}(2y'_n - y'_{n-1} + 2y'_{n-2}) \\
\text{modifier:} \quad & my_{n+1} = py_{n+1} + \frac{28}{29}(y_n - py_n) \\
\text{corrector:} \quad & y_{n+1} = y_{n-1} + \frac{h}{3}(my'_{n+1} + 4y'_n + y'_{n-1})
\end{aligned} \tag{35.2}$$

(See Problems 35.1 and 35.3.)

35.3 MODIFIED HAMMING'S METHOD

$$\begin{aligned}
\text{predictor:} \quad & py_{n+1} = y_{n-3} + \frac{4h}{3}(2y'_n - y'_{n-1} + 2y'_{n-2}) \\
\text{modifier:} \quad & my_{n+1} = py_{n+1} + \frac{112}{121}(y_n - py_n) \\
\text{corrector:} \quad & y_{n+1} = \frac{1}{8}(9y_n - y_{n-2}) + \frac{3h}{8}(my'_{n+1} + 2y'_n - y'_{n-1})
\end{aligned} \tag{35.3}$$

(See Problems 35.2 and 35.4.)

35.4 STARTING VALUES

Neither modified method given above can be started until y_0, y_1, y_2, y_3, and y_4 are known. The value of y_0 is prescribed by the initial condition. As was the case in Chapter 34, y_1, y_2, and y_3 are computed by the fourth-order Runge-Kutta method *(33.2)*. The value of y_4 is calculated from the corresponding unmodified predictor-corrector method given in Chapter 34. (Note that a modified predictor-corrector method cannot be used to generate y_4. At $n = 3$, y_4 is given in terms of my_4, which in turn depends on py_3, a value that is not known or obtainable.)

Solved Problems

35.1. Find $y(1)$ for $y' = y - x$; $y(0) = 2$, using the modified Milne's method with $h = 0.1$.

For this problem, $f(x, y) = y - x$ and $y_0 = 2$. Using Problem 33.2, we find $y_1 = 2.2051708$, $y_2 = 2.4214026$, and $y_3 = 2.6498585$. In addition, we have from Problem 34.3 that $py_4 = 2.8918208$ and $y_4 = 2.8918245$. Thus, from (*32.2*),

$$y_0' = y_0 - x_0 = 2 \qquad y_1' = y_1 - x_1 = 2.1051708 \qquad y_2' = y_2 - x_2 = 2.2214026$$

$$y_3' = y_3 - x_3 = 2.3498585 \qquad y_4' = y_4 - x_4 = 2.4918245$$

Then, using (*35.2*), we compute

$n = 4$:

$$\begin{aligned} py_5 &= y_1 + \frac{4h}{3}(2y_4' - y_3' + 2y_2') \\ &= 2.2051708 + \frac{4(0.1)}{3}[2(2.4918245) - 2.3498585 + 2(2.2214026)] \\ &= 3.1487169 \end{aligned}$$

$$\begin{aligned} my_5 &= py_5 + \frac{28}{29}(y_4 - py_4) \\ &= 3.1487169 + \frac{28}{29}(2.8918245 - 2.8918208) = 3.1487205 \end{aligned}$$

$$my_5' = my_5 - x_5 = 2.6487205$$

$$\begin{aligned} y_5 &= y_3 + \frac{h}{3}(my_5' + 4y_4' + y_3') \\ &= 2.6498585 + \frac{0.1}{3}[2.6487205 + 4(2.4918245) + 2.3498585] \\ &= 3.1487211 \end{aligned}$$

$n = 5$:

$$y_5' = y_5 - x_5 = 2.6487211$$

$$\begin{aligned} py_6 &= y_2 + \frac{4h}{3}(2y_5' - y_4' + 2y_3') \\ &= 2.4214026 + \frac{4(0.1)}{3}[2(2.6487211) - 2.4918245 + 2(2.3498585)] \\ &= 3.4221139 \end{aligned}$$

$$\begin{aligned} my_6 &= py_6 + \frac{28}{29}(y_5 - py_5) \\ &= 3.4221139 + \frac{28}{29}(3.1487211 - 3.1487169) = 3.4221180 \end{aligned}$$

$$my_6' = my_6 - x_6 = 2.8221180$$

$$\begin{aligned} y_6 &= y_4 + \frac{h}{3}(my_6' + 4y_5' + y_4') \\ &= 2.8918245 + \frac{0.1}{3}[2.8221180 + 4(2.6487211) + 2.4918245] \\ &= 3.4221187 \end{aligned}$$

Continuing in this manner, we obtain (Table 35-1) $y(1) = 4.7182822$, which differs from the true solution by only 0.0000004. Compare the results with Tables 34-3, 33-2, 32-1, 32-4, 32-6, and 32-8.

Table 35-1

Method: MODIFIED MILNE'S METHOD

Problem: $y' = y - x; \quad y(0) = 2$

x_n	$h = 0.1$			True solution
	py_n	my_n	y_n	$Y(x) = e^x + x + 1$
0.0	–	–	2.0000000	2.0000000
0.1	–	–	2.2051708	2.2051709
0.2	–	–	2.4214026	2.4214028
0.3	–	–	2.6498585	2.6498588
0.4	2.8918208	–	2.8918245	2.8918247
0.5	3.1487169	3.1487205	3.1487211	3.1487213
0.6	3.4221139	3.4221180	3.4221187	3.4221188
0.7	3.7137472	3.7137519	3.7137527	3.7137527
0.8	4.0255350	4.0255402	4.0255410	4.0255409
0.9	4.3595966	4.3596025	4.3596033	4.3596031
1.0	4.7182748	4.7182813	4.7182822	4.7182818

Table 35-2

Method: MODIFIED HAMMING'S METHOD

Problem: $y' = y - x; \quad y(0) = 2$

x_n	$h = 0.1$			True solution
	py_n	my_n	y_n	$Y(x) = e^x + x + 1$
0.0	–	–	2.0000000	2.0000000
0.1	–	–	2.2051708	2.2051709
0.2	–	–	2.4214026	2.4214028
0.3	–	–	2.6498585	2.6498588
0.4	2.8918208	–	2.8918245	2.8918247
0.5	3.1487169	3.1487204	3.1487214	3.1487213
0.6	3.4221140	3.4221182	3.4221194	3.4221188
0.7	3.7137474	3.7137524	3.7137539	3.7137527
0.8	4.0255354	4.0255414	4.0255428	4.0255409
0.9	4.3595975	4.3596044	4.3596059	4.3596031
1.0	4.7182763	4.7182840	4.7182856	4.7182818

35.2. Redo Problem 35.1 using the modified Hamming's method.

The starting values y_0, y_1, y_2, y_3, and their derivatives are identical to those given in Problem 35.1. From Problem 34.4 we find that $py_4 = 2.8918208$ and $y_4 = 2.8918245$; hence, $y_4' = y_4 - x_4 = 2.4918245$. Then, using (*35.3*), we compute

$$n = 4: \quad py_5 = y_1 + \frac{4h}{3}(2y_4' - y_3' + 2y_2')$$

$$= 2.2051708 + \frac{4(0.1)}{3}[2(2.4918245) - 2.3498585 + 2(2.2214026)]$$

$$= 3.1487169$$

$$my_5 = py_5 + \frac{112}{121}(y_4 - py_4)$$

$$= 3.1487169 + \frac{112}{121}(2.8918245 - 2.8918208) = 3.1487203$$

$$my_5' = my_5 - x_5 = 2.6487203$$

$$y_5 = \frac{1}{8}(9y_4 - y_2) + \frac{3h}{8}(my_5' + 2y_4' - y_3')$$

$$= \frac{1}{8}[9(2.8918245) - 2.4214026] + \frac{3(0.1)}{8}[2.6487203 + 2(2.4918245) - 2.3498585]$$

$$= 3.1487214$$

$$n = 5: \quad y_5' = y_5 - x_5 = 2.6487214$$

$$py_6 = y_2 + \frac{4h}{3}(2y_5' - y_4' + 2y_3')$$

$$= 2.4214026 + \frac{4(0.1)}{3}[2(2.6487214) - 2.4918245 + 2(2.3498585)]$$

$$= 3.4221140$$

$$my_6 = py_6 + \frac{112}{121}(y_5 - py_5)$$

$$= 3.4221140 + \frac{112}{121}(3.1487214 - 3.1487169) = 3.4221182$$

$$my_6' = my_6 - x_6 = 2.8221182$$

$$y_6 = \frac{1}{8}(9y_5 - y_3) + \frac{3h}{8}(my_6' + 2y_5' - y_4')$$

$$= \frac{1}{8}[9(3.1487214) - 2.6498585] + \frac{3(0.1)}{8}[2.8221182 + 2(2.6487214) - 2.4918245]$$

$$= 2.4221194$$

Continuing in this manner (Table 35-2), we obtain $y(1) = 4.7182856$. Compare the results with Table 35-1. Note that the statements in Section 34.4 concerning relative accuracy remain valid for modified methods.

35.3. Find $y(1)$ for $y' = y$; $y(0) = 1$, using the modified Milne's method with $h = 0.1$.

For this problem, $f(x, y) = y$ and $y_0 = 1$. Using Problem 33.3, we find $y_1 = 1.1051708$, $y_2 = 1.2214026$, and $y_3 = 1.3498585$. Also, from Problem 34.5, we have $py_4 = 1.4918208$ and $y_4 = 1.4918245$. Note that $y_1' = y_1$, $y_2' = y_2$, $y_3' = y_3$, and $y_4' = y_4$. Then, using (*35.2*), we compute

$$n = 4: \quad py_5 = y_1 + \frac{4h}{3}(2y_4' - y_3' + 2y_2')$$

$$= 1.1051708 + \frac{4(0.1)}{3}[2(1.4918245) - 1.3498585 + 2(1.2214026)]$$

$$= 1.6487169$$

$$my_5 = py_5 + \frac{28}{29}(y_4 - py_4) = 1.6487169 + \frac{28}{29}(1.4918245 - 1.4918208)$$
$$= 1.6487205$$
$$my_5' = my_5 = 1.6487205$$
$$y_5 = y_3 + \frac{h}{3}(my_5' + 4y_4' + y_3')$$
$$= 1.3498585 + \frac{0.1}{3}[1.6487205 + 4(1.4918245) + 1.3498585]$$
$$= 1.6487211$$

$n = 5$:

$$y_5' = y_5 = 1.6487211$$
$$py_6 = y_2 + \frac{4h}{3}(2y_5' - y_4' + 2y_3')$$
$$= 1.2214026 + \frac{4(0.1)}{3}[2(1.6487211) - 1.4918245 + 2(1.3498585)]$$
$$= 1.8221139$$
$$my_6 = py_6 + \frac{28}{29}(y_5 - py_5)$$
$$= 1.8221139 + \frac{28}{29}(1.6487211 - 1.6487169) = 1.8221180$$
$$my_6' = my_6 = 1.8221180$$
$$y_6 = y_4 + \frac{h}{3}(my_6' + 4y_5' + y_4')$$
$$= 1.4918245 + \frac{0.1}{3}[1.8221180 + 4(1.6487211) + 1.4918245]$$
$$= 1.8221187$$

Continuing in this manner (Table 35-3), we obtain $y(1) = 2.7182822$. Compare the results to Tables 34-5, 33-3, 32-2, 32-5, and 32-9.

35.4. Find $y(1)$ for $y' = y^2 + 1$; $y(0) = 0$, using the modified Hamming's method with $h = 0.1$.

For this problem, $f(x, y) = y^2 + 1$ and $y_0 = 0$. Using Problem 33.4, we find $y_1 = 0.1003346$, $y_2 = 0.2027099$, and $y_3 = 0.3093360$. Also, from Problem 34.8, we have $py_4 = 0.4227227$ and $y_4 = 0.4227980$. It now follows from (*32.2*) that

$$y_0' = y_0^2 + 1 = (0)^2 + 1 = 1$$
$$y_1' = y_1^2 + 1 = (0.1003346)^2 + 1 = 1.0100670$$
$$y_2' = y_2^2 + 1 = (0.2027099)^2 + 1 = 1.0410913$$
$$y_3' = y_3^2 + 1 = (0.3093360)^2 + 1 = 1.0956888$$
$$y_4' = y_4^2 + 1 = (0.4227980)^2 + 1 = 1.1787582$$

Then, using (*35.3*), we compute

$n = 4$:

$$py_5 = y_1 + \frac{4h}{3}(2y_4' - y_3' + 2y_2')$$
$$= 0.1003346 + \frac{4(0.1)}{3}[2(1.1787582) - 1.0956888 + 2(1.0410913)]$$
$$= 0.5462026$$
$$my_5 = py_5 + \frac{112}{121}(y_4 - py_4)$$
$$= 0.5462026 + \frac{112}{121}(0.4227980 - 0.4227227)$$
$$= 0.5462723$$

Table 35-3

Method: MODIFIED MILNE'S METHOD				
Problem: $y' = y$; $y(0) = 1$				
x_n	$h = 0.1$			True solution
	py_n	my_n	y_n	$Y(x) = e^x$
0.0	—	—	1.0000000	1.0000000
0.1	—	—	1.1051708	1.1051709
0.2	—	—	1.2214026	1.2214028
0.3	—	—	1.3498585	1.3498588
0.4	1.4918208	—	1.4918245	1.4918247
0.5	1.6487169	1.6487205	1.6487211	1.6487213
0.6	1.8221139	1.8221180	1.8221187	1.8221188
0.7	2.0137472	2.0137519	2.0137527	2.0137527
0.8	2.2255350	2.2255402	2.2255410	2.2255409
0.9	2.4595966	2.4596025	2.4596033	2.4596031
1.0	2.7182748	2.7182813	2.7182822	2.7182818

Table 35-4

Method: MODIFIED HAMMING'S METHOD				
Problem: $y' = y^2 + 1$; $y(0) = 0$				
x_n	$h = 0.1$			True solution
	py_n	my_n	y_n	$Y(x) = \tan x$
0.0	—	—	0.0000000	0.0000000
0.1	—	—	0.1003346	0.1003347
0.2	—	—	0.2027099	0.2027100
0.3	—	—	0.3093360	0.3093363
0.4	0.4227227	—	0.4227980	0.4227932
0.5	0.5462026	0.5462723	0.5463181	0.5463025
0.6	0.6839828	0.6840897	0.6841714	0.6841368
0.7	0.8420339	0.8422084	0.8423567	0.8422884
0.8	1.0291935	1.0294923	1.0297701	1.0296386
0.9	1.2593138	1.2598476	1.2604138	1.2601582
1.0	1.5556365	1.5566546	1.5579221	1.5574077

$$my'_5 = (my_5)^2 + 1 = (0.5462723)^2 + 1 = 1.2984134$$

$$\begin{aligned} y_5 &= \frac{1}{8}(9y_4 - y_2) + \frac{3h}{8}(my'_5 + 2y'_4 - y'_3) \\ &= \frac{1}{8}[9(0.4227980) - 0.2027099] + \frac{3(0.1)}{8}[1.2984134 + 2(1.1787582) - 1.0956888] \\ &= 0.5463181 \end{aligned}$$

$n = 5$: $$y'_5 = (y_5)^2 + 1 = (0.5463181)^2 + 1 = 1.2984635$$

$$\begin{aligned} py_6 &= y_2 + \frac{4h}{3}(2y'_5 - y'_4 + 2y'_3) \\ &= 0.2027099 + \frac{4(0.1)}{3}[2(1.2984635) - 1.1787582 + 2(1.0956888)] \\ &= 0.6839828 \end{aligned}$$

$$\begin{aligned} my_6 &= py_6 + \frac{112}{121}(y_5 - py_5) \\ &= 0.6839828 + \frac{112}{121}(0.5463181 - 0.5462026) = 0.6840897 \end{aligned}$$

$$my'_6 = (my_6)^2 + 1 = (0.6840897)^2 + 1 = 1.4679787$$

$$\begin{aligned} y_6 &= \frac{1}{8}(9y_5 - y_3) + \frac{3h}{8}(my'_6 + 2y'_5 - y'_4) \\ &= \frac{1}{8}[9(0.5463181) - 0.3093360] + \frac{3(0.1)}{8}[1.4679787 + 2(1.2984635) - 1.1787582] \\ &= 0.6841714 \end{aligned}$$

Continuing in this manner (Table 35-4), we obtain $y(1) = 1.5579221$. Compare the results to Tables 34-7, 34-8, 33-4, 32-3, 32-7, and 32-10.

Supplementary Problems

35.5. Redo Problem 35.3 using the modified Hamming's method.

35.6. Redo Problem 35.4 using the modified Milne's method.

35.7. Find $y(0.6)$ for $y' = -y$; $y(0) = 1$, using the modified Milne's method with $h = 0.1$.

35.8. Find $y(0.6)$ for $y' = -y + x + 2$; $y(0) = 2$, using the modified Hamming's method with $h = 0.1$.

35.9. Find $y(0.6)$ for $y' = 5x^4$; $y(0) = 0$, using the modified Milne's method with $h = 0.1$.

Answers to Supplementary Problems

For comparison with methods given in previous chapters, all answers are given through $x = 1.0$.

35.5.

Method: MODIFIED HAMMING'S METHOD				
Problem: $y' = y;\ y(0) = 1$				
x_n	$h = 0.1$			True solution
	py_n	my_n	y_n	$Y(x) = e^x$
0.0	—	—	1.0000000	1.0000000
0.1	—	—	1.1051708	1.1051709
0.2	—	—	1.2214026	1.2214028
0.3	—	—	1.3498585	1.3498588
0.4	1.4918208	—	1.4918245	1.4918247
0.5	1.6487169	1.6487203	1.6487214	1.6487213
0.6	1.8221140	1.8221181	1.8221194	1.8221188
0.7	2.0137474	2.0137524	2.0137539	2.0137527
0.8	2.2255354	2.2255414	2.2255428	2.2255409
0.9	2.4595975	2.4596044	2.4596059	2.4596031
1.0	2.7182763	2.7182840	2.7182856	2.7182818

35.6.

Method: MODIFIED MILNE'S METHOD				
Problem: $y' = y^2 + 1;\ y(0) = 0$				
x_n	$h = 0.1$			True solution
	py_n	my_n	y_n	$Y(x) = \tan x$
0.0	—	—	0.0000000	0.0000000
0.1	—	—	0.1003346	0.1003347
0.2	—	—	0.2027099	0.2027100
0.3	—	—	0.3093360	0.3093363
0.4	0.4227227	—	0.4227946	0.4227932
0.5	0.5462018	0.5462712	0.5463068	0.5463025
0.6	0.6839798	0.6840811	0.6841455	0.6841368
0.7	0.8420253	0.8421852	0.8423050	0.8422884
0.8	1.0291683	1.0294383	1.0296691	1.0296386
0.9	1.2592493	1.2597329	1.2602143	1.2601582
1.0	1.5554811	1.5564128	1.5575091	1.5574077

35.7.

Method: MODIFIED MILNE'S METHOD				
Problem: $y' = -y; \quad y(0) = 1$				
x_n	$h = 0.1$			True solution
	py_n	my_n	y_n	$Y(x) = e^{-x}$
0.0	–	–	1.0000000	1.0000000
0.1	–	–	0.9048375	0.9048374
0.2	–	–	0.8187309	0.8187308
0.3	–	–	0.7408184	0.7408182
0.4	0.6703225	–	0.6703200	0.6703201
0.5	0.6065331	0.6065306	0.6065308	0.6065307
0.6	0.5488138	0.5488116	0.5488115	0.5488116
0.7	0.4965875	0.4965858	0.4965854	0.4965853
0.8	0.4493306	0.4493286	0.4493288	0.4493290
0.9	0.4065714	0.4065697	0.4065697	0.4065697
1.0	0.3678807	0.3678790	0.3678792	0.3678794

35.8.

Method: MODIFIED HAMMING'S METHOD				
Problem: $y' = -y + x + 2; \quad y(0) = 2$				
x_n	$h = 0.1$			True solution
	py_n	my_n	y_n	$Y(x) = e^{-x} + x + 1$
0.0	–	–	2.000000	2.000000
0.1	–	–	2.004838	2.004837
0.2	–	–	2.018731	2.018731
0.3	–	–	2.040818	2.040818
0.4	2.070323	–	2.070320	2.070320
0.5	2.106533	2.106531	2.106530	2.106531
0.6	2.148814	2.148811	2.148811	2.148812
0.7	2.196588	2.196585	2.196585	2.196585
0.8	2.249331	2.249328	2.249328	2.249329
0.9	2.306571	2.306569	2.306569	2.306570
1.0	2.367881	2.367879	2.367879	2.367879

35.9.

Method: MODIFIED MILNE'S METHOD				
Problem: $y' = 5x^4;\ y(0) = 0$				
x_n	$h = 0.1$			True solution
	py_n	my_n	y_n	$Y(x) = x^5$
0.0	—	—	0.0000000	0.0000000
0.1	—	—	0.0000104	0.0000100
0.2	—	—	0.0003208	0.0003200
0.3	—	—	0.0024313	0.0024300
0.4	0.0098667	—	0.0102542	0.0102400
0.5	0.0308771	0.0312512	0.0312646	0.0312500
0.6	0.0773875	0.0777616	0.0777875	0.0777600
0.7	0.1676979	0.1680841	0.1680979	0.1680700
0.8	0.3273208	0.3277070	0.3277208	0.3276800
0.9	0.5901313	0.5905175	0.5905313	0.5904900
1.0	0.9996542	1.0000404	1.0000542	1.0000000

Chapter 36

Numerical Methods for Systems

36.1 GENERAL REMARKS

All methods for first-order initial-value problems given in Chapters 32 through 35 are easily extended to a *system* of first-order initial-value problems or to most *higher-order* initial-value problems. (Any higher-order problem can be reduced to a system of first-order problems by Steps 2 and 3 of Chapter 30 if Step 1 of Chapter 30 can be achieved.)

Below we give the generalizations of four numerical methods, for simplicity restricting ourselves to a system of only two equations:

$$\begin{aligned} y' &= f(x,y,z) \\ z' &= g(x,y,z); \\ y(x_0) &= y_0, \quad z(x_0) = z_0 \end{aligned} \tag{36.1}$$

We note that, with $y' = f(x,y,z) \equiv z$, system (*36.1*) represents the second-order initial-value problem

$$y'' = g(x,y,y'); \quad y(x_0) = y_0, \quad y'(x_0) = z_0$$

(See Problems 36.1 through 36.4.)

36.2 EULER'S METHOD

$$\begin{aligned} y_{n+1} &= y_n + hy'_n \\ z_{n+1} &= z_n + hz'_n \end{aligned} \tag{36.2}$$

(See Problems 36.5 and 36.6.)

36.3 A FOURTH-ORDER RUNGE-KUTTA METHOD

$$\begin{aligned} y_{n+1} &= y_n + \frac{1}{6}(k_1 + 2k_2 + 2k_3 + k_4) \\ z_{n+1} &= z_n + \frac{1}{6}(l_1 + 2l_2 + 2l_3 + l_4) \end{aligned} \tag{36.3}$$

where

$$\begin{aligned} k_1 &= hf(x_n,\ y_n,\ z_n) \\ l_1 &= hg(x_n,\ y_n,\ z_n) \\ k_2 &= hf(x_n + \tfrac{1}{2}h,\ y_n + \tfrac{1}{2}k_1,\ z_n + \tfrac{1}{2}l_1) \\ l_2 &= hg(x_n + \tfrac{1}{2}h,\ y_n + \tfrac{1}{2}k_1,\ z_n + \tfrac{1}{2}l_1) \\ k_3 &= hf(x_n + \tfrac{1}{2}h,\ y_n + \tfrac{1}{2}k_2,\ z_n + \tfrac{1}{2}l_2) \\ l_3 &= hg(x_n + \tfrac{1}{2}h,\ y_n + \tfrac{1}{2}k_2,\ z_n + \tfrac{1}{2}l_2) \\ k_4 &= hf(x_n + h,\ y_n + k_3,\ z_n + l_3) \\ l_4 &= hg(x_n + h,\ y_n + k_3,\ z_n + l_3) \end{aligned}$$

(See Problems 36.7 and 36.8.) This method may be used to provide additional starting values for Milne's and Hamming's methods below.

36.4 MILNE'S METHOD

$$\begin{aligned} py_{n+1} &= y_{n-3} + \frac{4h}{3}(2y'_n - y'_{n-1} + 2y'_{n-2}) \\ pz_{n+1} &= z_{n-3} + \frac{4h}{3}(2z'_n - z'_{n-1} + 2z'_{n-2}) \\ y_{n+1} &= y_{n-1} + \frac{h}{3}(py'_{n+1} + 4y'_n + y'_{n-1}) \\ z_{n+1} &= z_{n-1} + \frac{h}{3}(pz'_{n+1} + 4z'_n + z'_{n-1}) \end{aligned} \tag{36.4}$$

(See Problem 36.9.)

36.5 HAMMING'S METHOD

$$\begin{aligned} py_{n+1} &= y_{n-3} + \frac{4h}{3}(2y'_n - y'_{n-1} + 2y'_{n-2}) \\ pz_{n+1} &= z_{n-3} + \frac{4h}{3}(2z'_n - z'_{n-1} + 2z'_{n-2}) \\ y_{n+1} &= \frac{1}{8}(9y_n - y_{n-2}) + \frac{3h}{8}(py'_{n+1} + 2y'_n - y'_{n-1}) \\ z_{n+1} &= \frac{1}{8}(9z_n - z_{n-2}) + \frac{3h}{8}(pz'_{n+1} + 2z'_n - z'_{n-1}) \end{aligned} \tag{36.5}$$

(See Problem 36.10.)

Solved Problems

36.1. Reduce the initial-value problem $y'' - y = x$; $y(0) = 0$, $y'(0) = 1$ to system (*36.1*).

Defining $z = y'$, we have $z(0) = y'(0) = 1$ and $z' = y''$. The given differential equation can be rewritten as $y'' = y + x$, or $z' = y + x$. We thus obtain the first-order system

$$\begin{aligned} y' &= z \\ z' &= y + x; \\ y(0) &= 0, \quad z(0) = 1 \end{aligned}$$

36.2. Reduce the initial-value problem $y'' - 3y' + 2y = 0$; $y(0) = -1$, $y'(0) = 0$ to system (*36.1*).

Defining $z = y'$, we have $z(0) = y'(0) = 0$ and $z' = y''$. The given differential equation can be rewritten as $y'' = 3y' - 2y$, or $z' = 3z - 2y$. We thus obtain the first-order system

$$\begin{aligned} y' &= z \\ z' &= 3z - 2y; \\ y(0) &= -1, \quad z(0) = 0 \end{aligned}$$

36.3. Reduce the initial-value problem

$$2yy'' - 4xy^2y' + 2(\sin x)y^4 = 6; \quad y(1) = 0, \ y'(1) = 15$$

to system (*36.1*).

Defining $z = y'$, we have $z(1) = y'(1) = 15$ and $z' = y''$. The given differential equation can be rewritten as $y'' = 2xyy' - (\sin x)y^3 + (3/y)$, or $z' = 2xyz - (\sin x)y^3 + (3/y)$. We thus obtain the first-order system

$$\begin{aligned} y' &= z \\ z' &= 2xyz - (\sin x)y^3 + \frac{3}{y}; \\ y(1) &= 0, \quad z(1) = 15 \end{aligned}$$

36.4. Reduce the initial-value problem

$$y''' - 2xy'' + 4y' - x^2y = 1; \quad y(0) = 1, \ y'(0) = 2, \ y''(0) = 3$$

to a first-order system.

Following Steps 1 through 3 of Chapter 30, we obtain the system

$$\begin{aligned} y_1' &= y_2 \\ y_2' &= y_3 \\ y_3' &= x^2y_1 - 4y_2 + 2xy_3 + 1; \\ y_1(0) &= 1, \quad y_2(0) = 2, \quad y_3(0) = 3 \end{aligned}$$

To eliminate subscripting, we define $y = y_1$, $z = y_2$, and $w = y_3$. The system then becomes

$$\begin{aligned} y' &= z \\ z' &= w \\ w' &= x^2y - 4z + 2xw + 1; \\ y(0) &= 1, \quad z(0) = 2, \quad w(0) = 3 \end{aligned}$$

36.5. Find $y(1)$ for $y'' - y = x$; $y(0) = 0$, $y'(0) = 1$, using Euler's method with $h = 0.1$.

Using the results of Problem 36.1, we have $f(x, y, z) = z$, $g(x, y, z) = y + x$, $x_0 = 0$, $y_0 = 0$, and $z_0 = 1$. Then, using (*36.2*), we compute:

$n = 0$:

$$\begin{aligned} y_0' &= f(x_0, y_0, z_0) = z_0 = 1 \\ z_0' &= g(x_0, y_0, z_0) = y_0 + x_0 = 0 + 0 = 0 \\ y_1 &= y_0 + hy_0' = 0 + (0.1)(1) = 0.1 \\ z_1 &= z_0 + hz_0' = 1 + (0.1)(0) = 1 \end{aligned}$$

$n = 1$:

$$\begin{aligned} y_1' &= f(x_1, y_1, z_1) = z_1 = 1 \\ z_1' &= g(x_1, y_1, z_1) = y_1 + x_1 = 0.1 + 0.1 = 0.2 \\ y_2 &= y_1 + hy_1' = 0.1 + (0.1)(1) = 0.2 \\ z_2 &= z_1 + hz_1' = 1 + (0.1)(0.2) = 1.02 \end{aligned}$$

$n = 2$:

$$\begin{aligned} y_2' &= f(x_2, y_2, z_2) = z_2 = 1.02 \\ z_2' &= g(x_2, y_2, z_2) = y_2 + x_2 = 0.2 + 0.2 = 0.4 \\ y_3 &= y_2 + hy_2' = 0.2 + (0.1)(1.02) = 0.302 \\ z_3 &= z_2 + hz_2' = 1.02 + (0.1)(0.4) = 1.06 \end{aligned}$$

Continuing in this manner (Table 36-1), we obtain $y(1) = y_{10} = 1.2451$. Compare this table with Tables 36-3, 36-5, and 36-6, which contain results for the same differential equation using higher-order methods.

Table 36-1

Method: EULER'S METHOD			
Problem: $y'' - y = x; \quad y(0) = 0, \quad y'(0) = 1$			
x_n	$h = 0.1$		True solution
	y_n	z_n	$Y(x) = e^x - e^{-x} - x$
0.0	0.0000	1.0000	0.0000
0.1	0.1000	1.0000	0.1003
0.2	0.2000	1.0200	0.2027
0.3	0.3020	1.0600	0.3090
0.4	0.4080	1.1202	0.4215
0.5	0.5200	1.2010	0.5422
0.6	0.6401	1.3030	0.6733
0.7	0.7704	1.4270	0.8172
0.8	0.9131	1.5741	0.9762
0.9	1.0705	1.7454	1.1530
1.0	1.2451	1.9424	1.3504

Table 36-2

Method: EULER'S METHOD			
Problem: $y'' - 3y' + 2y = 0;$ $y(0) = -1, \quad y'(0) = 0$			
x_n	$h = 0.1$		True solution
	y_n	z_n	$Y(x) = e^{2x} - 2e^x$
0.0	−1.0000	0.0000	−1.0000
0.1	−1.0000	0.2000	−0.9889
0.2	−0.9800	0.4600	−0.9510
0.3	−0.9340	0.7940	−0.8776
0.4	−0.8546	1.2190	−0.7581
0.5	−0.7327	1.7556	−0.5792
0.6	−0.5571	2.4288	−0.3241
0.7	−0.3143	3.2689	0.0277
0.8	0.0126	4.3125	0.5020
0.9	0.4439	5.6037	1.1304
1.0	1.0043	7.1960	1.9525

36.6. Find $y(1)$ for $y''-3y'+2y=0$; $y(0)=-1$, $y'(0)=0$, using Euler's method with $h=0.1$.

Using the results of Problem 36.2, we have $f(x,y,z)=z$, $g(x,y,z)=3z-2y$, $x_0=0$, $y_0=-1$, and $z_0=0$. Then, using (*36.2*), we compute:

n = 0:

$$y_0' = f(x_0,y_0,z_0) = z_0 = 0$$

$$z_0' = g(x_0,y_0,z_0) = 3z_0-2y_0 = 3(0)-2(-1) = 2$$

$$y_1 = y_0+hy_0' = -1+(0.1)(0) = -1$$

$$z_1 = z_0+hz_0' = 0+(0.1)(2) = 0.2$$

n = 1:

$$y_1' = f(x_1,y_1,z_1) = z_1 = 0.2$$

$$z_1' = g(x_1,y_1,z_1) = 3z_1-2y_1 = 3(0.2)-2(-1) = 2.6$$

$$y_2 = y_1+hy_1' = -1+(0.1)(0.2) = -0.98$$

$$z_2 = z_1+hz_1' = 0.2+(0.1)(2.6) = 0.46$$

Continuing in this manner (Table 36-2), we obtain $y(1)=y_{10}=1.0043$. Compare this table with Table 36-4, which contains results for the same differential equation using a fourth-order Runge-Kutta method.

36.7. Find $y(1)$ for $y''-y=x$; $y(0)=0$, $y'(0)=1$, using a fourth-order Runge-Kutta method with $h=0.1$.

Using the results of Problem 36.1, we have $f(x,y,z)=z$, $g(x,y,z)=y+x$, $x_0=0$, $y_0=0$, and $z_0=1$. Then, using (*36.3*), we compute:

n = 0:

$$k_1 = hf(x_0,y_0,z_0) = hf(0,0,1) = (0.1)(1) = 0.1$$

$$l_1 = hg(x_0,y_0,z_0) = hg(0,0,1) = (0.1)(0+0) = 0$$

$$\begin{aligned} k_2 &= hf(x_0+\tfrac{1}{2}h,\ y_0+\tfrac{1}{2}k_1,\ z_0+\tfrac{1}{2}l_1) \\ &= hf[0+\tfrac{1}{2}(0.1),\ 0+\tfrac{1}{2}(0.1),\ 1+\tfrac{1}{2}(0)] \\ &= hf(0.05,0.05,1) = (0.1)(1) = 0.1 \end{aligned}$$

$$\begin{aligned} l_2 &= hg(x_0+\tfrac{1}{2}h,\ y_0+\tfrac{1}{2}k_1,\ z_0+\tfrac{1}{2}l_1) \\ &= hg(0.05,0.05,1) = (0.1)(0.05+0.05) = 0.01 \end{aligned}$$

$$\begin{aligned} k_3 &= hf(x_0+\tfrac{1}{2}h,\ y_0+\tfrac{1}{2}k_2,\ z_0+\tfrac{1}{2}l_2) \\ &= hf[0+\tfrac{1}{2}(0.1),\ 0+\tfrac{1}{2}(0.1),\ 1+\tfrac{1}{2}(0.01)] \\ &= hf(0.05,0.05,1.005) = (0.1)(1.005) = 0.101 \end{aligned}$$

$$\begin{aligned} l_3 &= hg(x_0+\tfrac{1}{2}h,\ y_0+\tfrac{1}{2}k_2,\ z_0+\tfrac{1}{2}l_2) \\ &= hg(0.05,0.05,1.005) = (0.1)(0.05+0.05) = 0.01 \end{aligned}$$

$$\begin{aligned} k_4 &= hf(x_0+h,\ y_0+k_3,\ z_0+l_3) \\ &= hf(0+0.1,\ 0+0.101,\ 1+0.01) \\ &= hf(0.1,0.101,1.01) = (0.1)(1.01) = 0.101 \end{aligned}$$

$$\begin{aligned} l_4 &= hg(x_0+h,\ y_0+k_3,\ z_0+l_3) \\ &= hg(0.1,0.101,1.01) = (0.1)(0.101+0.1) = 0.02 \end{aligned}$$

$$\begin{aligned} y_1 &= y_0+\tfrac{1}{6}(k_1+2k_2+2k_3+k_4) \\ &= 0+\tfrac{1}{6}[0.1+2(0.1)+2(0.101)+(0.101)] = 0.101 \end{aligned}$$

$$\begin{aligned} z_1 &= z_0+\tfrac{1}{6}(l_1+2l_2+2l_3+l_4) \\ &= 1+\tfrac{1}{6}[0+2(0.01)+2(0.01)+(0.02)] = 1.01 \end{aligned}$$

$n = 1$: $k_1 = hf(x_1, y_1, z_1) = hf(0.1, 0.101, 1.01)$
$= (0.1)(1.01) = 0.101$

$l_1 = hg(x_1, y_1, z_1) = hg(0.1, 0.101, 1.01)$
$= (0.1)(0.101 + 0.1) = 0.02$

$k_2 = hf(x_1 + \frac{1}{2}h,\ y_1 + \frac{1}{2}k_1,\ z_1 + \frac{1}{2}l_1)$
$= hf[0.1 + \frac{1}{2}(0.1),\ 0.101 + \frac{1}{2}(0.101),\ 1.01 + \frac{1}{2}(0.02)]$
$= hf(0.15, 0.152, 1.02) = (0.1)(1.02) = 0.102$

$l_2 = hg(x_1 + \frac{1}{2}h,\ y_1 + \frac{1}{2}k_1,\ z_1 + \frac{1}{2}l_1)$
$= hg(0.15, 0.152, 1.02) = (0.1)(0.152 + 0.15) = 0.03$

$k_3 = hf(x_1 + \frac{1}{2}h,\ y_1 + \frac{1}{2}k_2,\ z_1 + \frac{1}{2}l_2)$
$= hf[0.1 + \frac{1}{2}(0.1),\ 0.101 + \frac{1}{2}(0.102),\ 1.01 + \frac{1}{2}(0.03)]$
$= hf(0.15, 0.152, 1.025) = (0.1)(1.025) = 0.103$

$l_3 = hg(x_1 + \frac{1}{2}h,\ y_1 + \frac{1}{2}k_2,\ z_1 + \frac{1}{2}l_2)$
$= hg(0.15, 0.152, 1.025) = (0.1)(0.152 + 0.15) = 0.03$

$k_4 = hf(x_1 + h,\ y_1 + k_3,\ z_1 + l_3)$
$= hf(0.1 + 0.1,\ 0.101 + 0.103,\ 1.01 + 0.03)$
$= hf(0.2, 0.204, 1.04) = (0.1)(1.04) = 0.104$

$l_4 = hg(x_1 + h,\ y_1 + k_3,\ z_1 + l_3)$
$= hg(0.2, 0.204, 1.04) = (0.1)(0.204 + 0.2) = 0.04$

$y_2 = y_1 + \frac{1}{6}(k_1 + 2k_2 + 2k_3 + k_4)$
$= 0.101 + \frac{1}{6}[0.101 + 2(0.102) + 2(0.103) + (0.104)]$
$= 0.204$

$z_2 = z_1 + \frac{1}{6}(l_1 + 2l_2 + 2l_3 + l_4)$
$= 1.01 + \frac{1}{6}[0.02 + 2(0.03) + 2(0.03) + 0.04] = 1.04$

Continuing in this manner (Table 36-3), we obtain $y(1) = y_{10} = 1.350$. It is interesting to note that when $h = 0.01$, the computed solution y_n and the true solution at x_n are identical through the first seven decimal places for all $0 \leq x_n \leq 1$.

36.8. Find $y(1)$ for $y'' - 3y' + 2y = 0$; $y(0) = -1$, $y'(0) = 0$, using a fourth-order Runge-Kutta method with $h = 0.1$.

Using the results of Problem 36.2, we have $f(x, y, z) = z$, $g(x, y, z) = 3z - 2y$, $x_0 = 0$, $y_0 = -1$, and $z_0 = 0$. Then, using (*36.3*), we compute:

$n = 0$: $k_1 = hf(x_0, y_0, z_0) = hf(0, -1, 0) = (0.1)(0) = 0$
$l_1 = hg(x_0, y_0, z_0) = hg(0, -1, 0) = (0.1)[3(0) - 2(-1)] = 0.2$

$k_2 = hf(x_0 + \frac{1}{2}h,\ y_0 + \frac{1}{2}k_1,\ z_0 + \frac{1}{2}l_1)$
$= hf[0 + \frac{1}{2}(0.1),\ -1 + \frac{1}{2}(0),\ 0 + \frac{1}{2}(0.2)]$
$= hf(0.05, -1, 0.1) = (0.1)(0.1) = 0.01$

$l_2 = hg(x_0 + \frac{1}{2}h,\ y_0 + \frac{1}{2}k_1,\ z_0 + \frac{1}{2}l_1)$
$= hg(0.05, -1, 0.1) = (0.1)[3(0.1) - 2(-1)] = 0.23$

$k_3 = hf(x_0 + \frac{1}{2}h,\ y_0 + \frac{1}{2}k_2,\ z_0 + \frac{1}{2}l_2)$
$= hf[0 + \frac{1}{2}(0.1),\ -1 + \frac{1}{2}(0.01),\ 0 + \frac{1}{2}(0.23)]$
$= hf(0.05, -0.995, 0.115) = (0.1)(0.115) = 0.012$

Table 36-3

Method: FOURTH-ORDER RUNGE-KUTTA METHOD			
Problem: $y'' - y = x; \quad y(0) = 0, \quad y'(0) = 1$			
x_n	$h = 0.1$		True solution
	y_n	z_n	$Y(x) = e^x - e^{-x} - x$
0.0	0.0000000	1.0000000	0.0000000
0.1	0.1003333	1.0100083	0.1003335
0.2	0.2026717	1.0401335	0.2026720
0.3	0.3090401	1.0906769	0.3090406
0.4	0.4215040	1.1621445	0.4215047
0.5	0.5421897	1.2552516	0.5421906
0.6	0.6733060	1.3709300	0.6733072
0.7	0.8171660	1.5103373	0.8171674
0.8	0.9762103	1.6748689	0.9762120
0.9	1.1530314	1.8661714	1.1530335
1.0	1.3504000	2.0861595	1.3504024

Table 36-4

Method: FOURTH-ORDER RUNGE-KUTTA METHOD			
Problem: $y'' - 3y' + 2y = 0; \quad y(0) = -1, \quad y'(0) = 0$			
x_n	$h = 0.1$		True solution
	y_n	z_n	$Y(x) = e^{2x} - 2e^x$
0.0	−1.0000000	0.0000000	−1.0000000
0.1	−0.9889417	0.2324583	−0.9889391
0.2	−0.9509872	0.5408308	−0.9509808
0.3	−0.8776105	0.9444959	−0.8775988
0.4	−0.7581277	1.4673932	−0.7581085
0.5	−0.5791901	2.1390610	−0.5791607
0.6	−0.3241640	2.9959080	−0.3241207
0.7	0.0276326	4.0827685	0.0276946
0.8	0.5018638	5.4548068	0.5019506
0.9	1.1303217	7.1798462	1.1304412
1.0	1.9523298	9.3412190	1.9524924

$$
\begin{aligned}
l_3 &= hg(x_0 + \tfrac{1}{2}h,\ y_0 + \tfrac{1}{2}k_2,\ z_0 + \tfrac{1}{2}l_2) \\
&= hg(0.05, -0.995, 0.115) \;=\; (0.1)[3(0.115) - 2(-0.995)] \\
&= 0.234
\end{aligned}
$$

$$
\begin{aligned}
k_4 &= hf(x_0 + h,\ y_0 + k_3,\ z_0 + l_3) \\
&= hf(0 + 0.1,\ -1 + 0.012,\ 0 + 0.234) \\
&= hf(0.1, -0.988, 0.234) \;=\; (0.1)(0.234) \;=\; 0.023
\end{aligned}
$$

$$
\begin{aligned}
l_4 &= hg(x_0 + h,\ y_0 + k_3,\ z_0 + l_3) \\
&= hg(0.1, -0.988, 0.234) \;=\; (0.1)[3(0.234) - 2(-0.988)] \\
&= 0.268
\end{aligned}
$$

$$
\begin{aligned}
y_1 &= y_0 + \tfrac{1}{6}(k_1 + 2k_2 + 2k_3 + k_4) \\
&= -1 + \tfrac{1}{6}[0 + 2(0.01) + 2(0.012) + 0.023] \;=\; -0.989
\end{aligned}
$$

$$
\begin{aligned}
z_1 &= z_0 + \tfrac{1}{6}(l_1 + 2l_2 + 2l_3 + l_4) \\
&= 0 + \tfrac{1}{6}[0.2 + 2(0.23) + 2(0.234) + 0.268] \;=\; 0.233
\end{aligned}
$$

Continuing in this manner (Table 36-4), we obtain $y(1) = y_{10} = 1.952$.

36.9. Find $y(1)$ for $y'' - y = x;\ y(0) = 0,\ y'(0) = 1$, using Milne's method with $h = 0.1$.

Using the results of Table 36-3, we have $y_0 = 0$, $y_1 = 0.1003333$, $y_2 = 0.2026717$, $y_3 = 0.3090401$, $z_0 = 1$, $z_1 = 1.0100083$, $z_2 = 1.0401335$, and $z_3 = 1.0906769$. For this problem, $f(x,y,z) = z$ and $g(x,y,z) = y + x$, so $y'_0 = z_0 = 1$, $y'_1 = z_1 = 1.0100083$, $y'_2 = z_2 = 1.0401335$, $y'_3 = z_3 = 1.0906769$, $z'_0 = y_0 + x_0 = 0$, $z'_1 = y_1 + x_1 = 0.2003333$, $z'_2 = y_2 + x_2 = 0.4026717$, $z'_3 = y_3 + x_3 = 0.6090401$. Then, using (*36.4*), we compute:

$n = 3$:

$$
\begin{aligned}
py_4 &= y_0 + \frac{4h}{3}(2y'_3 - y'_2 + 2y'_1) \\
&= 0 + \frac{4(0.1)}{3}[2(1.0906769) - 1.0401335 + 2(1.0100083)] \\
&= 0.4214983
\end{aligned}
$$

$$
\begin{aligned}
pz_4 &= z_0 + \frac{4h}{3}(2z'_3 - z'_2 + 2z'_1) \\
&= 1 + \frac{4(0.1)}{3}[2(0.6090401) - 0.4026717 + 2(0.2003333)] \\
&= 1.1621433
\end{aligned}
$$

$$
py'_4 = pz_4 = 1.1621433
$$

$$
pz'_4 = py_4 + x_4 = 0.4214983 + 0.4 = 0.8214983
$$

$$
\begin{aligned}
y_4 &= y_2 + \frac{h}{3}(py'_4 + 4y'_3 + y'_2) \\
&= 0.2026717 + \frac{0.1}{3}[1.1621433 + 4(1.0906769) + 1.0401335] \\
&= 0.4215045
\end{aligned}
$$

$$
\begin{aligned}
z_4 &= z_2 + \frac{h}{3}(pz'_4 + 4z'_3 + z'_2) \\
&= 1.0401335 + \frac{0.1}{3}[0.8214983 + 4(0.6090401) + 0.4026717] \\
&= 1.1621445
\end{aligned}
$$

$n = 4$:

$$
y'_4 = z_4 = 1.1621445
$$

$$
z'_4 = y_4 + x_4 = 0.4215045 + 0.4 = 0.8215045
$$

$$
\begin{aligned}
py_5 &= y_1 + \frac{4h}{3}(2y_4' - y_3' + 2y_2') \\
&= 0.1003333 + \frac{4(0.1)}{3}[2(1.1621445) - 1.0906769 + 2(1.0401335)] \\
&= 0.5421838 \\
pz_5 &= z_1 + \frac{4h}{3}(2z_4' - z_3' + 2z_2') \\
&= 1.0100083 + \frac{4(0.1)}{3}[2(0.8215045) - 0.6090401 + 2(0.4026717)] \\
&= 1.2552500 \\
py_5' &= pz_5 = 1.2552500 \\
pz_5' &= py_5 + x_5 = 0.5421838 + 0.5 = 1.0421838 \\
y_5 &= y_3 + \frac{h}{3}(py_5' + 4y_4' + y_3') \\
&= 0.3090401 + \frac{0.1}{3}[1.2552500 + 4(1.1621445) + 1.0906769] \\
&= 0.5421903 \\
z_5 &= z_3 + \frac{h}{3}(pz_5' + 4z_4' + z_3') \\
&= 1.0906769 + \frac{0.1}{3}[1.0421838 + 4(0.8215045) + 0.6090401] \\
&= 1.2552517
\end{aligned}
$$

Continuing in this manner (Table 36-5), we obtain $y(1) = y_{10} = 1.3504024$. Compare the accuracy of $y(x_n)$ to $Y(x_n)$ in Table 36-5, especially at x_{10}.

36.10. Redo Problem 36.9 using Hamming's method.

All starting values and their derivatives are identical to those given in Problem 36.9. Using (*36.5*), we compute:

$n = 3$:

$$
\begin{aligned}
py_4 &= y_0 + \frac{4h}{3}(2y_3' - y_2' + 2y_1') \\
&= 0 + \frac{4(0.1)}{3}[2(1.0906769) - 1.0401335 + 2(1.0100083)] \\
&= 0.4214983 \\
pz_4 &= z_0 + \frac{4h}{3}(2z_3' - z_2' + 2z_1') \\
&= 1 + \frac{4(0.1)}{3}[2(0.6090401) - 0.4026717 + 2(0.200333)] \\
&= 1.1621433 \\
py_4' &= pz_4 = 1.1621433 \\
pz_4' &= py_4 + x_4 = 0.4214983 + 0.4 = 0.8214983 \\
y_4 &= \frac{1}{8}(9y_3 - y_1) + \frac{3h}{8}(py_4' + 2y_3' - y_2') \\
&= \frac{1}{8}[9(0.3090401) - 0.1003333] + \frac{3(0.1)}{8}[1.1621433 + 2(1.0906769) - 1.0401335] \\
&= 0.4215046 \\
z_4 &= \frac{1}{8}(9z_3 - z_1) + \frac{3h}{8}(pz_4' + 2z_3' - z_2') \\
&= \frac{1}{8}[9(1.0906769) - 1.0100083] + \frac{3(0.1)}{8}[0.8214983 + 2(0.6090401) - 0.4026717] \\
&= 1.1621445
\end{aligned}
$$

$n = 4$:

$$
\begin{aligned}
y_4' &= z_4 = 1.1621445 \\
z_4' &= y_4 + x_4 = 0.4215046 + 0.4 = 0.8215046
\end{aligned}
$$

Table 36-5

Method: MILNE'S METHOD					
Problem: $y'' - y = x; \quad y(0) = 0, \quad y'(0) = 1$					
x_n	$h = 0.1$				True solution
	py_n	pz_n	y_n	z_n	$Y(x) = e^x - e^{-x} - x$
0.0	—	—	0.0000000	1.0000000	0.0000000
0.1	—	—	0.1003333	1.0100083	0.1003335
0.2	—	—	0.2026717	1.0401335	0.2026720
0.3	—	—	0.3090401	1.0906769	0.3090406
0.4	0.4214983	1.1621433	0.4215045	1.1621445	0.4215047
0.5	0.5421838	1.2552500	0.5421903	1.2552517	0.5421906
0.6	0.6733000	1.3709276	0.6733071	1.3709300	0.6733072
0.7	0.8171597	1.5103347	0.8171671	1.5103376	0.8171674
0.8	0.9762043	1.6748655	0.9762120	1.6748693	0.9762120
0.9	1.1530250	1.8661678	1.1530332	1.8661723	1.1530335
1.0	1.3503938	2.0861552	1.3504024	2.0861606	1.3504024

Table 36-6

Method: HAMMING'S METHOD					
Problem: $y'' - y = x; \quad y(0) = 0, \quad y'(0) = 1$					
x_n	$h = 0.1$				True solution
	py_n	pz_n	y_n	z_n	$Y(x) = e^x - e^{-x} - x$
0.0	—	—	0.0000000	1.0000000	0.0000000
0.1	—	—	0.1003333	1.0100083	0.1003335
0.2	—	—	0.2026717	1.0401335	0.2026720
0.3	—	—	0.3090401	1.0906769	0.3090406
0.4	0.4214983	1.1621433	0.4215046	1.1621445	0.4215047
0.5	0.5421838	1.2552500	0.5421909	1.2552516	0.5421906
0.6	0.6733000	1.3709278	0.6733081	1.3709301	0.6733072
0.7	0.8171598	1.5103348	0.8171689	1.5103377	0.8171674
0.8	0.9762044	1.6748661	0.9762141	1.6748698	0.9762120
0.9	1.1530259	1.8661683	1.1530362	1.8661729	1.1530335
1.0	1.3503950	2.0861563	1.3504059	2.0861618	1.3504024

$$\begin{aligned}
py_5 &= y_1 + \frac{4h}{3}(2y_4' - y_3' + 2y_2') \\
&= 0.1003333 + \frac{4(0.1)}{3}[2(1.1621445) - 1.0906769 + 2(1.0401335)] \\
&= 0.5421838 \\
pz_5 &= z_1 + \frac{4h}{3}(2z_4' - z_3' + 2z_2') \\
&= 1.0100083 + \frac{4(0.1)}{3}[2(0.8215045) - 0.6090401 + 2(0.4026717)] \\
&= 1.2552500 \\
py_5' &= pz_5 = 1.2552500 \\
pz_5' &= py_5 + x_5 = 0.5421838 + 0.5 = 1.0421838 \\
y_5 &= \frac{1}{8}(9y_4 - y_2) + \frac{3h}{8}(py_5' + 2y_4' - y_3') \\
&= \frac{1}{8}[9(0.4215045) - 0.2026717] + \frac{3(0.1)}{8}[1.2552500 + 2(1.1621445) - 1.0906769] \\
&= 0.5421909 \\
z_5 &= \frac{1}{8}(9z_4 - z_2) + \frac{3h}{8}(pz_5' + 2z_4' - z_3') \\
&= \frac{1}{8}[9(1.1621445) - 1.0401335] + \frac{3(0.1)}{8}[1.0421838 + 2(0.8215045) - 0.6090401] \\
&= 1.2552516
\end{aligned}$$

Continuing in this manner (Table 36-6), we find that $y(1) = y_{10} = 1.3504059$.

Supplementary Problems

36.11. Reduce the initial-value problem $y'' + y = 0;\ y(0) = 1,\ y'(0) = 0$ to system (*36.1*).

36.12. Reduce the initial-value problem $y'' - y = x;\ y(0) = 0,\ y'(0) = -1$ to system (*36.1*).

36.13. Reduce the initial-value problem $xy''' - x^2y'' + (y')^2y = 0;\ y(0) = 1,\ y'(0) = 2,\ y''(0) = 3$ to a first-order system.

In Problems 36.14 through 36.21, round off all computations to three decimal places.

36.14. Find $y(0.5)$ for $y'' + y = 0;\ y(0) = 1,\ y'(0) = 0,$ using Euler's method with $h = 0.1$.

36.15. Find $y(0.5)$ for $y'' - y = x;\ y(0) = 0,\ y'(0) = -1,$ using Euler's method with $h = 0.1$.

36.16. Redo Problem 36.14 using a fourth-order Runge-Kutta method.

36.17. Redo Problem 36.15 using a fourth-order Runge-Kutta method.

36.18. Redo Problem 36.14 using Milne's method.

36.19. Redo Problem 36.14 using Hamming's method.

36.20. Find $y(0.5)$ for $y'' - 3y' + 2y = 0$; $y(0) = -1$, $y'(0) = 0$, using Milne's method with $h = 0.1$. For starting values use Table 36.4.

36.21. Redo Problem 36.20 using Hamming's method.

36.22. Formulate the modified Milne's method for system (*36.1*).

36.23. Formulate the modified Hamming's method for system (*36.1*).

36.24. Formulate Milne's method for the system

$$\begin{aligned} y' &= f(x, y, z, w) \\ z' &= g(x, y, z, w) \\ w' &= r(x, y, z, w); \\ y(x_0) = y_0, \quad & z(x_0) = z_0, \quad w(x_0) = w_0 \end{aligned}$$

36.25. Formulate a fourth-order Runge-Kutta method for the system of Problem 36.24.

Answers to Supplementary Problems

36.11. $y' = z, \quad z' = -y; \quad y(0) = 1, \quad z(0) = 0$

36.12. $y' = z, \quad z' = y + x; \quad y(0) = 0, \quad z(0) = -1$

36.13. $y' = z, \quad z' = w, \quad w' = xw - \dfrac{z^2 y}{x}; \quad y(0) = 1, \quad z(0) = 2, \quad w(0) = 3$

36.14.

Method: EULER'S METHOD

Problem: $y'' + y = 0; \quad y(0) = 1, \quad y'(0) = 0$

x_n	$h = 0.1$		True solution
	y_n	z_n	$Y(x) = \cos x$
0.0	1.0000	0.0000	1.0000
0.1	1.0000	−0.1000	0.9950
0.2	0.9900	−0.2000	0.9801
0.3	0.9700	−0.2990	0.9553
0.4	0.9401	−0.3960	0.9211
0.5	0.9005	−0.4900	0.8776
0.6	0.8515	−0.5801	0.8253
0.7	0.7935	−0.6652	0.7648
0.8	0.7270	−0.7446	0.6967
0.9	0.6525	−0.8173	0.6216
1.0	0.5708	−0.8825	0.5403

36.15. Since $Y(x) = -x$ is a first-degree polynomial, Euler's method is exact and $y(0.5) = Y(0.5) = -0.5$.

36.16.

Method: FOURTH-ORDER RUNGE-KUTTA METHOD			
Problem: $y'' + y = 0;\quad y(0) = 1,\quad y'(0) = 0$			
x_n	$h = 0.1$		True solution
	y_n	z_n	$Y(x) = \cos x$
0.0	1.0000000	0.0000000	1.0000000
0.1	0.9950042	−0.0998333	0.9950042
0.2	0.9800666	−0.1986692	0.9800666
0.3	0.9553365	−0.2955200	0.9553365
0.4	0.9210611	−0.3894180	0.9210610
0.5	0.8775827	−0.4794252	0.8775826
0.6	0.8253359	−0.5646420	0.8253356
0.7	0.7648425	−0.6442172	0.7648422
0.8	0.6967071	−0.7173556	0.6967067
0.9	0.6216105	−0.7833264	0.6216100
1.0	0.5403030	−0.8414705	0.5403023

36.17. Since $Y(x) = -x$ is a first-degree polynomial, a fourth-order Runge-Kutta method is exact and $y(0.5) = Y(0.5) = -0.5$.

36.18.

Method: MILNE'S METHOD					
Problem: $y'' + y = 0;\quad y(0) = 1,\quad y'(0) = 0$					
x_n	$h = 0.1$				True solution
	py_n	pz_n	y_n	z_n	$Y(x) = \cos x$
0.0	—	—	1.0000000	0.0000000	1.0000000
0.1	—	—	0.9950042	−0.0998333	0.9950042
0.2	—	—	0.9800666	−0.1986692	0.9800666
0.3	—	—	0.9553365	−0.2955200	0.9553365
0.4	0.9210617	−0.3894153	0.9210611	−0.3894183	0.9210610
0.5	0.8775835	−0.4794225	0.8775827	−0.4794254	0.8775826
0.6	0.8253369	−0.5646395	0.8253358	−0.5646426	0.8253356
0.7	0.7648437	−0.6442148	0.7648423	−0.6442178	0.7648422
0.8	0.6967086	−0.7173535	0.6967069	−0.7173564	0.6967067
0.9	0.6216120	−0.7833245	0.6216101	−0.7833272	0.6216100
1.0	0.5403047	−0.8414690	0.5403024	−0.8414715	0.5403023

36.19.

Method: HAMMING'S METHOD					
Problem: $y'' + y = 0;\quad y(0) = 1,\quad y'(0) = 0$					
x_n	$h = 0.1$				True solution $Y(x) = \cos x$
	py_n	pz_n	y_n	z_n	
0.0	—	—	1.0000000	0.0000000	1.0000000
0.1	—	—	0.9950042	−0.0998333	0.9950042
0.2	—	—	0.9800666	−0.1986692	0.9800666
0.3	—	—	0.9553365	−0.2955200	0.9553365
0.4	0.9210617	−0.3894153	0.9210611	−0.3894183	0.9210610
0.5	0.8775835	−0.4794225	0.8775827	−0.4794258	0.8775826
0.6	0.8253368	−0.5646395	0.8253357	−0.5646431	0.8253356
0.7	0.7648436	−0.6442148	0.7648423	−0.6442187	0.7648422
0.8	0.6967083	−0.7173536	0.6967067	−0.7173574	0.6967067
0.9	0.6216117	−0.7833248	0.6216097	−0.7833286	0.6216100
1.0	0.5403041	−0.8414694	0.5403018	−0.8414729	0.5403023

36.20.

Method: MILNE'S METHOD					
Problem: $y'' - 3y' + 2y = 0;\quad y(0) = -1,\quad y'(0) = 0$					
x_n	$h = 0.1$				True solution $Y(x) = e^{2x} - 2e^x$
	py_n	pz_n	y_n	z_n	
0.0	—	—	−1.0000000	0.0000000	−1.0000000
0.1	—	—	−0.9889417	0.2324583	−0.9889391
0.2	—	—	−0.9509872	0.5408308	−0.9509808
0.3	—	—	−0.8776105	0.9444959	−0.8775988
0.4	−0.7582563	1.4671290	−0.7581224	1.4674042	−0.7581085
0.5	−0.5793451	2.1387436	−0.5791820	2.1390779	−0.5791607
0.6	−0.3243547	2.9955182	−0.3241479	2.9959412	−0.3241207
0.7	0.0274045	4.0823034	0.0276562	4.0828171	0.0276946
0.8	0.5015908	5.4542513	0.5019008	5.4548828	0.5019506
0.9	1.1299955	7.1791838	1.1303739	7.1799534	1.1304412
1.0	1.9519398	9.3404286	1.9524049	9.3413729	1.9524924

36.21.

Method: HAMMING'S METHOD					
Problem: $y'' - 3y' + 2y = 0; \quad y(0) = -1, \quad y'(0) = 0$					
x_n	$h = 0.1$				True solution
	py_n	pz_n	y_n	z_n	$Y(x) = e^{2x} - 2e^x$
0.0	—	—	−1.0000000	0.0000000	−1.0000000
0.1	—	—	−0.9889417	0.2324583	−0.9889391
0.2	—	—	−0.9509872	0.5408308	−0.9509808
0.3	—	—	−0.8776105	0.9444959	−0.8775988
0.4	−0.7582563	1.4671290	−0.7581208	1.4674075	−0.7581085
0.5	−0.5793442	2.1387454	−0.5791725	2.1390975	−0.5791607
0.6	−0.3243499	2.9955280	−0.3241310	2.9959763	−0.3241207
0.7	0.0274121	4.0823189	0.0276868	4.0828802	0.0276946
0.8	0.5016098	5.4542900	0.5019470	5.4549778	0.5019506
0.9	1.1300312	7.1792566	1.1304442	7.1800975	1.1304412
1.0	1.9519994	9.3405500	1.9525051	9.3415779	1.9524924

36.22.

$$py_{n+1} = y_{n-3} + \frac{4h}{3}(2y'_n - y'_{n-1} + 2y'_{n-2})$$

$$pz_{n+1} = z_{n-3} + \frac{4h}{3}(2z'_n - z'_{n-1} + 2z'_{n-2})$$

$$my_{n+1} = py_{n+1} + \frac{28}{29}(y_n - py_n)$$

$$mz_{n+1} = pz_{n+1} + \frac{28}{29}(z_n - pz_n)$$

$$y_{n+1} = y_{n-1} + \frac{h}{3}(my'_{n+1} + 4y'_n + y'_{n-1})$$

$$z_{n+1} = z_{n-1} + \frac{h}{3}(mz'_{n+1} + 4z'_n + z'_{n-1})$$

36.23.

$$py_{n+1} = y_{n-3} + \frac{4h}{3}(2y'_n - y'_{n-1} + 2y'_{n-2})$$

$$pz_{n+1} = z_{n-3} + \frac{4h}{3}(2z'_n - z'_{n-1} + 2z'_{n-2})$$

$$my_{n+1} = py_{n+1} + \frac{112}{121}(y_n - py_n)$$

$$mz_{n+1} = pz_{n+1} + \frac{112}{121}(z_n - pz_n)$$

$$y_{n+1} = \frac{1}{8}(9y_n - y_{n-2}) + \frac{3h}{8}(my'_{n+1} + 2y'_n - y'_{n-1})$$

$$z_{n+1} = \frac{1}{8}(9z_n - z_{n-2}) + \frac{3h}{8}(mz'_{n+1} + 2z'_n - z'_{n-1})$$

36.24. Equations (*36.4*) with the addition of

$$pw_{n+1} = w_{n-3} + \frac{4h}{3}(2w'_n - w'_{n-1} + 2w'_{n-2})$$

$$w_{n+1} = w_{n-1} + \frac{h}{3}(pw'_{n+1} + 4w'_n + w'_{n-1})$$

36.25. $$y_{n+1} = y_n + \tfrac{1}{6}(k_1 + 2k_2 + 2k_3 + k_4)$$

$$z_{n+1} = z_n + \tfrac{1}{6}(l_1 + 2l_2 + 2l_3 + l_4)$$

$$w_{n+1} = w_n + \tfrac{1}{6}(m_1 + 2m_2 + 2m_3 + m_4)$$

where

$$\begin{aligned}
k_1 &= hf(x_n, y_n, z_n, w_n)\\
l_1 &= hg(x_n, y_n, z_n, w_n)\\
m_1 &= hr(x_n, y_n, z_n, w_n)\\
k_2 &= hf(x_n + \tfrac{1}{2}h,\ y_n + \tfrac{1}{2}k_1,\ z_n + \tfrac{1}{2}l_1,\ w_n + \tfrac{1}{2}m_1)\\
l_2 &= hg(x_n + \tfrac{1}{2}h,\ y_n + \tfrac{1}{2}k_1,\ z_n + \tfrac{1}{2}l_1,\ w_n + \tfrac{1}{2}m_1)\\
m_2 &= hr(x_n + \tfrac{1}{2}h,\ y_n + \tfrac{1}{2}k_1,\ z_n + \tfrac{1}{2}l_1,\ w_n + \tfrac{1}{2}m_1)\\
k_3 &= hf(x_n + \tfrac{1}{2}h,\ y_n + \tfrac{1}{2}k_2,\ z_n + \tfrac{1}{2}l_2,\ w_n + \tfrac{1}{2}m_2)\\
l_3 &= hg(x_n + \tfrac{1}{2}h,\ y_n + \tfrac{1}{2}k_2,\ z_n + \tfrac{1}{2}l_2,\ w_n + \tfrac{1}{2}m_2)\\
m_3 &= hr(x_n + \tfrac{1}{2}h,\ y_n + \tfrac{1}{2}k_2,\ z_n + \tfrac{1}{2}l_2,\ w_n + \tfrac{1}{2}m_2)\\
k_4 &= hf(x_n + h,\ y_n + k_3,\ z_n + l_3,\ w_n + m_3)\\
l_4 &= hg(x_n + h,\ y_n + k_3,\ z_n + l_3,\ w_n + m_3)\\
m_4 &= hr(x_n + h,\ y_n + k_3,\ z_n + l_3,\ w_n + m_3)
\end{aligned}$$

Chapter 37

Second-Order Boundary-Value Problems

37.1 HOMOGENEOUS AND NONHOMOGENEOUS PROBLEMS

We consider the boundary-value problem given by the second-order linear differential equation

$$y'' + P(x)y' + Q(x)y = \phi(x) \tag{37.1}$$

and the boundary conditions

$$\begin{aligned} \alpha_1 y(a) + \beta_1 y'(a) &= \gamma_1 \\ \alpha_2 y(b) + \beta_2 y'(b) &= \gamma_2 \end{aligned} \tag{37.2}$$

Here $P(x)$, $Q(x)$, and $\phi(x)$ are continuous in $[a, b]$ and α_1, α_2, β_1, β_2, γ_1, and γ_2 are all real constants. Furthermore, it is assumed that α_1 and β_1 are not both zero, and also that α_2 and β_2 are not both zero.

The boundary-value problem is said to be *homogeneous* if both the differential equation and the boundary conditions are homogeneous (i.e., $\phi(x) \equiv 0$ and $\gamma_1 = \gamma_2 = 0$). Otherwise the problem is *nonhomogeneous*. Thus a homogeneous boundary-value problem has the form

$$\begin{aligned} y'' + P(x)y' + Q(x)y &= 0; \\ \alpha_1 y(a) + \beta_1 y'(a) &= 0 \\ \alpha_2 y(b) + \beta_2 y'(b) &= 0 \end{aligned} \tag{37.3}$$

A somewhat more general homogeneous boundary-value problem than (*37.3*) is one where the coefficients $P(x)$ and $Q(x)$ also depend on an arbitrary constant λ. Such a problem has the form

$$\begin{aligned} y'' + P(x, \lambda)y' + Q(x, \lambda)y &= 0; \\ \alpha_1 y(a) + \beta_1 y'(a) &= 0 \\ \alpha_2 y(b) + \beta_2 y'(b) &= 0 \end{aligned} \tag{37.4}$$

Observe that (*37.3*) or (*37.4*) always admits the trivial solution $y(x) \equiv 0$.

37.2 UNIQUENESS OF SOLUTIONS

A boundary-value problem is solved by first obtaining the general solution to the differential equation, using any of the appropriate methods presented heretofore, and then applying the boundary conditions to evaluate the arbitrary constants. (See Problems 37.1 through 37.5.)

Conditions under which a given boundary-value problem has a *unique* solution are given by the following theorems.

Theorem 37.1. The system of linear algebraic equations $pc_1 + qc_2 = 0$, $rc_1 + sc_2 = 0$ in the unknowns c_1 and c_2 has at least one solution besides $c_1 = c_2 = 0$ if and only if the determinant

$$\begin{vmatrix} p & q \\ r & s \end{vmatrix}$$

equals zero.

Together with Theorem 11.1, page 57, Theorem 37.1 implies:

Theorem 37.2. Let $y_1(x)$ and $y_2(x)$ be two linearly independent solutions of

$$y'' + P(x)y' + Q(x)y = 0$$

Nontrivial solutions (i.e. solutions not identically equal to zero) to the homogeneous boundary-value problem *(37.3)* exist if and only if the determinant

$$\begin{vmatrix} \alpha_1 y_1(a) + \beta_1 y_1'(a) & \alpha_1 y_2(a) + \beta_1 y_2'(a) \\ \alpha_2 y_1(b) + \beta_2 y_1'(b) & \alpha_2 y_2(b) + \beta_2 y_2'(b) \end{vmatrix} \tag{37.5}$$

equals zero.

(See Problems 37.1 and 37.2.)

Theorem 37.3. The nonhomogeneous boundary-value problem defined by *(37.1)* and *(37.2)* has a unique solution if and only if the associated homogeneous problem *(37.3)* has only the trivial solution.

(See Problems 37.3 through 37.5.) In other words, *a nonhomogeneous problem has a unique solution when and only when the associated homogeneous problem has a unique solution.*

37.3. EIGENVALUE PROBLEMS

When applied to the boundary-value problem *(37.4)*, Theorem 37.2 shows that nontrivial solutions may exist for certain values of λ but not for other values of λ. Those values of λ for which nontrivial solutions do exist are called *eigenvalues*; the corresponding nontrivial solutions are called *eigenfunctions*.

Solved Problems

37.1. Solve $y'' + 2y' - 3y = 0$; $y(0) = 0$, $y'(1) = 0$.

This is a homogeneous boundary-value problem of the form *(37.3)*, with $P(x) \equiv 2$, $Q(x) \equiv -3$, $\alpha_1 = 1$, $\beta_1 = 0$, $\alpha_2 = 0$, $\beta_2 = 1$, $a = 0$, and $b = 1$. The general solution to the differential equation is $y = c_1e^{-3x} + c_2e^{x}$. Applying the boundary conditions, we find that $c_1 = c_2 = 0$; hence, the solution is $y \equiv 0$.

The same result follows from Theorem 37.2. Two linearly independent solutions are $y_1(x) = e^{-3x}$ and $y_2(x) = e^{x}$; hence, the determinant *(37.5)* becomes

$$\begin{vmatrix} 1 & 1 \\ -3e^{-3} & e \end{vmatrix} = e + 3e^{-3}$$

Since this determinant is not zero, the only solution is the trivial solution $y(x) \equiv 0$.

37.2. Solve $y'' = 0;\ \ y(-1) = 0,\ y(1) - 2y'(1) = 0.$

This is a homogeneous boundary-value problem, (*37.3*), where $P(x) = Q(x) \equiv 0$, $\alpha_1 = 1$, $\beta_1 = 0$, $\alpha_2 = 1$, $\beta_2 = -2$, $a = -1$, and $b = 1$. The general solution to the differential equation is $y = c_1 + c_2 x$. Applying the boundary conditions, we obtain the equations $c_1 - c_2 = 0$ and $c_1 - c_2 = 0$, which have the solution $c_1 = c_2$, c_2 arbitrary. Thus, the solution to the boundary-value problem is $y = c_2(1 + x)$, c_2 arbitrary. As a different solution is obtained for each value of c_2, the problem has infinitely many nontrivial solutions.

The existence of nontrivial solutions is also immediate from Theorem 37.2. Here $y_1(x) = 1$, $y_2(x) = x$, and determinant (*37.5*) becomes

$$\begin{vmatrix} 1 & -1 \\ 1 & -1 \end{vmatrix} = 0$$

37.3. Solve $y'' + 2y' - 3y = 9x;\ \ y(0) = 1,\ y'(1) = 2.$

This is a nonhomogeneous boundary-value problem, (*37.1*) and (*37.2*), where $\phi(x) = x$, $\gamma_1 = 1$, and $\gamma_2 = 2$. Since the associated homogeneous problem has only the trivial solution (Problem 37.1), it follows from Theorem 37.3 that the given problem has a unique solution. Solving the differential equation by the method of Chapter 14, we obtain

$$y = c_1 e^{-3x} + c_2 e^{x} - 3x - 2$$

Applying the boundary conditions, we find

$$c_1 + c_2 - 2 = 1 \qquad -3c_1 e^{-3} + c_2 e - 3 = 2$$

whence

$$c_1 = \frac{3e - 5}{e + 3e^{-3}} \qquad c_2 = \frac{5 + 9e^{-3}}{e + 3e^{-3}}$$

Finally,

$$y = \frac{(3e - 5)e^{-3x} + (5 + 9e^{-3})e^{x}}{e + 3e^{-3}} - 3x - 2$$

37.4. Solve $y'' = 2;\ \ y(-1) = 5,\ y(1) - 2y'(1) = 1.$

This is a nonhomogeneous boundary-value problem, (*37.1*) and (*37.2*), where $\phi(x) \equiv 2$, $\gamma_1 = 5$, and $\gamma_2 = 1$. Since the associated homogeneous problem has nontrivial solutions (Problem 37.2), this problem does not have a unique solution. There are, therefore, either no solutions or more than one solution. Solving the differential equation, we find that $y = c_1 + c_2 x + x^2$. Then, applying the boundary conditions, we obtain the equations $c_1 - c_2 = 4$ and $c_1 - c_2 = 4$; thus, $c_1 = 4 + c_2$, c_2 arbitrary. Finally, $y = c_2(1 + x) + 4 + x^2$; and this problem has infinitely many solutions, one for each value of the arbitrary constant c_2.

37.5. Solve $y'' = 2;\ \ y(-1) = 0,\ y(1) - 2y'(1) = 0.$

This is a nonhomogeneous boundary-value problem, (*37.1*) and (*37.2*), where $\phi(x) \equiv 2$ and $\gamma_1 = \gamma_2 = 0$. As in Problem 37.4, there are either no solutions or more than one solution. The solution to the differential equation is $y = c_1 + c_2 x + x^2$. Applying the boundary conditions, we obtain the equations $c_1 - c_2 = -1$ and $c_1 - c_2 = 3$. Since these equations have no solution, the boundary-value problem has no solution.

37.6. Find the eigenvalues and eigenfunctions of

$$y'' - 4\lambda y' + 4\lambda^2 y = 0; \quad y(0) = 0, \ y(1) + y'(1) = 0$$

The coefficients of the given differential equation are constants (with respect to x); hence, the general solution can be found by use of the characteristic equation. We write the characteristic equation in terms of the variable m, since λ (see Section 12.1) now has another meaning. Thus we have $m^2 - 4\lambda m + 4\lambda^2 = 0$, which has the double root $m = 2\lambda$; the solution to the differential equation is $y = c_1 e^{2\lambda x} + c_2 x e^{2\lambda x}$. Applying the boundary conditions and simplifying, we obtain

$$c_1 = 0 \qquad c_1(1+2\lambda) + c_2(2+2\lambda) = 0$$

It now follows that $c_1 = 0$ and either $c_2 = 0$ or $\lambda = -1$. The choice $c_2 = 0$ results in the trivial solution $y \equiv 0$; the choice $\lambda = -1$ results in the nontrivial solution $y = c_2 x e^{-2x}$, c_2 arbitrary. Thus, the boundary-value problem has the eigenvalue $\lambda = -1$ and the eigenfunction $y = c_2 x e^{-2x}$.

37.7. Find the eigenvalues and eigenfunctions of

$$y'' - 4\lambda y' + 4\lambda^2 y = 0; \quad y'(1) = 0, \quad y(2) + 2y'(2) = 0$$

As in Problem 37.6, the solution to the differential equation is $y = c_1 e^{2\lambda x} + c_2 x e^{2\lambda x}$. Applying the boundary conditions and simplifying, we obtain the equations

$$\begin{aligned} (2\lambda)c_1 + (1+2\lambda)c_2 &= 0 \\ (1+4\lambda)c_1 + (4+8\lambda)c_2 &= 0 \end{aligned} \tag{1}$$

By Theorem 37.1, (1) has a nontrivial solution for c_1 and c_2 if and only if the determinant

$$\begin{vmatrix} 2\lambda & 1+2\lambda \\ 1+4\lambda & 4+8\lambda \end{vmatrix} = (1+2\lambda)(4\lambda - 1)$$

is zero; that is, if and only if either $\lambda = -\frac{1}{2}$ or $\lambda = \frac{1}{4}$. When $\lambda = -\frac{1}{2}$, (1) has the solution $c_1 = 0$, c_2 arbitrary; when $\lambda = \frac{1}{4}$, (1) has the solution $c_1 = -3c_2$, c_2 arbitrary. It follows that the eigenvalues are $\lambda_1 = -\frac{1}{2}$ and $\lambda_2 = \frac{1}{4}$ and the corresponding eigenfunctions are $y_1 = c_2 x e^{-x}$ and $y_2 = c_2(-3 + x)e^{x/2}$.

37.8. Find the eigenvalues and eigenfunctions of

$$y'' + \lambda y' = 0; \quad y(0) + y'(0) = 0, \quad y'(1) = 0$$

In terms of the variable m, the characteristic equation is $m^2 + \lambda m = 0$. We consider the cases $\lambda = 0$ and $\lambda \neq 0$ separately, since they result in different solutions (see Chapter 12).

$\lambda = 0$: The solution to the differential equation is $y = c_1 + c_2 x$. Applying the boundary conditions, we obtain the equations $c_1 + c_2 = 0$ and $c_2 = 0$. It follows that $c_1 = c_2 = 0$, and $y \equiv 0$. Therefore, $\lambda = 0$ is not an eigenvalue.

$\lambda \neq 0$: The solution to the differential equation is $y = c_1 + c_2 e^{-\lambda x}$. Applying the boundary conditions, we obtain

$$\begin{aligned} c_1 + (1-\lambda)c_2 &= 0 \\ (-\lambda e^{-\lambda})c_2 &= 0 \end{aligned}$$

These equations have a nontriival solution for c_1 and c_2 if and only if (Theorem 37.1)

$$\begin{vmatrix} 1 & 1-\lambda \\ 0 & -\lambda e^{-\lambda} \end{vmatrix} = -\lambda e^{-\lambda} = 0$$

which is an impossibility, since $\lambda \neq 0$.

Since we obtain only the trivial solution for $\lambda = 0$ and $\lambda \neq 0$, we can conclude that the problem does not have any eigenvalues.

37.9. Find the eigenvalues and eigenfunctions of

$$y'' - 4\lambda y' + 4\lambda^2 y = 0; \quad y(0) + y'(0) = 0, \quad y(1) - y'(1) = 0$$

As in Problem 37.6, the solution to the differential equation is $y = c_1 e^{2\lambda x} + c_2 x e^{2\lambda x}$. Applying the boundary conditions and simplifying, we obtain the equations

$$\begin{aligned}(1+2\lambda)c_1 + \quad\quad c_2 &= 0\\(1-2\lambda)c_1 + (-2\lambda)c_2 &= 0\end{aligned} \tag{1}$$

Equations (*1*) have a nontrivial solution for c_1 and c_2 if and only if (see Theorem 37.1) the determinant

$$\begin{vmatrix} 1+2\lambda & 1 \\ 1-2\lambda & -2\lambda \end{vmatrix} = -4\lambda^2 - 1$$

is zero; that is, if and only if $\lambda = \pm\frac{1}{2}i$. These eigenvalues are complex. In order to keep the differential equation under consideration real, we require that λ be real. Therefore this problem has no (real) eigenvalues and the only (real) solution is the trivial one: $y(x) \equiv 0$.

37.10. Find the eigenvalues and eigenfunctions of

$$y'' + \lambda y = 0; \quad y(0) = 0, \quad y(1) = 0.$$

The characteristic equation is $m^2 + \lambda = 0$. We consider the cases $\lambda = 0$, $\lambda < 0$, and $\lambda > 0$ separately, since they lead to different solutions (see Chapter 12).

$\lambda = 0$: The solution is $y = c_1 + c_2 x$. Applying the boundary conditions, we obtain $c_1 = c_2 = 0$, which results in the trivial solution.

$\lambda < 0$: The solution is $y = c_1 e^{\sqrt{-\lambda}x} + c_2 e^{-\sqrt{-\lambda}x}$, where $-\lambda$ and $\sqrt{-\lambda}$ are positive. Applying the boundary conditions, we obtain

$$c_1 + c_2 = 0 \qquad c_1 e^{\sqrt{-\lambda}} + c_2 e^{-\sqrt{-\lambda}} = 0$$

Using Theorem 37.1, we find

$$\begin{vmatrix} 1 & 1 \\ e^{\sqrt{-\lambda}} & e^{-\sqrt{-\lambda}} \end{vmatrix} = e^{-\sqrt{-\lambda}} - e^{\sqrt{-\lambda}}$$

which is never zero for any value of $\lambda < 0$. Hence, $c_1 = c_2 = 0$ and $y \equiv 0$.

$\lambda > 0$: The solution is $A \sin\sqrt{\lambda}\,x + B\cos\sqrt{\lambda}\,x$. Applying the boundary conditions, we obtain $B = 0$ and $A\sin\sqrt{\lambda} = 0$. Note that $\sin\theta = 0$ if and only if $\theta = n\pi$, where $n = 0, \pm1, \pm2, \ldots$. Furthermore, if $\theta > 0$, then n must be positive. To satisfy the boundary conditions, $B = 0$ and either $A = 0$ or $\sin\sqrt{\lambda} = 0$. This last equation is equivalent to $\sqrt{\lambda} = n\pi$ where $n = 1, 2, 3, \ldots$. The choice $A = 0$ results in the trivial solution; the choice $\sqrt{\lambda} = n\pi$ results in the nontrivial solution $y_n = A_n \sin n\pi x$. Here the notation A_n signifies that the arbitrary constant A_n can be different for different values of n.

Collecting the results of all three cases, we conclude that the eigenvalues are $\lambda_n = n^2\pi^2$ and the corresponding eigenfunctions are $y_n = A_n \sin n\pi x$, for $n = 1, 2, 3, \ldots$.

37.11. Find the eigenvalues and eigenfunctions of

$$y'' + \lambda y = 0; \quad y(0) = 0, \quad y'(\pi) = 0.$$

As in Problem 37.10, the cases $\lambda = 0$, $\lambda < 0$, and $\lambda > 0$ must be considered separately.

$\lambda = 0$: The solution is $y = c_1 + c_2 x$. Applying the boundary conditions, we obtain $c_1 = c_2 = 0$, hence $y \equiv 0$.

$\lambda < 0$: The solution is $y = c_1 e^{\sqrt{-\lambda}x} + c_2 e^{-\sqrt{-\lambda}x}$, where $-\lambda$ and $\sqrt{-\lambda}$ are positive. Applying the boundary conditions, we obtain

$$c_1 + c_2 = 0 \qquad c_1\sqrt{-\lambda}\, e^{\sqrt{-\lambda}\pi} - c_2\sqrt{-\lambda}\, e^{-\sqrt{-\lambda}\pi} = 0$$

Using Theorem 37.1, we find that the only solution of these equations is $c_1 = c_2 = 0$, hence $y \equiv 0$.

$\lambda > 0$: The solution is $y = A \sin\sqrt{\lambda}\,x + B\cos\sqrt{\lambda}\,x$. Applying the boundary conditions, we obtain $B = 0$ and $A\sqrt{\lambda}\cos\sqrt{\lambda}\,\pi = 0$. For $\theta > 0$, $\cos\theta = 0$ if and only if θ is a positive odd multiple of $\pi/2$; that is, when $\theta = (2n-1)\frac{\pi}{2} = (n-\frac{1}{2})\pi$, where $n = 1, 2, 3, \ldots$. Therefore, to satisfy the boundary conditions, we must have $B = 0$ and either $A = 0$ or $\cos\sqrt{\lambda}\,\pi = 0$. This last equation is equivalent to $\sqrt{\lambda} = n - \frac{1}{2}$. The choice $A = 0$ results in the trivial solution; the choice $\sqrt{\lambda} = n - \frac{1}{2}$ results in the nontrivial solution $y_n = A_n \sin(n-\frac{1}{2})x$.

Collecting all three cases, we conclude that the eigenvalues are $\lambda_n = (n-\frac{1}{2})^2$ and the corresponding eigenfunctions are $y_n = A_n \sin(n-\frac{1}{2})x$, where $n = 1, 2, 3, \ldots$.

Supplementary Problems

In Problems 37.12 through 37.19, find all solutions, if solutions exist, to the given boundary-value problem.

37.12. $y'' + y = 0;\quad y(0) = 0,\quad y(\pi/2) = 0.$

37.13. $y'' + y = x;\quad y(0) = 0,\quad y(\pi/2) = 0.$

37.14. $y'' + y = 0;\quad y(0) = 0,\quad y(\pi/2) = 1.$

37.15. $y'' + y = x;\quad y(0) = -1,\quad y(\pi/2) = 1.$

37.16. $y'' + y = 0;\quad y'(0) = 0,\quad y(\pi/2) = 0.$

37.17. $y'' + y = 0;\quad y'(0) = 1,\quad y(\pi/2) = 0.$

37.18. $y'' + y = x;\quad y'(0) = 1,\quad y(\pi/2) = 0.$

37.19. $y'' + y = x;\quad y'(0) = 1,\quad y(\pi/2) = \pi/2.$

In Problems 37.20 through 37.26, find the eigenvalues and eigenfunctions, if any, of the given boundary-value problems.

37.20. $y'' + 2\lambda y' + \lambda^2 y = 0;\quad y(0) + y'(0) = 0,\quad y(1) + y'(1) = 0.$

37.21. $y'' + 2\lambda y' + \lambda^2 y = 0;\quad y(0) = 0,\quad y(1) = 0.$

37.22. $y'' + 2\lambda y' + \lambda^2 y = 0;\quad y(1) + y'(1) = 0,\quad 3y(2) + 2y'(2) = 0.$

37.23. $y'' + \lambda y' = 0;\quad y(0) + y'(0) = 0;\quad y(2) + y'(2) = 0.$

37.24. $y'' - \lambda y = 0;\quad y(0) = 0,\quad y(1) = 0.$

37.25. $y'' + \lambda y = 0;\quad y'(0) = 0,\quad y(5) = 0.$

37.26. $y'' + \lambda y = 0;\quad y'(0) = 0,\quad y'(\pi) = 0.$

Answers to Supplementary Problems

37.12. $y \equiv 0$

37.13. $y = x - \frac{\pi}{2} \sin x$

37.14. $y = \sin x$

37.15. $y = x + (1 - \frac{1}{2}\pi) \sin x - \cos x$

37.16. $y = B \cos x$, B arbitrary

37.17. no solution

37.18. no solution

37.19. $y = x + B \cos x$, B arbitrary

37.20. $\lambda = 1$, $y = c_1 e^{-x}$

37.21. no eigenvalues or eigenfunctions

37.22. $\lambda = 2$, $y = c_2 x e^{-2x}$ and $\lambda = \frac{1}{2}$, $y = c_2(-3 + x)e^{-x/2}$

37.23. $\lambda = 1$, $y = c_2 e^{-x}$ (c_2 arbitrary)

37.24. $\lambda_n = -n^2\pi^2$, $y_n = A_n \sin n\pi x$, for $n = 1, 2, \ldots$ (A_n arbitrary)

37.25. $\lambda_n = (\frac{1}{5}n - \frac{1}{10})^2\pi^2$, $y_n = B_n \cos(\frac{1}{5}n - \frac{1}{10})\pi x$, for $n = 1, 2, \ldots$ (B_n arbitrary)

37.26. $\lambda_n = n^2$, $y_n = B_n \cos nx$, for $n = 0, 1, 2, \ldots$ (B_n arbitrary)

Chapter 38

Sturm-Liouville Problems

38.1 DEFINITION

A second-order *Sturm-Liouville problem* is a homogeneous boundary-value problem of the form

$$[p(x)y']' + q(x)y + \lambda w(x)y = 0; \tag{38.1}$$

$$\begin{aligned} \alpha_1 y(a) + \beta_1 y'(a) &= 0 \\ \alpha_2 y(b) + \beta_2 y'(b) &= 0 \end{aligned} \tag{38.2}$$

where $p(x)$, $p'(x)$, $q(x)$, and $w(x)$ are continuous on $[a, b]$, and both $p(x)$ and $w(x)$ are positive on $[a, b]$.

Example 38.1. Problem 37.10 is a Sturm-Liouville problem with $p(x) \equiv 1$, $q(x) \equiv 0$ and $w(x) \equiv 1$. The problem

$$(xy')' + [x^2 + 1 + \lambda e^x]y = 0; \quad y(1) + 2y'(1) = 0, \quad y(2) - 3y'(2) = 0$$

is a Sturm-Liouville problem with $p(x) = x$, $q(x) = x^2 + 1$, and $w(x) = e^x$. Note that on $[1, 2]$, which is the interval of interest, both $p(x)$ and $w(x)$ are positive.

The second-order differential equation

$$a_2(x)y'' + a_1(x)y' + a_0(x)y + \lambda r(x)y = 0 \tag{38.3}$$

where $a_2(x)$ does not vanish on $[a, b]$, is equivalent to *(38.1)* if and only if $a_2'(x) = a_1(x)$. (See Problem 38.2.) This condition can always be forced by multiplying *(38.3)* by a suitable factor. (See Problem 38.3.)

38.2 PROPERTIES OF STURM-LIOUVILLE PROBLEMS

Sturm-Liouville problems have desirable features not shared by more general eigenvalue problems.

Theorem 38.1. The eigenvalues of a Sturm-Liouville problem are all real and nonnegative.

(Compare this theorem to the result of Problem 37.9.)

Theorem 38.2. The eigenvalues of a Sturm-Liouville problem can be arranged to form a strictly increasing infinite sequence; that is, $0 \leq \lambda_1 < \lambda_2 < \lambda_3 < \cdots$. Furthermore, $\lambda_n \to \infty$ as $n \to \infty$.

(Compare this theorem to the results of Problems 37.6 through 37.8.)

Theorem 38.3. For each eigenvalue of a Sturm-Liouville problem, there exists one and only one linearly independent eigenfunction.

(By this theorem, there corresponds to each eigenvalue λ_n a unique eigenfunction with lead coefficient unity; we denote this eigenfunction by $e_n(x)$.)

Theorem 38.4. The set of eigenfunctions $\{e_1(x), e_2(x), \ldots\}$ of a Sturm-Liouville problem satisfies the relation

$$\int_a^b w(x)e_n(x)e_m(x)\,dx = 0 \tag{38.4}$$

for $n \neq m$, where $w(x)$ is given in (*38.1*).

Solved Problems

38.1. Determine which of the following differential equations with the boundary conditions $y(0) = 0$, $y'(1) = 0$ form Sturm-Liouville problems:

(*a*) $e^x y'' + e^x y' + \lambda y = 0$

(*b*) $xy'' + y' + (x^2 + 1 + \lambda)y = 0$

(*c*) $\left(\frac{1}{x}y'\right)' + (x + \lambda)y = 0$

(*d*) $y'' + \lambda(1 + x)y = 0$

(*e*) $e^{2x}y'' + e^{2x}y' + \lambda y = 0$

(*a*) The equation can be rewritten as $(e^x y')' + \lambda y = 0$; hence $p(x) = e^x$, $q(x) \equiv 0$, and $w(x) \equiv 1$. This is a Sturm-Liouville problem.

(*b*) The equation is equivalent to $(xy')' + (x^2 + 1)y + \lambda y = 0$; hence $p(x) = x$, $q(x) = x^2 + 1$, and $w(x) \equiv 1$. Since $p(x)$ is zero at a point in the interval $[0, 1]$, this is not a Sturm-Liouville problem.

(*c*) Here $p(x) = 1/x$, $q(x) = x$, and $w(x) \equiv 1$. Since $p(x)$ is not continuous in $[0, 1]$, in particular at $x = 0$, this is not a Sturm-Liouville problem.

(*d*) The equation can be rewritten as $(y')' + \lambda(1 + x)y = 0$; hence $p(x) \equiv 1$, $q(x) \equiv 0$, and $w(x) = 1 + x$. This is a Sturm-Liouville problem.

(*e*) The equation, in its present form, is not equivalent to (*38.1*); this is not a Sturm-Liouville problem. However, if we first multiply the equation by e^{-x}, we obtain $(e^x y')' + \lambda e^{-x} y = 0$; this is a Sturm-Liouville problem with $p(x) = e^x$, $q(x) \equiv 0$, and $w(x) = e^{-x}$.

38.2. Prove that (*38.1*) is equivalent to (*38.3*) if and only if $a_2'(x) = a_1(x)$.

Applying the product rule of differentiation to (*38.1*), we find that

$$p(x)y'' + p'(x)y' + q(x)y + \lambda w(x)y = 0 \tag{1}$$

Setting $a_2(x) = p(x)$, $a_1(x) = p'(x)$, $a_0(x) = q(x)$, and $r(x) = w(x)$, it follows that (*1*), which is (*38.1*) rewritten, is precisely (*38.3*) with $a_2'(x) = p'(x) = a_1(x)$.

Conversely, if $a_2'(x) = a_1(x)$, then (*38.3*) has the form

$$a_2(x)y'' + a_2'(x)y' + a_0(x)y + \lambda r(x)y = 0$$

which is equivalent to $[a_2(x)y']' + a_0(x)y + \lambda r(x)y = 0$. This last equation is precisely (*38.1*) with $p(x) = a_2(x)$, $q(x) = a_0(x)$, and $w(x) = r(x)$.

38.3. Show that if *(38.3)* is multiplied by $I(x) = e^{\int [a_1(x)/a_2(x)]\,dx}$, the resulting equation is equivalent to *(38.1)*.

Multiplying *(38.3)* by $I(x)$, we obtain

$$I(x)a_2(x)y'' + I(x)a_1(x)y' + I(x)a_0(x)y + \lambda I(x)r(x)y = 0$$

which can be rewritten as

$$a_2(x)[I(x)y']' + I(x)a_0(x)y + \lambda I(x)r(x)y = 0 \tag{1}$$

Divide *(1)* by $a_2(x)$ and then set $p(x) = I(x)$, $q(x) = I(x)a_0(x)/a_2(x)$ and $w(x) = I(x)r(x)/a_2(x)$; the resulting equation is precisely *(38.1)*. Note that since $I(x)$ is an exponential and since $a_2(x)$ does not vanish, $I(x)$ is positive.

38.4. Transform $y'' + 2xy' + (x+\lambda)y = 0$ into *(38.1)* by means of the procedure outlined in Problem 38.3.

Here $a_2(x) \equiv 1$ and $a_1(x) = 2x$; hence $a_1(x)/a_2(x) = 2x$ and $I(x) = e^{\int 2x\,dx} = e^{x^2}$. Multiplying the given differential equation by $I(x)$, we obtain

$$e^{x^2}y'' + 2xe^{x^2}y' + xe^{x^2}y + \lambda e^{x^2}y = 0$$

which can be rewritten as

$$(e^{x^2}y')' + xe^{x^2}y + \lambda e^{x^2}y = 0$$

This last equation is precisely *(38.1)* with $p(x) = e^{x^2}$, $q(x) = xe^{x^2}$, and $w(x) = e^{x^2}$.

38.5. Transform $(x+2)y'' + 4y' + xy + \lambda e^x y = 0$ into *(38.1)* by means of the procedure outlined in Problem 38.3.

Here $a_2(x) = x+2$ and $a_1(x) \equiv 4$; hence $a_1(x)/a_2(x) = 4/(x+2)$ and

$$I(x) = e^{\int [4/(x+2)]\,dx} = e^{4\ln|x+2|} = e^{\ln(x+2)^4} = (x+2)^4$$

Multiplying the given differential equation by $I(x)$, we obtain

$$(x+2)^5y'' + 4(x+2)^4y' + (x+2)^4xy + \lambda(x+2)^4e^xy = 0$$

which can be rewritten as

$$(x+2)[(x+2)^4y']' + (x+2)^4xy + \lambda(x+2)^4e^xy = 0$$

or

$$[(x+2)^4y']' + (x+2)^3y + \lambda(x+2)^3e^xy = 0$$

This last equation is precisely *(38.1)* with $p(x) = (x+2)^4$, $q(x) = (x+2)^3$, and $w(x) = (x+2)^3e^x$. Note that since we divided by $a_2(x)$, it is necessary to restrict $x \neq -2$. Furthermore, in order that both $p(x)$ and $w(x)$ be positive, we must require $x > -2$.

38.6. Verify Theorems 38.1 through 38.4 for the Sturm-Liouville problem

$$y'' + \lambda y = 0; \quad y(0) = 0, \quad y(1) = 0$$

Using the results of Problem 37.10, we have that the eigenvalues are $\lambda_n = n^2\pi^2$ and the corresponding eigenfunctions are $y_n(x) = A_n \sin n\pi x$, for $n = 1, 2, 3, \ldots$. The eigenvalues are obviously real and nonnegative, and they can be ordered as $\lambda_1 = \pi^2 < \lambda_2 = 4\pi^2 < \lambda_3 = 9\pi^2 < \cdots$. Each eigenvalue has a single linearly independent eigenfunction $e_n(x) = \sin n\pi x$ associated with it. Finally, since

$$\sin n\pi x \sin m\pi x = \frac{1}{2}\cos(n-m)\pi x - \frac{1}{2}\cos(n+m)\pi x$$

we have for $n \neq m$ and $w(x) \equiv 1$:

$$\begin{aligned}
\int_a^b w(x)e_n(x)e_m(x)\,dx &= \int_0^1 \left[\tfrac{1}{2}\cos(n-m)\pi x - \tfrac{1}{2}\cos(n+m)\pi x\right] dx \\
&= \left[\frac{1}{2(n-m)\pi}\sin(n-m)\pi x - \frac{1}{2(n+m)\pi}\sin(n+m)\pi x\right]_{x=0}^{x=1} \\
&= 0
\end{aligned}$$

38.7. Verify Theorems 38.1 through 38.4 for the Sturm-Liouville problem

$$y'' + \lambda y = 0; \quad y'(0) = 0, \quad y(\pi) = 0$$

For this problem, we calculate the eigenvalues $\lambda_n = (n-\frac{1}{2})^2$ and the corresponding eigenfunctions $y_n(x) = A_n \cos(n-\frac{1}{2})x$, for $n = 1, 2, \ldots$. The eigenvalues are real and positive, and can be ordered as

$$\lambda_1 = \frac{1}{4} < \lambda_2 = \frac{9}{4} < \lambda_3 = \frac{25}{4} < \cdots$$

Each eigenvalue has only one linearly independent eigenfunction $e_n(x) = \cos(n-\frac{1}{2})x$ associated with it. Also, for $n \neq m$ and $w(x) \equiv 1$,

$$\begin{aligned} \int_a^b w(x)e_n(x)e_m(x)\,dx &= \int_0^\pi \cos(n-\tfrac{1}{2})x \cos(m-\tfrac{1}{2})x\,dx \\ &= \int_0^\pi [\tfrac{1}{2}\cos(n+m-1)x + \tfrac{1}{2}\cos(n-m)x]\,dx \\ &= \left[\frac{1}{2(n+m-1)}\sin(n+m-1)x + \frac{1}{2(n-m)}\sin(n-m)x\right]_{x=0}^{x=\pi} \\ &= 0 \end{aligned}$$

38.8. Prove that if the set of nonzero functions $\{y_1(x), y_2(x), \ldots, y_p(x)\}$ satisfies (*38.4*), then the set is linearly independent on $[a, b]$.

Using Section 11.2, we consider the equation

$$c_1y_1(x) + c_2y_2(x) + \cdots + c_ky_k(x) + \cdots + c_py_p(x) \equiv 0 \tag{1}$$

Multiplying this equation by $w(x)y_k(x)$ and then integrating from a to b, we obtain

$$\begin{aligned} c_1\int_a^b w(x)y_k(x)y_1(x)\,dx + c_2\int_a^b w(x)y_k(x)y_2(x)\,dx + \cdots \\ + c_k\int_a^b w(x)y_k(x)y_k(x)\,dx + \cdots + c_p\int_a^b w(x)y_k(x)y_p(x)\,dx = 0 \end{aligned}$$

From (*38.4*) we conclude that for $i \neq k$,

$$c_k\int_a^b w(x)y_k(x)y_i(x)\,dx = 0$$

But since $y_k(x)$ is a nonzero function and $w(x)$ is positive on $[a, b]$, it follows that

$$\int_a^b w(x)[y_k(x)]^2\,dx \neq 0$$

hence, $c_k = 0$. Since $c_k = 0$, $k = 1, 2, \ldots, p$, is the only solution to (*1*), the given set of functions is linearly independent on $[a, b]$.

Supplementary Problems

In Problems 38.9 through 38.15, determine whether each of the given differential equations with the boundary conditions $y(-1) + 2y'(-1) = 0$, $y(1) + 2y'(1) = 0$ is a Sturm-Liouville problem.

38.9. $(2 + \sin x)y'' + (\cos x)y' + (1 + \lambda)y = 0.$

38.10. $(\sin \pi x)y'' + (\pi \cos \pi x)y' + (x + \lambda)y = 0.$

38.11. $(\sin x)y'' + (\cos x)y' + (1 + \lambda)y = 0.$

38.12. $(x+2)^2y'' + 2(x+2)y' + (e^x + \lambda e^{2x})y = 0.$

38.13. $(x+2)^2y'' + (x+2)y' + (e^x + \lambda e^{2x})y = 0.$

38.14. $y'' + \dfrac{3}{x^2}\lambda y = 0.$

38.15. $y'' + \dfrac{3}{(x-4)^2}\lambda y = 0.$

38.16. Transform $e^{2x}y'' + e^{2x}y' + (x+\lambda)y = 0$ into *(38.1)* by means of the procedure outlined in Problem 38.3.

38.17. Transform $x^2y'' + xy' + \lambda xy = 0$ into *(38.1)* by means of the procedure outlined in Problem 38.3.

38.18. Verify Theorems 38.1 through 38.4 for the Sturm-Liouville problem

$$y'' + \lambda y = 0; \qquad y'(0) = 0, \quad y'(\pi) = 0$$

38.19. Verify Theorems 38.1 through 38.4 for the Sturm-Liouville problem

$$y'' + \lambda y = 0; \qquad y(0) = 0, \quad y(2\pi) = 0$$

Answers to Supplementary Problems

38.9. yes

38.10. no, $p(x) = \sin \pi x$ is zero at $x = \pm 1, 0$

38.11. no, $p(x) = \sin x$ is zero at $x = 0$

38.12. yes

38.13. no, the equation is not equivalent to *(38.1)*

38.14. no, $w(x) = \dfrac{3}{x^2}$ is not continuous at $x = 0$

38.15. yes

38.16. $I(x) = e^x; \quad (e^xy')' + xe^{-x}y + \lambda e^{-x}y = 0$

38.17. $I(x) = x; \quad (xy')' + \lambda y = 0$

38.18. $\lambda_n = n^2, \quad e_n(x) = \cos nx \qquad (n = 0, 1, 2, \ldots)$

38.19. $\lambda_n = \dfrac{n^2}{4}, \quad e_n(x) = \sin \dfrac{nx}{2} \qquad (n = 1, 2, \ldots)$

Chapter 39

Eigenfunction Expansions

39.1 PIECEWISE SMOOTH FUNCTIONS

A wide class of functions can be represented by infinite series of eigenfunctions of a Sturm-Liouville problem. We say that a function $f(x)$ is *piecewise smooth* on $[a, b]$ if both $f(x)$ and its derivative $f'(x)$ are piecewise continuous (Section 22.3) on $[a, b]$.

Theorem 39.1. If $f(x)$ is piecewise smooth on $[a, b]$ and if $\{e_n(x)\}$ is the set of all eigenfunctions of a Sturm-Liouville problem (Theorem 38.3), then

$$f(x) = \sum_{n=1}^{\infty} c_n e_n(x) \tag{39.1}$$

where

$$c_n = \frac{\int_a^b w(x)\, f(x)\, e_n(x)\, dx}{\int_a^b w(x)\, e_n^2(x)\, dx} \tag{39.2}$$

The representation (*39.1*) is valid at all points in the open interval (a, b) where $f(x)$ is continuous. The function $w(x)$ in (*39.2*) is given by (*38.1*).

Because different Sturm-Liouville problems usually generate different sets of eigenfunctions, a given piecewise smooth function will have many expansions of the form (*39.1*). The basic features of all such expansions are exhibited by the trigonometric series discussed below.

39.2 FOURIER SINE SERIES

The eigenfunctions of the Sturm-Liouville problem $y'' + \lambda y = 0$; $y(0) = 0$, $y(L) = 0$, where L is a real positive number, are $e_n(x) = \sin \frac{n\pi x}{L}$ $(n = 1, 2, 3, \ldots)$. Substituting these functions into (*39.1*), we obtain

$$f(x) = \sum_{n=1}^{\infty} c_n \sin \frac{n\pi x}{L} \tag{39.3}$$

For this Sturm-Liouville problem, $w(x) \equiv 1$, $a = 0$, and $b = L$; so that

$$\int_a^b w(x)\, e_n^2(x)\, dx = \int_0^L \sin^2 \frac{n\pi x}{L}\, dx = \frac{L}{2}$$

and (*39.2*) becomes

$$c_n = \frac{2}{L} \int_0^L f(x) \sin \frac{n\pi x}{L}\, dx \tag{39.4}$$

The expansion (*39.3*) with coefficients given by (*39.4*) is the *Fourier sine series* for $f(x)$ on $(0, L)$. (See Problems 39.3, 39.5, and 39.6.)

39.3 FOURIER COSINE SERIES

The eigenfunctions of the Sturm-Liouville problem $y'' + \lambda y = 0$; $y'(0) = 0$, $y'(L) = 0$, where L is a real positive number, are $e_0(x) = 1$ and $e_n(x) = \cos\frac{n\pi x}{L}$ $(n = 1, 2, 3, \ldots)$. Here $\lambda = 0$ is an eigenvalue with corresponding eigenfunction $e_0(x) = 1$. Substituting these functions into (*39.1*), where because of the additional eigenfunction $e_0(x)$ the summation now begins at $n = 0$, we obtain

$$f(x) = c_0 + \sum_{n=1}^{\infty} c_n \cos\frac{n\pi x}{L} \tag{39.5}$$

For this Sturm-Liouville problem, $w(x) \equiv 1$, $a = 0$, and $b = L$; so that

$$\int_a^b w(x)\, e_0^2(x)\, dx = \int_0^L dx = L \qquad \int_a^b w(x)\, e_n^2(x)\, dx = \int_0^L \cos^2\frac{n\pi x}{L}\, dx = \frac{L}{2}$$

Thus (*39.2*) becomes

$$c_0 = \frac{1}{L}\int_0^L f(x)\, dx \qquad c_n = \frac{2}{L}\int_0^L f(x)\cos\frac{n\pi x}{L}\, dx \quad (n = 1, 2, \ldots) \tag{39.6}$$

The expansion (*39.5*) with coefficients given by (*39.6*) is the *Fourier cosine series* for $f(x)$ on $(0, L)$. (See Problems 39.4 and 39.7.)

Solved Problems

39.1. Is the function

$$f(x) = \begin{cases} x^2 + 1 & x < 0 \\ 1 & 0 \le x \le 1 \\ 2x + 1 & x > 1 \end{cases}$$

piecewise smooth on $[-2, 2]$?

The function is continuous everywhere on $[-2, 2]$ except at $x_1 = 1$. Since the required limits exist at x_1, $f(x)$ is piecewise continuous. Differentiating $f(x)$, we obtain

$$f'(x) = \begin{cases} 2x & x < 0 \\ 0 & 0 \le x < 1 \\ 2 & x > 1 \end{cases}$$

The derivative does not exist at $x_1 = 1$ but is continuous at all other points in $[-2, 2]$. At x_1 the required limits exist, hence $f'(x)$ is piecewise continuous. It follows that $f(x)$ is piecewise smooth on $[-2, 2]$.

39.2. Is the function

$$f(x) = \begin{cases} 1 & x < 0 \\ \sqrt{x} & 0 \le x \le 1 \\ x^3 & x > 1 \end{cases}$$

piecewise smooth on $[-1, 3]$?

The function $f(x)$ is continuous everywhere on $[-1,3]$ except at $x_1 = 0$. Since the required limits exist at x_1, $f(x)$ is piecewise continuous. Differentiating $f(x)$, we obtain

$$f'(x) = \begin{cases} 0 & x < 0 \\ \dfrac{1}{2\sqrt{x}} & 0 < x < 1 \\ 3x^2 & x > 1 \end{cases}$$

which is continuous everywhere on $[-1,3]$ except at the two points $x_1 = 0$ and $x_2 = 1$ where the derivative does not exist. At x_1,

$$\lim_{\substack{x \to x_1 \\ x > x_1}} f'(x) = \lim_{\substack{x \to 0 \\ x > 0}} \frac{1}{2\sqrt{x}} = \infty$$

hence, one of the required limits does not exist. It follows that $f'(x)$ is not piecewise continuous, and therefore that $f(x)$ is not piecewise smooth, on $[-1,3]$.

39.3. Find a Fourier sine series for $f(x) = 1$ on $(0,5)$.

Using (*39.4*) with $L = 5$, we have

$$c_n = \frac{2}{L}\int_0^L f(x) \sin\frac{n\pi x}{L}\,dx = \frac{2}{5}\int_0^5 (1) \sin\frac{n\pi x}{5}\,dx$$

$$= \frac{2}{5}\left[-\frac{5}{n\pi}\cos\frac{n\pi x}{5}\right]_{x=0}^{x=5} = \frac{2}{n\pi}[1 - \cos n\pi] = \frac{2}{n\pi}[1 - (-1)^n]$$

Thus (*39.3*) becomes

$$1 = \sum_{n=1}^{\infty} \frac{2}{n\pi}[1-(-1)^n] \sin\frac{n\pi x}{5}$$

$$= \frac{4}{\pi}\left(\sin\frac{\pi x}{5} + \frac{1}{3}\sin\frac{3\pi x}{5} + \frac{1}{5}\sin\frac{5\pi x}{5} + \cdots\right) \quad (1)$$

Since $f(x) = 1$ is piecewise smooth on $[0,5]$ and continuous everywhere in the open interval $(0,5)$, it follows from Theorem 39.1 that (*1*) is valid for all x in $(0,5)$.

39.4. Find a Fourier cosine series for $f(x) = x$ on $(0,3)$.

Using (*39.6*) with $L = 3$, we have

$$c_0 = \frac{1}{L}\int_0^L f(x)\,dx = \frac{1}{3}\int_0^3 x\,dx = \frac{3}{2}$$

$$c_n = \frac{2}{L}\int_0^L f(x)\cos\frac{n\pi x}{L}\,dx = \frac{2}{3}\int_0^3 x\cos\frac{n\pi x}{3}\,dx$$

$$= \frac{2}{3}\left[\frac{3x}{n\pi}\sin\frac{n\pi x}{3} + \frac{9}{n^2\pi^2}\cos\frac{n\pi x}{3}\right]_{x=0}^{x=3}$$

$$= \frac{2}{3}\left(\frac{9}{n^2\pi^2}\cos n\pi - \frac{9}{n^2\pi^2}\right) = \frac{6}{n^2\pi^2}[(-1)^n - 1]$$

Thus (*39.5*) becomes

$$x = \frac{3}{2} + \sum_{n=1}^{\infty}\frac{6}{n^2\pi^2}[(-1)^n - 1]\cos\frac{n\pi x}{3}$$

$$= \frac{3}{2} - \frac{12}{\pi^2}\left(\cos\frac{\pi x}{3} + \frac{1}{9}\cos\frac{3\pi x}{3} + \frac{1}{25}\cos\frac{5\pi x}{3} + \cdots\right) \quad (1)$$

Since $f(x) = x$ is piecewise smooth on $[0,3]$ and continuous everywhere in the open interval $(0,3)$, it follows from Theorem 39.1 that (*1*) is valid for all x in $(0,3)$.

39.5. Find a Fourier sine series for $f(x) = \begin{cases} 0 & x \leq 2 \\ 2 & x > 2 \end{cases}$ on $(0, 3)$.

Using (*39.4*) with $L = 3$, we obtain

$$c_n = \frac{2}{3}\int_0^3 f(x) \sin\frac{n\pi x}{3}\,dx$$

$$= \frac{2}{3}\int_0^2 (0) \sin\frac{n\pi x}{3}\,dx + \frac{2}{3}\int_2^3 (2) \sin\frac{n\pi x}{3}\,dx$$

$$= 0 + \frac{4}{3}\left[-\frac{3}{n\pi}\cos\frac{n\pi x}{3}\right]_{x=2}^{x=3} = \frac{4}{n\pi}\left[\cos\frac{2n\pi}{3} - \cos n\pi\right]$$

Thus (*39.3*) becomes

$$f(x) = \sum_{n=1}^{\infty} \frac{4}{n\pi}\left[\cos\frac{2n\pi}{3} - (-1)^n\right] \sin\frac{n\pi x}{3}$$

Furthermore, $\cos\frac{2\pi}{3} = -\frac{1}{2}$, $\cos\frac{4\pi}{3} = -\frac{1}{2}$, $\cos\frac{6\pi}{3} = 1, \ldots;$ hence,

$$f(x) = \frac{4}{\pi}\left(\frac{1}{2}\sin\frac{\pi x}{3} - \frac{3}{4}\sin\frac{2\pi x}{3} + \frac{2}{3}\sin\frac{3\pi x}{3} - \cdots\right) \tag{1}$$

Since $f(x)$ is piecewise smooth on $[0, 3]$ and continuous everywhere in $(0, 3)$ except at $x = 2$, it follows from Theorem 39.1 that (*1*) is valid everywhere in $(0, 3)$ except at $x = 2$.

39.6. Find a Fourier sine series for $f(x) = e^x$ on $(0, \pi)$.

Using (*39.4*) with $L = \pi$, we obtain

$$c_n = \frac{2}{\pi}\int_0^{\pi} e^x \sin\frac{n\pi x}{\pi}\,dx = \frac{2}{\pi}\left[\frac{e^x}{1+n^2}(\sin nx - n\cos nx)\right]_{x=0}^{x=\pi}$$

$$= \frac{2}{\pi}\left(\frac{n}{1+n^2}\right)(1 - e^{\pi}\cos n\pi)$$

Thus (*39.3*) becomes

$$e^x = \frac{2}{\pi}\sum_{n=1}^{\infty} \frac{n}{1+n^2}[1 - e^{\pi}(-1)^n] \sin nx$$

It follows from Theorem 39.1 that this last equation is valid for all x in $(0, \pi)$.

39.7. Find a Fourier cosine series for $f(x) = e^x$ on $(0, \pi)$.

Using (*39.6*) with $L = \pi$, we have

$$c_0 = \frac{1}{\pi}\int_0^{\pi} e^x\,dx = \frac{1}{\pi}(e^{\pi} - 1)$$

$$c_n = \frac{2}{\pi}\int_0^{\pi} e^x \cos\frac{n\pi x}{\pi}\,dx = \frac{2}{\pi}\left[\frac{e^x}{1+n^2}(\cos nx + n\sin nx)\right]_{x=0}^{x=\pi}$$

$$= \frac{2}{\pi}\left(\frac{1}{1+n^2}\right)(e^{\pi}\cos n\pi - 1)$$

Thus (*39.5*) becomes

$$e^x = \frac{1}{\pi}(e^{\pi} - 1) + \frac{2}{\pi}\sum_{n=1}^{\infty} \frac{1}{1+n^2}[(-1)^n e^{\pi} - 1] \cos nx$$

As in Problem 39.6, this last equation is valid for all x in $(0, \pi)$.

39.8. Find an expansion for $f(x) = e^x$ in terms of the eigenfunctions of the Sturm-Liouville problem $y'' + \lambda y = 0$; $y'(0) = 0$, $y(\pi) = 0$.

From Problem 38.7, we have $e_n(x) = \cos(n - \frac{1}{2})x$ for $n = 1, 2, \ldots$. Substituting these functions and $w(x) \equiv 1$, $a = 0$, and $b = \pi$ into (*39.2*), we obtain for the numerator:

$$\int_a^b w(x) f(x) e_n(x)\, dx = \int_0^\pi e^x \cos(n - \tfrac{1}{2})x\, dx$$

$$= \frac{e^x}{1 + (n - \frac{1}{2})^2}\left[\cos(n - \tfrac{1}{2})x + (n - \tfrac{1}{2})\sin(n - \tfrac{1}{2})x\right]\Bigg|_{x=0}^{x=\pi}$$

$$= \frac{-1}{1 + (n - \frac{1}{2})^2}\left[e^\pi(n - \tfrac{1}{2})(-1)^n + 1\right]$$

and for the denominator:

$$\int_a^b w(x) e_n^2(x)\, dx = \int_0^\pi \cos^2(n - \tfrac{1}{2})x\, dx$$

$$= \left[\frac{x}{2} + \frac{\sin(2n - 1)x}{4(n - \frac{1}{2})}\right]_{x=0}^{x=\pi} = \frac{\pi}{2}$$

Thus

$$c_n = \frac{2}{\pi}\left[\frac{-1}{1 + (n - \frac{1}{2})^2}\right]\left[e^\pi(n - \tfrac{1}{2})(-1)^n + 1\right]$$

and (*39.1*) becomes

$$e^x = \frac{-2}{\pi}\sum_{n=1}^{\infty}\frac{1 + (-1)^n e^\pi(n - \frac{1}{2})}{1 + (n - \frac{1}{2})^2}\cos(n - \tfrac{1}{2})x$$

By Theorem 39.1 this last equation is valid for all x in $(0, \pi)$.

39.9. Find an expansion for $f(x) = 1$ in terms of the eigenfunctions of the Sturm-Liouville problem $y'' + \lambda y = 0$; $y(0) = 0$, $y'(1) = 0$.

We can show that the eigenfunctions are $e_n(x) = \sin(n - \frac{1}{2})\pi x$ $(n = 1, 2, \ldots)$. Substituting these functions and $w(x) \equiv 1$, $a = 0$, $b = 1$ into (*39.2*), we obtain for the numerator:

$$\int_a^b w(x) f(x) e_n(x)\, dx = \int_0^1 \sin(n - \tfrac{1}{2})\pi x\, dx$$

$$= \frac{-1}{(n - \frac{1}{2})\pi}\cos(n - \tfrac{1}{2})\pi x\Bigg|_0^1 = \frac{1}{(n - \frac{1}{2})\pi}$$

and for the denominator:

$$\int_a^b w(x) e_n^2(x)\, dx = \int_0^1 \sin^2(n - \tfrac{1}{2})\pi x\, dx$$

$$= \left[\frac{x}{2} - \frac{\sin(2n - 1)\pi x}{4(n - \frac{1}{2})}\right]_{x=0}^{x=1} = \frac{1}{2}$$

Thus

$$c_n = \frac{2}{(n - \frac{1}{2})\pi}$$

and (*39.1*) becomes

$$1 = \frac{2}{\pi}\sum_{n=1}^{\infty}\frac{\sin(n - \frac{1}{2})\pi x}{n - \frac{1}{2}}$$

By Theorem 39.1 this last equation is valid for all x in $(0, 1)$.

Supplementary Problems

39.10. Find a Fourier sine series for $f(x) = 1$ on $(0, 1)$.

39.11. Find a Fourier sine series for $f(x) = x$ on $(0, 3)$.

39.12. Find a Fourier cosine series for $f(x) = x^2$ on $(0, \pi)$.

39.13. Find a Fourier cosine series for $f(x) = \begin{cases} 0 & x \leqq 2 \\ 2 & x > 2 \end{cases}$ on $(0, 3)$.

39.14. Find a Fourier cosine series for $f(x) = 1$ on $(0, 7)$.

39.15. Find a Fourier sine series for $f(x) = \begin{cases} x & x \leqq 1 \\ 2 & x > 1 \end{cases}$ on $(0, 2)$.

39.16. Find an expansion for $f(x) = 1$ in terms of the eigenfunctions of the Sturm-Liouville problem $y'' + \lambda y = 0$; $y'(0) = 0$, $y(\pi) = 0$.

39.17. Find an expansion for $f(x) = x$ in terms of the eigenfunctions of the Sturm-Liouville problem $y'' + \lambda y = 0$; $y(0) = 0$, $y'(\pi) = 0$.

39.18. Which of the following functions are piecewise smooth on $[-2, 3]$?

(*a*) $f(x) = \begin{cases} x^3 & x < 0 \\ \sin \pi x & 0 \leqq x \leqq 1 \\ x^2 - 5x & x > 1 \end{cases}$ (*c*) $f(x) = \ln |x|$

(*b*) $f(x) = \begin{cases} e^x & x < 1 \\ \sqrt{x} & x \geqq 1 \end{cases}$ (*d*) $f(x) = \begin{cases} (x-1)^2 & x \leqq 1 \\ (x-1)^{1/3} & x > 1 \end{cases}$

Answers to Supplementary Problems

39.10. $\dfrac{2}{\pi} \sum_{n=1}^{\infty} \dfrac{1}{n}[1 - (-1)^n] \sin n\pi x$ **39.11.** $-\dfrac{6}{\pi} \sum_{n=1}^{\infty} \dfrac{(-1)^n}{n} \sin \dfrac{n\pi x}{3}$ **39.12.** $\dfrac{1}{3}\pi^2 + 4 \sum_{n=1}^{\infty} \dfrac{(-1)^n}{n^2} \cos nx$

39.13. $\dfrac{2}{3} - \dfrac{4}{\pi} \sum_{n=1}^{\infty} \dfrac{1}{n} \sin \dfrac{2n\pi}{3} \cos \dfrac{n\pi x}{3}$ **39.14.** 1

39.15. $\sum_{n=1}^{\infty} \left(\dfrac{4}{n^2\pi^2} \sin \dfrac{n\pi}{2} + \dfrac{2}{n\pi} \cos \dfrac{n\pi}{2} - \dfrac{4}{n\pi} \cos n\pi \right) \sin \dfrac{n\pi x}{2}$

39.16. $-\dfrac{2}{\pi} \sum_{n=1}^{\infty} \dfrac{(-1)^n}{n - \frac{1}{2}} \cos (n - \tfrac{1}{2})x$ **39.17.** $-\dfrac{2}{\pi} \sum_{n=1}^{\infty} \dfrac{(-1)^n}{(n - \frac{1}{2})^2} \sin (n - \tfrac{1}{2})x$

39.18. (*a*) yes (*c*) no, since $\lim_{\substack{x \to 0 \\ x > 0}} \ln |x| = -\infty$

(*b*) yes (*d*) no, since $\lim_{\substack{x \to 1 \\ x > 1}} \dfrac{1}{3(x-1)^{2/3}} = \infty$

Appendix A

THE GAMMA FUNCTION $(1.00 \leqq x \leqq 1.99)$

x	$\Gamma(x)$	x	$\Gamma(x)$
1.00	1.0000 0000	1.50	0.8862 2693
1.01	0.9943 2585	1.51	0.8865 9169
1.02	0.9888 4420	1.52	0.8870 3878
1.03	0.9835 4995	1.53	0.8875 6763
1.04	0.9784 3820	1.54	0.8881 7766
1.05	0.9735 0427	1.55	0.8888 6835
1.06	0.9687 4365	1.56	0.8896 3920
1.07	0.9641 5204	1.57	0.8904 8975
1.08	0.9597 2531	1.58	0.8914 1955
1.09	0.9554 5949	1.59	0.8924 2821
1.10	0.9513 5077	1.60	0.8935 1535
1.11	0.9473 9550	1.61	0.8946 8061
1.12	0.9435 9019	1.62	0.8959 2367
1.13	0.9399 3145	1.63	0.8972 4423
1.14	0.9364 1607	1.64	0.8986 4203
1.15	0.9330 4093	1.65	0.9001 1682
1.16	0.9298 0307	1.66	0.9016 6837
1.17	0.9266 9961	1.67	0.9032 9650
1.18	0.9237 2781	1.68	0.9050 0103
1.19	0.9208 8504	1.69	0.9067 8182
1.20	0.9181 6874	1.70	0.9086 3873
1.21	0.9155 7649	1.71	0.9105 7168
1.22	0.9131 0595	1.72	0.9125 8058
1.23	0.9107 5486	1.73	0.9146 6537
1.24	0.9085 2106	1.74	0.9168 2603
1.25	0.9064 0248	1.75	0.9190 6253
1.26	0.9043 9712	1.76	0.9213 7488
1.27	0.9025 0306	1.77	0.9237 6313
1.28	0.9007 1848	1.78	0.9262 2731
1.29	0.8990 4159	1.79	0.9287 6749
1.30	0.8974 7070	1.80	0.9313 8377
1.31	0.8960 0418	1.81	0.9340 7626
1.32	0.8946 4046	1.82	0.9368 4508
1.33	0.8933 7805	1.83	0.9396 9040
1.34	0.8922 1551	1.84	0.9426 1236
1.35	0.8911 5144	1.85	0.9456 1118
1.36	0.8901 8453	1.86	0.9486 8704
1.37	0.8893 1351	1.87	0.9518 4019
1.38	0.8885 3715	1.88	0.9550 7085
1.39	0.8878 5429	1.89	0.9583 7931
1.40	0.8872 6382	1.90	0.9617 6583
1.41	0.8867 6466	1.91	0.9652 3073
1.42	0.8863 5579	1.92	0.9787 7431
1.43	0.8860 3624	1.93	0.9723 9692
1.44	0.8858 0506	1.94	0.9760 9891
1.45	0.8856 6138	1.95	0.9798 8065
1.46	0.8856 0434	1.96	0.9837 4254
1.47	0.8856 3312	1.97	0.9876 8498
1.48	0.8857 4696	1.98	0.9917 0841
1.49	0.8859 4513	1.99	0.9958 1326

Appendix B

BESSEL FUNCTIONS $(0.0 \leq x \leq 14.9)$

x	$J_0(x)$	$J_1(x)$	$Y_0(x)$	$Y_1(x)$
0.0	1.0000 0000	0.0000 0000	$-\infty$	$-\infty$
0.1	0.9975 0156	0.0499 3753	−1.5342 3865	−6.4589 5109
0.2	0.9900 2497	0.0995 0083	−1.0811 0532	−3.3238 2499
0.3	0.9776 2625	0.1483 1882	−0.8072 7358	−2.2931 0514
0.4	0.9603 9823	0.1960 2658	−0.6060 2457	−1.7808 7204
0.5	0.9384 6981	0.2422 6846	−0.4445 1873	−1.4714 7239
0.6	0.9120 0486	0.2867 0099	−0.3085 0987	−1.2603 9135
0.7	0.8812 0089	0.3289 9574	−0.1906 6493	−1.1032 4987
0.8	0.8462 8735	0.3688 4205	−0.0868 0228	−0.9781 4418
0.9	0.8075 2380	0.4059 4955	+0.0056 2831	−0.8731 2658
1.0	0.7651 9769	0.4400 5059	0.0882 5696	−0.7812 1282
1.1	0.7196 2202	0.4709 0239	0.1621 6320	−0.6981 1956
1.2	0.6711 3274	0.4982 8906	0.2280 8350	−0.6211 3638
1.3	0.6200 8599	0.5220 2325	0.2865 3536	−0.5485 1973
1.4	0.5668 5512	0.5419 4771	0.3378 9513	−0.4791 4697
1.5	0.5118 2767	0.5579 3651	0.3824 4892	−0.4123 0863
1.6	0.4554 0217	0.5698 9594	0.4204 2690	−0.3475 7801
1.7	0.3979 8486	0.5777 6523	0.4520 2700	−0.2847 2625
1.8	0.3399 8641	0.5815 1695	0.4774 3171	−0.2236 6487
1.9	0.2818 1856	0.5811 5707	0.4968 1997	−0.1644 0577
2.0	0.2238 9078	0.5767 2481	0.5103 7567	−0.1070 3243
2.1	0.1666 0698	0.5682 9214	0.5182 9374	−0.0516 7861
2.2	0.1103 6227	0.5559 6305	0.5207 8429	+0.0014 8779
2.3	0.0555 3978	0.5398 7253	0.5180 7540	0.0522 7732
2.4	+0.0025 0768	0.5201 8527	0.5104 1475	0.1004 8894
2.5	−0.0483 8378	0.4970 9410	0.4980 7036	0.1459 1814
2.6	−0.0968 0495	0.4708 1827	0.4813 3059	0.1883 6354
2.7	−0.1424 4937	0.4416 0138	0.4605 0355	0.2276 3245
2.8	−0.1850 3603	0.4097 0925	0.4359 1599	0.2635 4539
2.9	−0.2243 1155	0.3754 2748	0.4079 1177	0.2959 4005
3.0	−0.2600 5195	0.3390 5896	0.3768 5001	0.3246 7442
3.1	−0.2920 6435	0.3009 2113	0.3431 0289	0.3496 2948
3.2	−0.3201 8817	0.2613 4325	0.3070 5325	0.3707 1134
3.3	−0.3442 9626	0.2206 6345	0.2690 9200	0.3878 5293
3.4	−0.3642 9560	0.1792 2585	0.2296 1534	0.4010 1529
3.5	−0.3801 2774	0.1373 7753	0.1890 2194	0.4101 8842
3.6	−0.3917 6898	0.0954 6555	0.1477 1001	0.4153 9176
3.7	−0.3992 3020	0.0538 3399	0.1060 7432	0.4166 7437
3.8	−0.4025 5641	+0.0128 2100	0.0645 0325	0.4141 1469
3.9	−0.4018 2601	−0.0272 4404	+0.0233 7591	0.4078 2002
4.0	−0.3971 4981	−0.0660 4333	−0.0169 4074	0.3979 2571
4.1	−0.3886 6968	−0.1032 7326	−0.0560 9463	0.3845 9403
4.2	−0.3765 5705	−0.1386 4694	−0.0937 5120	0.3680 1281
4.3	−0.3610 1112	−0.1718 9656	−0.1295 9590	0.3483 9376
4.4	−0.3422 5679	−0.2027 7552	−0.1633 3646	0.3259 7067
4.5	−0.3205 4251	−0.2310 6043	−0.1947 0501	0.3009 9732
4.6	−0.2961 3782	−0.2565 5284	−0.2234 5995	0.2737 4524
4.7	−0.2693 3079	−0.2790 8074	−0.2493 8765	0.2445 0130
4.8	−0.2404 2533	−0.2984 9986	−0.2723 0379	0.2135 6517
4.9	−0.2097 3833	−0.3146 9467	−0.2920 5459	0.1812 4669

x	$J_0(x)$	$J_1(x)$	$Y_0(x)$	$Y_1(x)$
5.0	−0.1775 9677	−0.3275 7914	−0.3085 1763	0.1478 6314
5.1	−0.1443 3475	−0.3370 9720	−0.3216 0245	0.1137 3644
5.2	−0.1102 9044	−0.3432 2301	−0.3312 5093	0.0791 9034
5.3	−0.0758 0311	−0.3459 6083	−0.3374 3730	0.0445 4762
5.4	−0.0412 1010	−0.3453 4479	−0.3401 6788	+0.0101 2727
5.5	−0.0068 4387	−0.3414 3822	−0.3394 8059	−0.0237 5824
5.6	+0.0269 7088	−0.3343 3284	−0.3354 4418	−0.0568 0561
5.7	0.0599 2001	−0.3241 4768	−0.3281 5714	−0.0887 2334
5.8	0.0917 0257	−0.3110 2774	−0.3177 4643	−0.1192 3411
5.9	0.1220 3335	−0.2951 4244	−0.3043 6593	−0.1480 7715
6.0	0.1506 4526	−0.2766 8386	−0.2881 9468	−0.1750 1034
6.1	0.1772 9142	−0.2558 6477	−0.2694 3493	−0.1998 1220
6.2	0.2017 4722	−0.2329 1657	−0.2483 0995	−0.2222 8364
6.3	0.2238 1201	−0.2080 8694	−0.2250 6175	−0.2422 4950
6.4	0.2433 1060	−0.1816 3751	−0.1999 4860	−0.2595 5989
6.5	0.2600 9461	−0.1538 4130	−0.1732 4243	−0.2740 9127
6.6	0.2740 4336	−0.1249 8017	−0.1452 2622	−0.2857 4728
6.7	0.2850 6474	−0.0953 4212	−0.1161 9114	−0.2944 5931
6.8	0.2930 9560	−0.0652 1866	−0.0864 3387	−0.3001 8688
6.9	0.2981 0204	−0.0349 0210	−0.0562 5369	−0.3029 1763
7.0	0.3000 7927	−0.0046 8282	−0.0259 4974	−0.3026 6724
7.1	0.2990 5138	+0.0251 5327	+0.0041 8179	−0.2994 7887
7.2	0.2950 7069	0.0543 2742	0.0338 5040	−0.2934 2259
7.3	0.2882 1695	0.0825 7043	0.0627 7389	−0.2845 9437
7.4	0.2785 9623	0.1096 2509	0.0906 8088	−0.2731 1496
7.5	0.2663 3966	0.1352 4843	0.1173 1329	−0.2591 2851
7.6	0.2516 0183	0.1592 1377	0.1424 2852	−0.2428 0100
7.7	0.2345 5914	0.1813 1272	0.1658 0163	−0.2243 1847
7.8	0.2154 0781	0.2013 5687	0.1872 2717	−0.2038 8510
7.9	0.1943 6184	0.2191 7940	0.2065 2095	−0.1817 2108
8.0	0.1716 5081	0.2346 3635	0.2235 2149	−0.1580 6046
8.1	0.1475 1745	0.2476 0777	0.2380 9133	−0.1331 4880
8.2	0.1222 1530	0.2579 9860	0.2501 1803	−0.1072 4072
8.3	0.0960 0610	0.2657 3930	0.2595 1496	−0.0805 9750
8.4	0.0691 5726	0.2707 8627	0.2662 2187	−0.0534 8451
8.5	0.0419 3925	0.2731 2196	0.2702 0511	−0.0261 6868
8.6	+0.0146 2299	0.2727 5484	0.2714 5771	+0.0010 8399
8.7	−0.0125 2273	0.2697 1902	0.2699 9917	0.0280 1096
8.8	−0.0392 3380	0.2640 7370	0.2658 7494	0.0543 5556
8.9	−0.0652 5325	0.2559 0237	0.2591 5576	0.0798 6940
9.0	−0.0903 3361	0.2453 1179	0.2499 3670	0.1043 1458
9.1	−0.1142 3923	0.2324 3075	0.2383 3599	0.1274 6588
9.2	−0.1367 4837	0.2174 0866	0.2244 9369	0.1491 1279
9.3	−0.1576 5519	0.2004 1393	0.2085 7007	0.1690 6131
9.4	−0.1767 7157	0.1816 3220	0.1907 4392	0.1871 3568
9.5	−0.1939 2875	0.1612 6443	0.1712 1063	0.2031 7990
9.6	−0.2089 7872	0.1395 2481	0.1501 8014	0.2170 5897
9.7	−0.2217 9548	0.1166 3865	0.1278 7479	0.2286 6003
9.8	−0.2322 7603	0.0928 4009	0.1045 2708	0.2378 9324
9.9	−0.2403 4111	0.0683 6983	0.0803 7731	0.2446 9241

x	$J_0(x)$	$J_1(x)$	$Y_0(x)$	$Y_1(x)$
10.0	−0.2459 3576	0.0434 7275	0.0556 7117	0.2490 1542
10.1	−0.2490 2965	+0.0183 9552	0.0306 5738	0.2508 4444
10.2	−0.2496 1707	−0.0066 1574	+0.0055 8523	0.2501 8583
10.3	−0.2477 1681	−0.0313 1783	−0.0192 9785	0.2470 6994
10.4	−0.2433 7175	−0.0554 7276	−0.0437 4862	0.2415 5056
10.5	−0.2366 4819	−0.0788 5001	−0.0675 3037	0.2337 0423
10.6	−0.2276 3505	−0.1012 2866	−0.0904 1515	0.2236 2929
10.7	−0.2164 4274	−0.1223 9942	−0.1121 8589	0.2114 4478
10.8	−0.2032 0197	−0.1421 6657	−0.1326 3838	0.1972 8909
10.9	−0.1880 6225	−0.1603 4969	−0.1515 8319	0.1813 1851
11.0	−0.1711 9030	−0.1767 8530	−0.1688 4732	0.1637 0554
11.1	−0.1527 6830	−0.1913 2829	−0.1842 7577	0.1446 3711
11.2	−0.1329 9194	−0.2038 5315	−0.1977 3287	0.1243 1268
11.3	−0.1120 6846	−0.2142 5503	−0.2091 0343	0.1029 4219
11.4	−0.0902 1450	−0.2224 5059	−0.2182 9371	0.0807 4397
11.5	−0.0676 5395	−0.2283 7862	−0.2252 3211	0.0579 4255
11.6	−0.0446 1567	−0.2320 0047	−0.2298 6973	0.0347 6647
11.7	−0.0213 3128	−0.2333 0024	−0.2321 8059	+0.0114 4601
11.8	+0.0019 6717	−0.2322 8473	−0.2321 6178	−0.0117 8901
11.9	0.0250 4944	−0.2289 8325	−0.2298 3321	−0.0347 1150
12.0	0.0476 8931	−0.2234 4710	−0.2252 3731	−0.0570 9922
12.1	0.0696 6677	−0.2157 4897	−0.2184 3838	−0.0787 3693
12.2	0.0907 7012	−0.2059 8202	−0.2095 2181	−0.0994 1842
12.3	0.1107 9795	−0.1942 5885	−0.1985 9309	−0.1189 4840
12.4	0.1295 6103	−0.1807 1025	−0.1857 7662	−0.1371 4438
12.5	0.1468 8405	−0.1654 8380	−0.1712 1431	−0.1538 3826
12.6	0.1626 0727	−0.1487 4234	−0.1550 6412	−0.1688 7792
12.7	0.1765 8789	−0.1306 6223	−0.1374 9838	−0.1821 2855
12.8	0.1887 0135	−0.1114 3156	−0.1187 0195	−0.1934 7385
12.9	0.1988 4244	−0.0912 4825	−0.0988 7037	−0.2028 1697
13.0	0.2069 2610	−0.0703 1805	−0.0782 0786	−0.2100 8141
13.1	0.2128 8820	−0.0488 5247	−0.0569 2526	−0.2152 1151
13.2	0.2166 8592	−0.0270 6670	−0.0352 3788	−0.2181 7291
13.3	0.2182 9809	−0.0051 7748	−0.0133 6342	−0.2189 5271
13.4	0.2177 2518	+0.0165 9902	+0.0084 8021	−0.2175 5947
13.5	0.2149 8917	0.0380 4929	0.0300 7701	−0.2140 2293
13.6	0.2101 3316	0.0589 6456	0.0512 1501	−0.2083 9360
13.7	0.2032 2083	0.0791 4277	0.0716 8830	−0.2007 4215
13.8	0.1943 3564	0.0983 9052	0.0912 9901	−0.1911 5851
13.9	0.1835 7986	0.1165 2489	0.1098 5919	−0.1797 5095
14.0	0.1710 7348	0.1333 7515	0.1271 9257	−0.1666 4484
14.1	0.1569 5288	0.1487 8435	0.1431 3623	−0.1519 8133
14.2	0.1413 6938	0.1626 1073	0.1575 4209	−0.1359 1587
14.3	0.1244 8769	0.1747 2905	0.1702 7826	−0.1186 1660
14.4	0.1064 8412	0.1850 3166	0.1812 3024	−0.1002 6259
14.5	0.0875 4487	0.1934 2946	0.1903 0189	−0.0810 4209
14.6	0.0678 6407	0.1998 5265	0.1974 1629	−0.0611 5056
14.7	0.0476 4185	0.2042 5127	0.2025 1632	−0.0407 8875
14.8	0.0270 8231	0.2065 9557	0.2055 6516	−0.0201 6071
14.9	0.0063 9154	0.2068 7617	0.2065 4643	+0.0005 2828

Appendix C

ADDITIONAL LAPLACE TRANSFORMS

The following entries supplement Table 22-1, page 132.

	$f(x)$	$F(s) = \mathcal{L}\{f(x)\}$
18.	$\frac{1}{a} e^{-x/a}$	$\frac{1}{1+as}$
19.	$\frac{1}{a}(e^{ax}-1)$	$\frac{1}{s(s-a)}$
20.	$1 - e^{-x/a}$	$\frac{1}{s(1+as)}$
21.	$\frac{1}{a^2} x e^{-x/a}$	$\frac{1}{(1+as)^2}$
22.	$\frac{e^{ax}-e^{bx}}{a-b}$	$\frac{1}{(s-a)(s-b)}$
23.	$\frac{e^{-x/a}-e^{-x/b}}{a-b}$	$\frac{1}{(1+as)(1+bs)}$
24.	$(1+ax)e^{ax}$	$\frac{s}{(s-a)^2}$
25.	$\frac{1}{a^3}(a-x)e^{-x/a}$	$\frac{s}{(1+as)^2}$
26.	$\frac{ae^{ax}-be^{bx}}{a-b}$	$\frac{s}{(s-a)(s-b)}$
27.	$\frac{ae^{-x/b}-be^{-x/a}}{ab(a-b)}$	$\frac{s}{(1+as)(1+bs)}$
28.	$\frac{1}{a^2}(e^{ax}-1-ax)$	$\frac{1}{s^2(s-a)}$
29.	$\sin^2 ax$	$\frac{2a^2}{s(s^2+4a^2)}$
30.	$\sinh^2 ax$	$\frac{2a^2}{s(s^2-4a^2)}$
31.	$\frac{1}{\sqrt{2}}\left(\cosh\frac{ax}{\sqrt{2}}\sin\frac{ax}{\sqrt{2}} - \sinh\frac{ax}{\sqrt{2}}\cos\frac{ax}{\sqrt{2}}\right)$	$\frac{a^3}{s^4+a^4}$

	$f(x)$	$F(s) = \mathcal{L}\{f(x)\}$
32.	$\sin \frac{ax}{\sqrt{2}} \sinh \frac{ax}{\sqrt{2}}$	$\frac{a^2 s}{s^4 + a^4}$
33.	$\frac{1}{\sqrt{2}}\left(\cos \frac{ax}{\sqrt{2}} \sinh \frac{ax}{\sqrt{2}} + \sin \frac{ax}{\sqrt{2}} \cosh \frac{ax}{\sqrt{2}}\right)$	$\frac{as^2}{s^4 + a^4}$
34.	$\cos \frac{ax}{\sqrt{2}} \cosh \frac{ax}{\sqrt{2}}$	$\frac{s^3}{s^4 + a^4}$
35.	$\frac{1}{2}(\sinh ax - \sin ax)$	$\frac{a^3}{s^4 - a^4}$
36.	$\frac{1}{2}(\cosh ax - \cos ax)$	$\frac{a^2 s}{s^4 - a^4}$
37.	$\frac{1}{2}(\sinh ax + \sin ax)$	$\frac{as^2}{s^4 - a^4}$
38.	$\frac{1}{2}(\cosh ax + \cos ax)$	$\frac{s^3}{s^4 - a^4}$
39.	$\sin ax \sinh ax$	$\frac{2a^2 s}{s^4 + 4a^4}$
40.	$\cos ax \sinh ax$	$\frac{a(s^2 - 2a^2)}{s^4 + 4a^4}$
41.	$\sin ax \cosh ax$	$\frac{a(s^2 + 2a^2)}{s^4 + 4a^4}$
42.	$\cos ax \cosh ax$	$\frac{s^3}{s^4 + 4a^4}$
43.	$\frac{1}{2}(\sin ax + ax \cos ax)$	$\frac{as^2}{(s^2 + a^2)^2}$
44.	$\cos ax - \frac{ax}{2} \sin ax$	$\frac{s^3}{(s^2 + a^2)^2}$
45.	$\frac{1}{2}(ax \cosh ax - \sinh ax)$	$\frac{a^3}{(s^2 - a^2)^2}$
46.	$\frac{x}{2} \sinh ax$	$\frac{as}{(s^2 - a^2)^2}$
47.	$\frac{1}{2}(\sinh ax + ax \cosh ax)$	$\frac{as^2}{(s^2 - a^2)^2}$
48.	$\cosh ax + \frac{ax}{2} \sinh ax$	$\frac{s^3}{(s^2 - a^2)^2}$

	$f(x)$	$F(s) = \mathcal{L}\{f(x)\}$
49.	$\dfrac{a \sin bx - b \sin ax}{a^2 - b^2}$	$\dfrac{ab}{(s^2+a^2)(s^2+b^2)}$
50.	$\dfrac{\cos bx - \cos ax}{a^2 - b^2}$	$\dfrac{s}{(s^2+a^2)(s^2+b^2)}$
51.	$\dfrac{a \sin ax - b \sin bx}{a^2 - b^2}$	$\dfrac{s^2}{(s^2+a^2)(s^2+b^2)}$
52.	$\dfrac{a^2 \cos ax - b^2 \cos bx}{a^2 - b^2}$	$\dfrac{s^3}{(s^2+a^2)(s^2+b^2)}$
53.	$\dfrac{b \sinh ax - a \sinh bx}{a^2 - b^2}$	$\dfrac{ab}{(s^2-a^2)(s^2-b^2)}$
54.	$\dfrac{\cosh ax - \cosh bx}{a^2 - b^2}$	$\dfrac{s}{(s^2-a^2)(s^2-b^2)}$
55.	$\dfrac{a \sinh ax - b \sinh bx}{a^2 - b^2}$	$\dfrac{s^2}{(s^2-a^2)(s^2-b^2)}$
56.	$\dfrac{a^2 \cosh ax - b^2 \cosh bx}{a^2 - b^2}$	$\dfrac{s^3}{(s^2-a^2)(s^2-b^2)}$
57.	$x - \dfrac{1}{a} \sin ax$	$\dfrac{a^2}{s^2(s^2+a^2)}$
58.	$\dfrac{1}{a} \sinh ax - x$	$\dfrac{a^2}{s^2(s^2-a^2)}$
59.	$1 - \cos ax - \dfrac{ax}{2} \sin ax$	$\dfrac{a^4}{s(s^2+a^2)^2}$
60.	$1 - \cosh ax + \dfrac{ax}{2} \sinh ax$	$\dfrac{a^4}{s(s^2-a^2)^2}$
61.	$1 + \dfrac{b^2 \cos ax - a^2 \cos bx}{a^2 - b^2}$	$\dfrac{a^2b^2}{s(s^2+a^2)(s^2+b^2)}$
62.	$1 + \dfrac{b^2 \cosh ax - a^2 \cosh bx}{a^2 - b^2}$	$\dfrac{a^2b^2}{s(s^2-a^2)(s^2-b^2)}$
63.	$\dfrac{1}{8}[(3 - a^2x^2) \sin ax - 3ax \cos ax]$	$\dfrac{a^5}{(s^2+a^2)^3}$
64.	$\dfrac{x}{8}[\sin ax - ax \cos ax]$	$\dfrac{a^3 s}{(s^2+a^2)^3}$
65.	$\dfrac{1}{8}[(1 + a^2x^2) \sin ax - ax \cos ax]$	$\dfrac{a^3 s^2}{(s^2+a^2)^3}$

	$f(x)$	$F(s) = \mathcal{L}\{f(x)\}$
66.	$\frac{1}{8}[(3 + a^2x^2)\sinh ax - 3ax\cosh ax]$	$\frac{a^5}{(s^2 - a^2)^3}$
67.	$\frac{x}{8}(ax\cosh ax - \sinh ax)$	$\frac{a^3 s}{(s^2 - a^2)^3}$
68.	$\frac{1}{8}[ax\cosh ax - (1 - a^2x^2)\sinh ax]$	$\frac{a^3 s^2}{(s^2 - a^2)^3}$
69.	$\frac{1}{n!}(1 - e^{-x/a})^n$	$\frac{1}{s(as+1)(as+2)\cdots(as+n)}$
70.	$\sin(ax + b)$	$\frac{s\sin b + a\cos b}{s^2 + a^2}$
71.	$\cos(ax + b)$	$\frac{s\cos b - a\sin b}{s^2 + a^2}$
72.	$e^{-ax} - e^{ax/2}\left(\cos\frac{ax\sqrt{3}}{2} - \sqrt{3}\sin\frac{ax\sqrt{3}}{2}\right)$	$\frac{3a^2}{s^3 + a^3}$
73.	$\frac{1 + 2ax}{\sqrt{\pi x}}$	$\frac{s + a}{s\sqrt{s}}$
74.	$e^{-ax}/\sqrt{\pi x}$	$\frac{1}{\sqrt{s + a}}$
75.	$\frac{1}{2x\sqrt{\pi x}}(e^{bx} - e^{ax})$	$\sqrt{s - a} - \sqrt{s - b}$
76.	$\frac{1}{\sqrt{\pi x}}\cos 2\sqrt{ax}$	$\frac{1}{\sqrt{s}}e^{-a/s}$
77.	$\frac{1}{\sqrt{\pi x}}\cosh 2\sqrt{ax}$	$\frac{1}{\sqrt{s}}e^{a/s}$
78.	$\frac{1}{\sqrt{a\pi}}\sin 2\sqrt{ax}$	$s^{-3/2}e^{-a/s}$
79.	$\frac{1}{\sqrt{a\pi}}\sinh 2\sqrt{ax}$	$s^{-3/2}e^{a/s}$
80.	$J_0(2\sqrt{ax})$	$\frac{1}{s}e^{-a/s}$
81.	$\sqrt{x/a}\,J_1(2\sqrt{ax})$	$\frac{1}{s^2}e^{-a/s}$
82.	$(x/a)^{(p-1)/2}J_{p-1}(2\sqrt{ax})$ $(p > 0)$	$s^{-p}e^{-a/s}$

	$f(x)$	$F(s) = \mathcal{L}\{f(x)\}$
83.	$J_0(x)$	$\dfrac{1}{\sqrt{s^2+1}}$
84.	$J_1(x)$	$\dfrac{\sqrt{s^2+1}-s}{\sqrt{s^2+1}}$
85.	$J_p(x) \qquad (p > -1)$	$\dfrac{(\sqrt{s^2+1}-s)^p}{\sqrt{s^2+1}}$
86.	$x^p J_p(ax) \qquad (p > -\frac{1}{2})$	$\dfrac{(2a)^p\,\Gamma(p+\frac{1}{2})}{\sqrt{\pi}\,(s^2+a^2)^{p+(1/2)}}$
87.	$\dfrac{x^{p-1}}{\Gamma(p)} \qquad (p > 0)$	$\dfrac{1}{s^p}$
88.	$\dfrac{4^n\,n!}{(2n)!\,\sqrt{\pi}}\,x^{n-(1/2)}$	$\dfrac{1}{s^n\sqrt{s}}$
89.	$\dfrac{x^{p-1}}{\Gamma(p)}\,e^{-ax} \qquad (p > 0)$	$\dfrac{1}{(s+a)^p}$
90.	$\dfrac{1-e^{ax}}{x}$	$\ln\dfrac{s-a}{s}$
91.	$\dfrac{e^{bx}-e^{ax}}{x}$	$\ln\dfrac{s-a}{s-b}$
92.	$\dfrac{2}{x}\sinh ax$	$\ln\dfrac{s+a}{s-a}$
93.	$\dfrac{2}{x}(1-\cos ax)$	$\ln\dfrac{s^2+a^2}{s^2}$
94.	$\dfrac{2}{x}(\cos bx - \cos ax)$	$\ln\dfrac{s^2+a^2}{s^2+b^2}$
95.	$\dfrac{\sin ax}{x}$	$\arctan\dfrac{a}{s}$
96.	$\dfrac{2}{x}\sin ax \cos bx$	$\arctan\dfrac{2as}{s^2-a^2+b^2}$
97.	$\sin\lvert ax\rvert$	$\left(\dfrac{a}{s^2+a^2}\right)\left(\dfrac{1+e^{-(\pi/a)s}}{1-e^{-(\pi/a)s}}\right)$

INDEX